There Is No B Planet: Relevant Experiences and Case Studies on Future Challenges in Sustainable Urban Planning & Territorial Management

There Is No B Planet: Relevant Experiences and Case Studies on Future Challenges in Sustainable Urban Planning & Territorial Management

Guest Editors

Salvador García-Ayllón Veintimilla

Josep Lluís Miralles Garcia

MDPI

Basel • Beijing • Wuhan • Barcelona • Belgrade • Novi Sad • Cluj • Manchester

Guest Editors
Salvador García-Ayllón Veintimilla
Department of Mining and Civil Engineering
Technical University of Cartagena
Cartagena
Spain

Josep Lluís Miralles Garcia
Department of Urban Planning
Universitat Politècnica de València
Valencia
Spain

Editorial Office
MDPI AG
Grosspeteranlage 5
4052 Basel, Switzerland

This is a reprint of the Special Issue, published open access by the journal *Land* (ISSN 2073-445X), freely accessible at: https://www.mdpi.com/journal/land/special_issues/29QIUE6QR3.

For citation purposes, cite each article independently as indicated on the article page online and as indicated below:

Lastname, A.A.; Lastname, B.B. Article Title. *Journal Name* **Year**, *Volume Number*, Page Range.

ISBN 978-3-7258-4273-5 (Hbk)
ISBN 978-3-7258-4274-2 (PDF)
https://doi.org/10.3390/books978-3-7258-4274-2

Cover image courtesy of Salvador García-Ayllón Veintimilla

Contents

About the Editors

Salvador García-Ayllón Veintimilla

Salvador García-Ayllón Veintimilla is an Architect and Civil Engineer. He holds a PhD in Urban Planning and Territorial Management. Since 2010, he has been a professor responsible for various courses related to land use policy and environmental and infrastructure planning at the Faculty Civil Engineering of the Technical University of Cartagena (UPCT). He has also been a visiting professor at the universities of California Berkeley (USA), Columbia New York (USA), Alcalá de Henares (Spain), CEPES (Mexico), Cyprus University of Technology, and Massachusetts Institute of Technology (MIT) through a Fulbright grant from the United States government. He is currently the Director of the department of Civil and Mining Engineering at UPCT.

In the scientific field, he is principal investigator of the R&D group of Territorial Policy, Environmental and Infrastructure Planning. He is also Director of the Biyectiva GIS & AI Business Professorship and Chair of the Cartagena City Council sustainable urban mobility professorship of the UPCT. He is the author of more than 50 publications in international research journals and 15 books, being part of the top 2% most cited authors in the world according to the list published by Stanford University. He is also a member of the Spanish Ministry of Universities' Committee for the Recognition of Foreign University Degrees Granting Access to Regulated Professions and regularly participates as a project evaluator for the European Commission's European Research Agency, as well as for public research institutions in several countries around the world.

At a professional level, he has held positions of diverse responsibility in the private sector and public administration. He is currently an advisor to various public institutions and has also been an elected member of the governing board of the College of Civil Engineers, and of the Professional Ethics Committee of the College of Architects.

Josep Lluís Miralles Garcia

Josep Lluís Miralles Garcia is a professor at the Department of Urbanism and the Civil Engineering School at the Universitat Politècnica de València (Spain) where he has been since 1979. He received his PhD in Civil Engineering in 1988 from the same university. He has been an Associate Professor since 1989. Since 2024, he has been a professor "Ad Honorem".

His research interest is focused on economic, social, and environmental sustainability of urban development. More specifically, he works on the introduction of sustainability in urban and territorial planning with special attention to transport infrastructures. He pays special attention to green infrastructure and ecosystem services, environmental management or peri-urban natural resources like L'Horta de València (Spain), the environmental and territorial impacts of high-speed train infrastructures, conditions of sustainability within urban development and urban tourism development in coastal areas with flooding risks, real estates, and speculative processes.

He has directed 12 research projects. In addition, he has participated both as a team member and director in 18 contracts with companies or public administration for urban studies, urban and territorial planning, and territory management projects. He is the author or coauthor of more than 50 articles in research journals, more than 20 chapters of various books and more than 58 papers in national and international conferences. He is the author of four research books and he has directed nine doctoral research programs.

Between 1991 and 2012, he was the Director of the Master's in Territorial, Environmental and Urban Planning course. From 2008 to 2017, Miralles-Garcia was the Director of Master's in Transport,

Territory and Urbanism course. Between 2012 and 2020, he was the Director of Department of Urbanism. Between 2015 and 2023, he was a member of the Urban Territorial Commission of the Generalitat Valenciana, which approves urban plans in the Valencian territory. He has been an editor of different publications and as of 2025, he is coordinator of 'Professor Association of urbanism and territorial planning'.

Preface

We scientists always enjoy coming across future challenges in our own field of research because investigating technologies, tools, or management methods to improve our knowledge and, consequently, generating solutions to the problems of our society is an essential part of science—problems that society itself increasingly generates through its social and economic activities. In fact, urban and territorial planning has been and continues to be a tool developed to rationalize urbanization processes in the territory. This rationalization allows society to act preemptively to avoid unaffordable damage and costs of all kinds, not just economic ones. Damage that can be caused by natural phenomena or conditioned by the effects of climate change (floods, forest fires, melting ice, etc.) or by human activities. Science, by definition, investigates reality as it is and seeks the rules that explain it as a basis for making proposals that lead to improvements. This is precisely the opposite of what denialism advocates: denying science as an objective source of knowledge. The international conferences on future challenges in Sustainable Urban Planning and Territorial Management (SUPTM) focus on urban and territorial planning, with the ultimate goal of achieving an urban environment based on management rules that guarantee quality human living conditions, at a minimum, maintained over time or, if possible, progressively improved. That is, an understanding of reality to be obtained scientifically, just as the authors who present their studies at conferences do.

This reprint presents a selection of relevant experiences and case studies presented at the 2nd International Conference on Future Challenges in Sustainable Urban Planning and Territorial Management, held in 2024. This selection reflects the scientific areas of preferred scientific research production and the current trends of investigation from a methodological point of view. Some articles have focused on different issues relative to the process of urban growth in China, which is currently one of the epicenters in the study of urban and especially territorial analysis due to the high rates of urban growth that its cities have experienced during the last decade.

Huiguang Chen et al. have studied the response of rural–urban migration to land endowment while recognizing the heterogeneity of land tenure security perceptions. China is in the process of urban consolidation after an era of exponential economic expansion that has created numerous megalopolises of several million inhabitants on the east coast of the country from the rural interior, and it is important to identify the causes that explain the decision-making of migrant people. Changjun Jiang and Huiguang Chen explore the laws and trends of urban–rural land use in highly urbanized areas that have a great significance in promoting rural transformation. This one is another consequence of rural–urban migration. Tianxiang Long, Yuxin Liu, and Qikang Zhong present an innovative study that aims to analyze the patterns of system coupling coordination development in the Yangtze River Basin under the interacting influences of population growth, ecological conservation, energy utilization, and digital economic development. This way authors show a new approach to analyze the relationship between urban growth and sustainability in a regional zone. By the other hand, Siwen Xia et al. focus their study on the development of a Free Trade Zone in China, and the effects on urban land use efficiency.

How to make urban development processes environmentally sustainable is also a recurring research question in this field of research. There are several approaches to achieving this goal, but this issue will require increasing attention because of climate change anthropic derivatives. Sprawling land development patterns have exacerbated ecological degradation, social fragmentation, and public health problems. Dakota B. Walker et al. propose an ecological impact-weighted land value taxation policy (ELVT) which incorporates the ecological footprint of land use into one's land value tax burden as a possible way to socialize ecological cost.

Broadening the perspective to a more general approach to the problem, Hongpeng Fu et al. have conducted a study on the implementation of Sustainable Development Goals in China. The results obtained allow us to understand the progress of its implementation, bringing surprising results to light.

Sustainable development implies maintaining and increasing, if possible, the environmental services generated by natural ecosystems. These environmental services are the necessary input to social and economic activities and include natural resource production and waste recycling. Under this framework, the role of territorial vegetation and forests is essential. Benxu Wang at al. present a study that analyzes the spatiotemporal patterns of economic density and vegetation coverage in the Yellow River Basin based on forest resource inventory and socio-economic data from 448 counties in 2008, 2013, and 2018. Over the 2008–2018 period, a significant negative correlation between economic density and vegetation cover is observed. This study delineates a practical pathway for high-quality economic development and high-level ecological protection in the Yellow River Basin. On the other hand, Xiaobing Gu et al. present a study that delves into the impact of climate, specifically hydrothermal conditions, on the complex interplay between urbanization and forest ecological function in the Yellow River Basin.

In urban areas, public spaces are essential to sustainable urban forms according to Sustainable Development Goal 11.7. Kan Wang, Xing Dang, and Jianjun Bai develop a system to characterize public spaces from the point of view of safety, inclusiveness, accessibility, and greenness through 11 indicators. They consider these four characteristics as a significance of three integrated scales: apartment complexes, street blocks, and counties. It is a possible model to monitor the urban public spaces by public administration to achieve urban sustainable goals. Regarding the public spaces of streets, Maria Konstantinidou and Josep Maria Salanova Grau present an interesting study that applies 3D technology to the analysis of urban mobility. With this methodological approach, they show how digital twins can be used as a tool for vulnerable citizen engagement, improving efficiency of co-creation in mobility planning.

All these proposals on relevant cases and methodologically innovative experiences in the complex field of urban planning and territorial planning have been considered worthy of appearing in this volume by the scientific committee of the event. As Editors of this reprint, we hope that the contributions made by the authors of the selected published articles will serve as a reference for researchers in their own fields of scientific interest, and that the science will lead to more efficient sustainable urban and territorial planning.

Salvador García-Ayll6n Veintimilla and Josep Lluís Miralles Garcia
Guest Editors

land

MDPI

Article

Relationship between Climate-Shaped Urbanization and Forest Ecological Function: A Case Study of the Yellow River Basin, China

Xiaobing Gu [1,2], Guangyu Wang [3], Shunli Zhang [1,2], Linyan Feng [4], Ram P. Sharma [5], Huoyan Zhou [6], Liyong Fu [4], Qingjun Wu [1,2], Yaquan Dou [1] and Xiaodi Zhao [1,3,*]

1 Research Institute of Forestry Policy and Information, Chinese Academy of Forestry, Beijing 100091, China; guxiaobing@cau.edu.cn (X.G.); bs20203110778@cau.edu.cn (S.Z.); wqj2020@cau.edu.cn (Q.W.); douyq@caf.ac.cn (Y.D.)
2 College of Economics and Management, China Agricultural University, Beijing 100083, China
3 Faculty of Forestry, The University of British Columbia, Vancouver, BC V6T 1Z4, Canada; guangyu.wang@ubc.ca
4 Research Institute of Forest Resource Information Techniques, Chinese Academy of Forestry, Beijing 100091, China; linyan_feng@caf.ac.cn (L.F.); fuly@ifrit.ac.cn (L.F.)
5 Institute of Forestry, Tribhuvan University, Kathmandu 44600, Nepal; ramsharm1@gmail.com
6 School of Ecology and Environment Science, Yunnan University, Kunming 650031, China; zhouhuoyan@itc.ynu.edu.cn
* Correspondence: zhaoxiaodi@caf.ac.cn; Tel.: +86-136-7105-3260

Abstract: Amidst the compounded challenges posed by global climate change and urbanization on forest ecosystems, the integration of urbanization control measures within a climate-focused framework may offer an avenue for breakthroughs. This study delves into the impact of climate, specifically hydrothermal conditions, on the complex interplay between urbanization (Urb) and forest ecological function (Eco) in the Yellow River Basin (YRB) in China. Our findings reveal: (1) The application of a coupled coordination model reveals a stronger alignment between urbanization and forest ecological function in the warm and humid regions in the YRB. (2) Through the cross-sectional threshold regression model, we elucidate the diverse responses of Urb to Eco across varying climate gradients. Among them, annual precipitation shows a double-threshold effect at 532.34 mm and 694.18 mm. As precipitation increases, the impact of Urb shifts from negative to positive on Eco. Moreover, in regions with precipitation below 532.34 mm and above 694.18 mm, the absolute value of response coefficients of Eco to Urb is amplified. Annual average temperature displays a single-threshold effect at 10.11 °C, leading to a transition from negative to positive impact as temperature rises. This study establishes the climate-based threshold system that governs the urbanization–forest ecological function relationship.

Keywords: urbanization; forest ecological function; climate; coupled coordination model; threshold regression model

1. Introduction

Climate plays a paramount role in shaping the characteristics and geographic distribution of forest ecosystems [1]. Over the past century, the Earth's climate has been undergoing a significant transformation, primarily characterized by global warming [2]. Compounded by the effects of urban development, the tension between preserving forest resources and the growing land requirements for urban expansion is becoming increasingly conspicuous [3]. Extent-wise, with urban expansion, the substantial reduction in forest cover becomes highly evident, thereby resulting in the diminishing ecological functions of regional forest ecosystems [4]. Demographically, the concentration of people in urban areas significantly amplifies the demand for forest resources, such as timber and water, thus intensifying the ecological pressure on forests [5]. Industrially, non-agricultural sectors further elevate the ongoing demand for forest resources [6]. Spatially, despite alterations in land use due to urban spatial expansion, it is plausible that urban green coverage may not adequately correspond to these changes [7]. Therefore, it is imperative to promptly implement the corresponding environmental and social governance measures to mitigate the adverse impacts that climate change and urbanization may exert on forests.

Compared to other ecosystems, the forest ecosystem exhibits a reduced capacity for adaptation and a slower response to the impacts of climate change [8]. Managing the urbanization process in line with climate conditions may offer a potential solution to the ongoing conflicts between urbanization and forest ecology. However, the influence of urbanization on forest ecology is intricate and multifaceted. Studies in this area can be categorized into two primary groups. The first category of research contends that urbanization has had detrimental effects on forest ecology. The expansion of urban areas, along with high-intensity human activities and irrational land use practices, have resulted in several adverse consequences for forest ecology. These include a decline in ecological function, a reduction in biodiversity, and the destruction of habitats for various plant and animal species [9]. For instance, the study by Deng et al. (2021), which examined natural forests in the Liaohe River Basin in China, revealed that urbanization led to a decrease in the water-holding capacity of the soil layer, the apomictic layer, and the forest canopy [10]. Additionally, Qian et al. (2022) found that key drivers of the spatial and temporal shifts in forest ecological function in Guangzhou, China, between 1979 and 2012 included alterations in forestry policies, urbanization, industrialization, and human disturbances. Of these, excessive logging due to the division of mountains into household plots, forest land encroachment driven by economic overexpansion, and forest degradation all contributed to a decline in the quality of regional forest ecological functions [6].

An alternative body of research believes that urbanization yields a positive impact on forest ecology. The factors contributing to this shift in the impact direction can be summarized as follows: Firstly, the development of urbanization is often accompanied by technological innovations and the adoption of green technologies, reducing cities' reliance on natural resources [11]. This fosters a more intensive development mode, leading to a substantial increase in production efficiency. Ehrhardt-Martinez et al. (2002) found that during the initial stages of urbanization, the deforestation rate continued to rise. However, as urbanization progresses, labor productivity and the environmental consciousness of residents improve, leading to a decline in the rate of destruction and consumption of natural resources, such as forest land [12]. Secondly, the growing global environmental awareness has the potential to instigate more stringent environmental measures. Through the agglomeration effect, urbanization fosters the dissemination of knowledge and skill, as well as the sharing of public resources, thereby enhancing the efficacy of centralized environmental governance [7]. Song et al. (2018) pointed out that environmental education and public engagement could prompt cities to actively invest in their ecological environments [13]. Thirdly, the positive impact of urbanization is interconnected with climate conditions [14]. Climate comfort significantly influences gross regional product growth and population expansion [15], albeit across different stages. Wen et al. (2018) discovered that precipitation plays a substantial positive regulatory role in the process of urbanization affecting vegetation cover [16]. However, Li et al. (2021), based on the Beijing–Tianjin–Hebei region, found that the level of urbanization had no significant impact on the degree of forest fragmentation [17].

The aforementioned studies indicate that there is no consistent consensus regarding the relationship between urbanization and forest ecology, and the influence of climate in this context still remains ambiguous. To illustrate this, this study utilizes the Yellow River Basin (YRB) in China as a case study. It explores the relationship between urbanization (Urb) and forest ecological function (Eco) by examining the coupling coordination level and impact coefficient, respectively, while considering the influence of climate. The specific research ideas are outlined as follows: (1) Employing a coupled coordination model to investigate the developmental status of the "Urb-Eco" system in the YRB. This model enables an exploration of the degree of coordination between urbanization and forest ecological function. (2) Utilizing the cross-sectional threshold model to assess the nonlinear impact of urbanization on forest ecological function. This analysis aims to identify the changing patterns of urbanization's effect on forest ecological function under varying climate gradients, providing valuable insights into the dynamic nature of this relationship.

At the county level, this study conducts a quantitative analysis of the interplay between forest ecology and urbanization in the YRB, encompassing different regions within the basin. By doing so, the research elucidates the existing conflicts between urbanization development and ecological environmental preservation in the YRB and its adjacent areas. This analysis serves as a foundation for understanding and addressing the challenges arising from the intersection of urbanization and forest ecology in the context of global climate change. Additionally, the findings of this study are expected to inform the formulation of pertinent environmental and socio-economic governance policies, offering practical solutions to the complex dilemma between urbanization and forest ecology.

2. Materials and Methods

2.1. Study Area

The YRB (Figure 1) spans from 95°53′ E to 119°12′ E in longitude and from 32°9′ N to 41°50′ N in latitude, covering an area of 7.52 × 105 km^2 [18]. This basin encompasses nine provinces and autonomous regions, including Qinghai, Sichuan, Gansu, Ningxia, Inner Mongolia, Shaanxi, Shanxi, Henan, and Shandong. The predominant land use types are grasslands and agricultural land, followed by unused land, forests, and water bodies [19]. Due to the influence of atmospheric circulation and monsoon patterns, there are significant climate variations across different regions within the basin, leading to uneven seasonal distribution [20]. The northwest experiences arid climate conditions, the central region has a semi-arid climate, while the southeast enjoys a semi-humid climate [21]. The annual average temperature ranges from −4 °C to 14 °C, with notable variations and west-to-east gradient from cold to warm [20]. Precipitation is concentrated and unevenly distributed, with an average annual rainfall of 470 mm [22]. Low and uneven rainfall, low humidity, and high evaporation rates are the main climatic characteristics of this region, with frequent occurrences of hail and sandstorms.

2.2. Research Methods

2.2.1. Coupled Coordination Model

To analyze the coordinated development status of the urbanization and forest ecological function subsystems, and explore their preliminary relationship with climate factors, we attempted to consider Urb and Eco as two subsystems and introduced a coupling coordination model. In this paper, we referred to the practices of Wang Jintao (2019) to calculate the comprehensive score, coupling degree, and coupling coordination degree [23], which involves the following steps:

First, calculating the comprehensive score, $T_i = \alpha Eco_i + \beta Urb_i$, where α and β represent the weights of the comprehensive evaluation indexes for the forest ecological function subsystem and urbanization, respectively. Both α and β were set to 0.5. Second, calculating the coupling degree, $C_i = \sqrt{Eco_i \times Urb_i / (\frac{Eco_i + Urb_i}{2})^2}$, where C_i represents the overall coupling degree of the two subsystems in county i. The coupling degree values range between 0 and 1. C equals 1 when the coupling degree is at its maximum, indicating the highest level of interaction and mutual influence between the systems and moving towards new and more advanced functional states. On the other hand, C equals 0 indicating no interaction or mutual influence between the systems, tending towards disordered development. Third, calculating the coupling coordination degree, $D_i = \sqrt{C_i \times T_i}$. The larger the value of D, the higher the coupling coordination degree, indicating not only an overall improvement in the composite system, but also a more coordinated coupling relationship. We classified the sample counties based on the Urb-Eco coupling coordination degree according to the classification criteria [24] in Table 1.

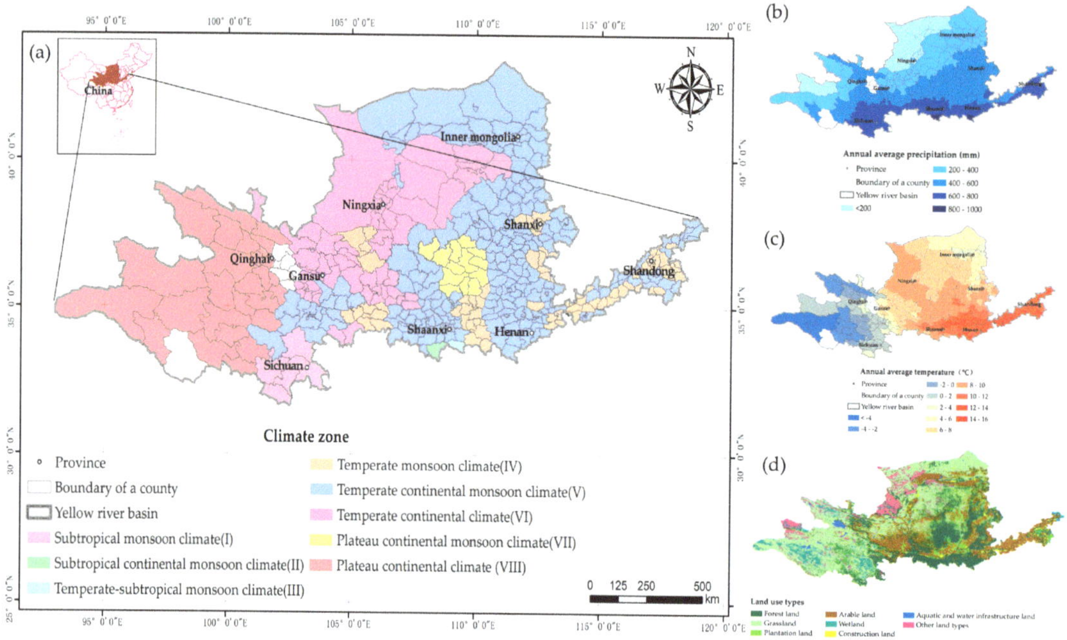

Figure 1. The overview of the Yellow River Basin (YRB): (**a**) climate zone; (**b**) annual average precipitation; (**c**) annual average temperature; (**d**) land use types. Note: climate type data for each county were collected from the local government's official websites; see Table S1 for more specific information. Annual average precipitation and annual average temperature data were obtained from the "WorldClim 2.1 Climate Data (1970–2000)" published by the WorldClim database (https://www.worldclim.org, accessed on 29 September 2023) and land use type data were provided by the National Forestry and Grassland Administration Planning Institute.

Table 1. Grade criteria of coupling degree and coupling coordination degree.

C Grade Criteria	[0.3,0.5)		[0.5,0.8)	[0.8,1.0]	
	Antagonistic Stage		**Adjustment Stage**	**High Coupling Stage**	
D Grade Criteria	[0,0.1) Extremely imbalanced	[0.1,0.2) Seriously imbalanced	[0.2,0.3) Moderately imbalanced	[0.3,0.4) Slightly imbalanced	[0.4,0.5) Nearly imbalanced
	[0.5,0.6) Barely coordinated	[0.6,0.7) Primary coordinated	[0.7,0.8) Intermediate coordinated	[0.8,0.9) Good coordinated	[0.9,1.0] High-quality coordinated

C refers to the coupling degree and D refers to the coupling coordination standard degree.

2.2.2. Cross-Sectional Threshold Model

To investigate the moderating effect of climate on the impact of urbanization on forest ecological function, this study employed Hansen's cross-sectional threshold model. Precipitation and temperature were used as threshold variables separately. This study explored the nonlinear influence of urbanization on the forest ecological function and further determined the relationship between these factors and the role played by climate variables.

Artificially defining the gradients and introducing interaction or quadratic terms for explanatory variables are common methods for performing grouping heterogeneity analyses. However, most of these methods had shortcomings, such as inconsistent grouping criteria, subjectivity, unclear interpretability, and an inability to test significance [25,26]. Hansen's threshold model was developed to address these issues. This model selects the number and values of thresholds "endogenously", thus substantially reducing the subjectivity associated with artificial grouping [27]. Moreover, the existence of the threshold effect and the authenticity of threshold values can be verified [27]. Without loss of generality, the construction of the cross-sectional multi-threshold model is as follows:

$$\begin{aligned} \text{Eco}_i = {} & \beta_0 + \beta_1 \cdot Urb_i \cdot I_i(Thres_i \leq \lambda_1) + \beta_2 \cdot Urb_i \cdot I_i(\lambda_2 < Thres_i \leq \lambda_2) + \cdots \\ & + \beta_k \cdot Urb_i \cdot I_i(\lambda_k < Thres_i \leq \lambda_k) + \gamma_i Control_i + \varepsilon_i (k < 436) \end{aligned} \quad (1)$$

where: i represents county-level variables; Eco represents forest ecological function scores; Urb represents the level of urbanization; $Control$ represents a series of control variables; $Thres$ represents threshold variables, specifically set as the annual average precipitation (Pre) and the annual average temperature (Temp); I is an indicator function, essentially a segmented function related to the threshold variable $Thres$. It takes the value 1 when the threshold variable is within that interval; otherwise, it takes 0. λ is the estimated threshold value, β is the coefficient estimated using the nonlinear least square method, k is the number of thresholds, γ is the coefficient of each control variable, and ε is the random disturbance term in the model.

The testing of the threshold model primarily consists of two parts [27–29]. First is the F-test for the existence of threshold effects, that is, the significance test of threshold effects, with the null hypothesis $\text{H}_0 : \beta_1 = \beta_2 = \cdots = \beta_k$. The F-statistic is $\text{F} = \left(S_0 - \text{S}_k\left(\hat{\lambda}\right)\right)/\hat{\sigma}^2$, where S_0 is the sum of squared residuals when there is no threshold value. Since the threshold value λ cannot be identified under the null hypothesis H_0, the distribution of the F-statistic is non-standard [28]. Therefore, Hansen (1999) proposed the "Bootstrap" resampling method to obtain the asymptotic distribution of the F-statistic and construct the corresponding p-value for significance testing [30]. This test assesses whether there would be significant differences in the parameters among the $k+1$ sample groups divided by k threshold values. Second, the LR-test for the validity of threshold values, that is, using maximum likelihood estimation to test whether the estimated threshold value $\hat{\lambda}$ significantly matches the true value λ, with the null hypothesis $\text{H}_0 : \hat{\lambda} = \lambda$. LR-statistic is $\text{LR}_k(\lambda) = \left(S_k(\lambda) - \text{S}_k\left(\hat{\lambda}\right)\right)/\hat{\sigma}^2$. When $\text{LR}_k(\lambda) \leq -2\ln\left(1 - \sqrt{1-\alpha}\right)$, the null hypothesis cannot be rejected in a condition when α is the significance level for the test of validity.

2.3. Variables

2.3.1. Dependent Variable

Forest Ecological Function Index (Eco): Forest ecological function refers to the ecological environment and benefits provided by a forest ecosystem through its inherent ecological characteristics and processes [31–34]. It is conducive to human survival and progress and encompasses a wide range of aspects, including water conservation, soil and water preservation, climate regulation, water purification, and the protection of biodiversity [35]. We constructed Eco based on the existing practices [31–36], following the guidelines outlined in "Technical Regulations for National Forest Resource Continuous Inventory" (GB/T 38590-2020) [37]. The Eco subsystem includes eight components: forest biomass, naturalness, community structure, tree species composition, total vegetation cover, canopy closure, average tree height, and dead leaves thickness. Using the predefined weights, a comprehensive score for this subsystem was calculated. The specific steps are: (1) Each factor is assigned with different scores for the three types (I, II, and III) based on its relevance. These scores for each factor are denoted as X_i. (2) Weights (w_i) were determined based on their relative importance. (3) The forest ecological function score (s_F) for each site was calculated using the formula $s_F = 1/\sum_{i=1}^{8} X_i w_i$. (4) By weighting the site data according to the site area, the site-level data was transformed into the county-level data, resulting in the comprehensive Eco scores for each sample county in the YRB. For assessment factors of forest ecological function and their classification, please see Table S2 for details.

2.3.2. Threshold-Dependent Variable

Urbanization (Urb): Urbanization takes on the multidimensional form of socio-economic change, and a single dimension of the population urbanization is insufficient to reflect the complex characteristics of land use types, migration, industrial structure, residents' lifestyles, and ecological environments [23,38]. Additionally, different dimensions of urbanization representation have varying impacts on ecological efficiency [38]. Therefore, referring to the previous studies [38–42], we evaluated the comprehensive level of urbanization (Urb) in the YRB using indicators from four dimensions: land urbanization (URL), population urbanization (URP), industrial urbanization (URI), and spatial urbanization (URS). We determined the weights of each evaluation indicator using the entropy method. This method calculates the information entropy of the indicator items to determine their weights, making the weight determination more objective [43]. Specifically, we used the proportion of built-up area as a proxy variable for URL, the proportion of the urban population to the total resident population for URP, the proportion of value added of non-agricultural industries for URI, and the green coverage rate of built-up areas for URS.

2.3.3. Threshold Variables

The annual average precipitation (Pre) and the annual average temperature (Temp): Hydrothermal conditions exert a significant influence on both the urban and forest ecosystems. Particularly in the region located north of China's Qinling Mountains and Huai River, precipitation has consistently served as a crucial limiting factor for vegetation growth [44–46]. Indeed, climate intricately governs the distribution of forest tree species [47], influences forest productivity [48], shapes forest structure [49], modulates plant functional traits [50], and ultimately impacts the forest ecological functions and services [51,52]. In the previous studies, key explanatory variables related to human socio-economic activities, such as gross domestic product per capita (GDP) [17], degree of openness to foreign trade [53], and urbanization level [38] have been used as threshold variables. However, one of the main objectives of our study was to explore the changes in the relationship between urbanization and forest ecological function under different climatic conditions. Therefore, natural climatic factors were distinguished from the socio-economic activity factors, and hydrothermal conditions were used here as thresholds in our analyses.

2.3.4. Control Variables

Within the cross-sectional threshold model, we incorporated controls for pertinent variables that have impacted the forest ecological functions. These variables encompass natural factors, such as altitude, slope, and contiguity level, alongside socio-economic factors including the regional economic scale (GDP), economic density, population density, disposable income of urban residents, and disposable income of rural residents. Moreover, in the domain of human activity factors affecting forest ecological function, it would be crucial not to disregard the policy-related factors. Notably, following the implementation of China's Grain for Green Program, the rate of vegetation recovery has surpassed that of the pre-implementation period by over sixfold [20]. Consequently, our study, in alignment with the actual circumstances in the YRB, considered introducing the three policy-related virtual variables: reforestation intensity, grassland restoration intensity, and wetland restoration intensity. These variables are considered very instrumental in representing the policy impact of significant initiatives in different counties [54,55].

2.4. Data Sources and Processing

2.4.1. Data Sources

The data for all variables were obtained from the year 2018, which covers 448 counties in the YRB. The data sources with specific indicators are shown in Table 2.

Table 2. Overview of data sources.

Criteria	Sub-Criteria	Original Data Source
Natural indicators	Forest biomass; Naturalness; Community structure; Tree species composition; Total vegetation cover; Canopy closure; Average tree height; Litter layer thickness; Altitude; Slope; Contiguous area level; Land use types	The 9th National Forest Resources Inventory in China results, the original data were provided by the Planning Institute of the National Forestry and Grassland Administration.
	Climatic zones	Official website of each county's government.
	Annual average temperature; Annual average precipitation data	The "WorldClim 2.1 Climate Data (1970–2000)" published by the WorldClim database (https://www.worldclim.org, accessed on 29 September 2023).
Socio-economic indicators	Total area by county; Built-up area; Urban population; Year-end resident population; Value added of non-agricultural industry; Regional gross domestic product; Green coverage; Economic density; Disposable income of urban residents; Disposable income of rural residents	The CNKI Big Data Platform (https://data.cnki.net, accessed on 29 September 2023) and national economic and social development statistical bulletins of various counties.
	Reforestation intensity; Grassland restoration intensity; Wetland restoration intensity	China Forestry and Grassland Statistical Yearbook 2018 (http://www.forestry.gov.cn, accessed on 29 September 2023) and China Forestry Statistical Yearbook 2002–2017 (https://data.cnki.net, accessed on 29 September 2023).

2.4.2. Data Processing

For the natural indicators, the county-level data were obtained by applying the area-weighted coefficients based on the sample areas. In cases where certain counties had missing variable values, they were replaced with the averages of the same variable from other counties within the same city. For counties where replacement was not possible, the samples were excluded. After data processing, a total of 436 valid samples were retained, which accounts for 97.32% of the total sample size. Descriptive statistics for the relevant variables are presented in Table 3. Data processing and model analysis were conducted using Microsoft Excel 2010 and StataMP 17, while the visualization of map data was carried out using ArcGIS 10.2.

Table 3. Descriptive statistics of variables.

Variable	Definition	N	Mean	Std. Dev.	Min	Max
Eco	Forest ecological function index.	436	0.438	0.065	0.342	0.691
Urb	Comprehensive urbanization score.	436	0.108	0.158	0.010	1.826
Pre	The annual average precipitation (mm).	436	514.639	144.314	115.708	813.681
Temp	The annual average temperature (°C).	436	8.947	4.216	−4.557	15.140
Altitude	Average altitude by county (m). Here, altitude has been log-transformed.	436	6.637	1.283	0.898	8.434
Slope	Average slope by county (°).	436	10.886	8.099	0.000	37.022
GDP	Regional gross domestic product (10^8 CNY). Here, GDP has been log-transformed.	436	4.790	1.149	1.176	7.490
Economic density	The ratio of regional gross domestic product to the total area (10^8 CNY/km^2).	436	0.652	2.718	0.000	41.303

Table 3. *Cont.*

Variable	Definition	N	Mean	Std. Dev.	Min	Max
Population density	The ratio of the permanent resident population to the total area (10^4 person/km^2).	436	0.084	0.135	0.000	0.686
Disposable income of urban residents	(10^4 CNY)	436	3.052	0.572	0.956	4.594
Disposable income of rural residents	(10^4 CNY)	436	1.268	0.431	0.377	3.352
Contiguous area level	The size of the contiguous area, evaluated according to the forest cover type and takes values from 0 to 7.	436	4.313	1.171	1.000	6.779
Reforestation intensity	A summation dummy variable equals 1 when a county has implemented only one of the projects. including "natural forest resource protection", "grain for green", "construction of key protective forest systems in the Three-North region", "control of wind and sand sources in Beijing and Tianjin", and "construction of fast-growing and high-yield timber forest bases in key areas"; 0 when none of the above projects have been implemented; and the highest value is taken to be 5.	436	3.323	0.878	1.000	4.000
Grassland restoration intensity	A dummy variable equals 1 when a county has implemented the "grazing prohibition and grassland restoration project" and 0 otherwise.	436	0.312	0.464	0.000	1.000
Wetland restoration intensity	A summation dummy variable equals 1 when a county has implemented only one of the projects including the "converting croplands to wetlands" and "wetland protection and restoration"; 0 when none of the above projects have been implemented; and the highest value is taken to be 2.	436	0.789	0.750	0.000	2.000

2.5. Study Flowchart

Based on the above analyses, we drew a study flowchart (Figure 2).

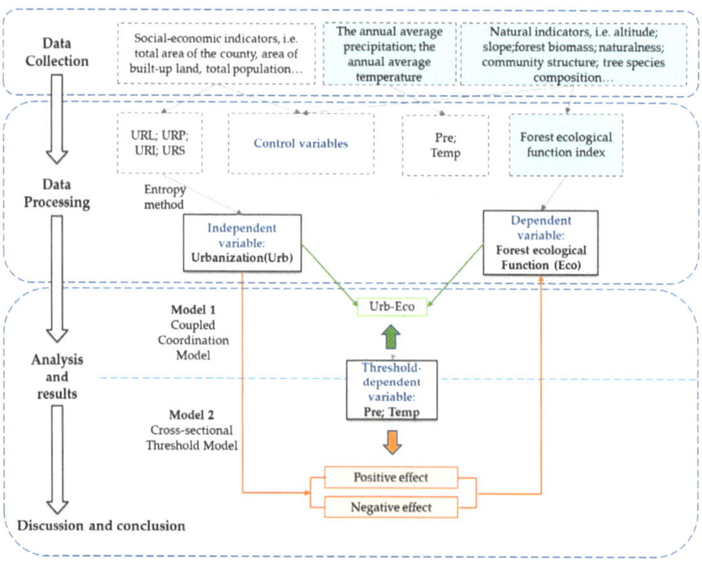

Figure 2. A study flowchart. Note: variables in blue boxes are site-level data and variables in white boxes are county-level data, which were weighted and aggregated based on the sample areas.

3. Results

3.1. Coupling and Coordination Results

3.1.1. Subsystem Scores

The scores of two subsystems, namely the comprehensive urbanization level (Urb) and the forest ecological function index (Eco), are shown in Figure 3.

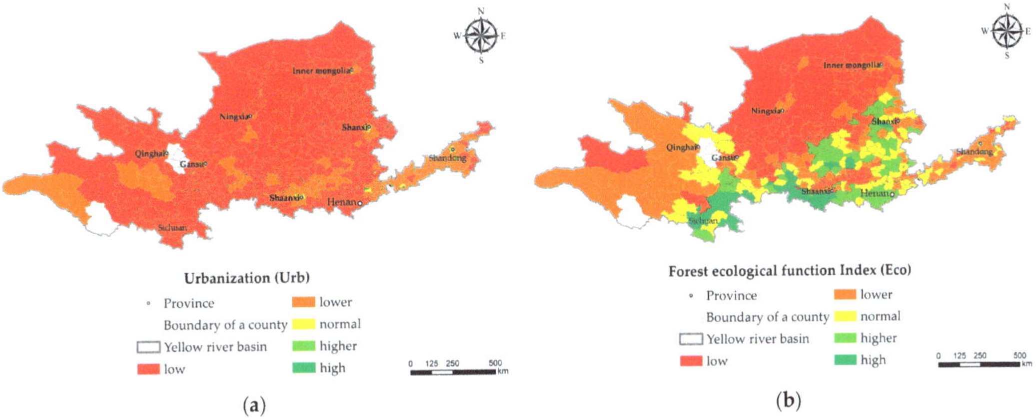

Figure 3. Subsystem scores in counties of the YRB calculated by coupled coordination model. (**a**) Subsystem scores for urbanization (Urb); (**b**) subsystem scores for forest ecological function (Eco). Note: each subsystem score is sliced into 5 parts: low, lower, normal, higher, and high, using the natural breakpoint method, respectively.

Figure 3a illustrates the comprehensive scores of the Urb subsystem calculated using the entropy method. In terms of regional distribution, Urb in the YRB shows relatively low overall spatial differentiation, with 75.51% of county-level samples below the average level. Counties with higher Urb scores were only sporadically distributed in the southern part of the YRB. In a county-by-county comparison, Beilin District in Shaanxi province had the highest Urb score, while Alashan Zuoqi in Inner Mongolia had the lowest Urb score.

Figure 3b shows that the forest ecological function scores in the YRB exhibit significant spatial variations. Taking the 400 mm annual precipitation line as a dividing criterion, a basic pattern of "higher in the southeast, lower in the northwest" was observed. Moving from the southern part of the YRB towards the surrounding areas, especially to the north, the forest ecological function gradually weakened. Counties with a "high" score of Eco are located in the southern part of the YRB, where annual precipitation exceeds 600 mm, including southern Shaanxi and Sichuan provinces, with Zhouzhi County in Shaanxi province having the highest Eco score. Counties with a "low" level of forest ecological function had a broader distribution, covering almost the entire YRB region in Inner Mongolia, with the poorest Eco found in Honggu District, Gansu Province. Counties with Eco scores between "high" and "low" show significant differences in the spatial distribution. The counties categorized as "normal" are scattered across various provinces and cities, while those classified as "lower" are mainly found in the upper and lower reaches of the YRB. Counties with a "higher" level are predominantly located in the central-right part of the YRB, primarily in Shaanxi and Shanxi provinces.

Arranging the average scores of each province and municipality, the Urb scores in the YRB can be ranked as follows: Henan > Shaanxi > Shandong > Overall (0.1093) > Qinghai > Gansu > Shanxi > Ningxia > Inner Mongolia > Sichuan. As for the Eco scores, the ranking is Sichuan > Shaanxi > Henan > Shanxi > Gansu > Overall (0.4383) > Qinghai > Shandong > Ningxia > Inner Mongolia. More than half of the provinces scored higher than the YRB's average Eco score. Sichuan province exhibited a significant spatial discrepancy between Urb and Eco: while its Eco score was high, its Urb score was relatively low.

3.1.2. Coupling Results

To further explore the relationship between urbanization and forest ecological function, the coupling degree and coupling coordination results of the two were calculated and presented (Figure 4).

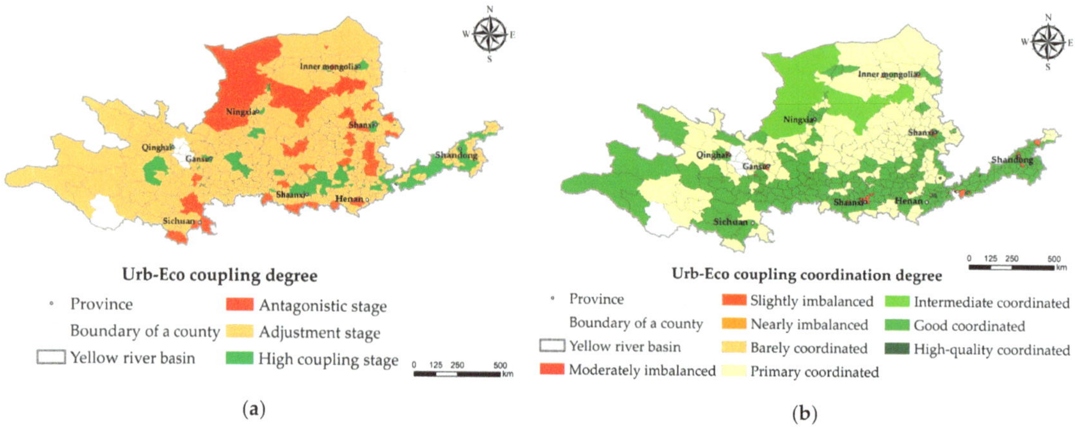

Figure 4. Coupling degree and coupling coordination degree of Urb-Eco system in the YRB calculated by coupled coordination model. (**a**) Urb-Eco coupling degree; (**b**) Urb-Eco coupling coordination degree.

Figure 4a shows that most counties are still in the "adjustment stage" of the Urb-Eco continuum. In counties located in the northern part of the YRB, which receives the least annual precipitation, the overall coupling degree was at a relatively low "antagonistic stage". This could be characterized by competition and resource competition between the two subsystems, with alternating dynamics. Counties in the "high coupling stage" are primarily situated in the regions with an annual average precipitation exceeding 600 mm and an annual average temperature above 10 °C. In these areas, there was a high level of interaction between urbanization and the forest ecological function. However, this proportion of counties was relatively low. While coupling degree is important for assessing the strength of coupling interactions and predicting the development order between the Urb and Eco subsystems, it might sometimes fail to reflect the overall functionality or development level of the systems. Therefore, it was necessary to construct a coupling coordination degree model. As evident from Figure 4b, counties with high-quality Urb-Eco coupling coordination are relatively scarce and dispersed, while the majority of counties are still in the "primary coordinated" stage, indicating that there is significant room for improvement in the overall coupling coordination of the system.

Scatter plots and linear regression lines illustrating the relationships between the coupling degree and coupling coordination of the two systems with changes in annual precipitation and annual average temperature are shown in Figure 5. The coordinated development level of urbanization and forest ecological function was positively correlated with both categories of climate factors, indicating that a warm and humid climate favors the coordinated development of urbanization and forest ecological function.

3.2. Threshold Model Results

3.2.1. Test of Threshold Effect Existence

Before identifying the threshold effects of the annual temperature and the annual precipitation in the impacts of urbanization on forest ecological function in the YRB, a threshold effect existence test was conducted and the results are presented in Table 4.

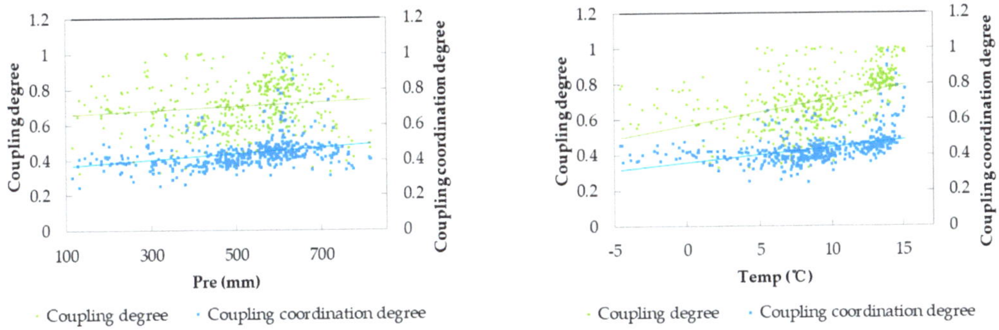

Figure 5. Scatter plot and fitting trend line of Urb-Eco coupling degree and coupling coordination degree with the annual average precipitation and the annual average temperature.

Table 4. Test results of threshold effect existence.

Threshold Variable	Threshold-Dependent Variable	Threshold Model Selection	F-Value	*p*-Value	Critical Value		
					1%	5%	10%
Pre	Urb	Single	26.613 ***	0.000	7.443	4.400	3.140
		Double	6.206 **	0.013	6.685	3.294	2.339
		Triple	5.671 **	0.047	9.914	5.591	3.745
Temp	Urb	Single	9.730 ***	0.002	7.515	4.193	2.892
		Double	5.578 **	0.023	8.654	4.339	3.017
		Triple	2.599	0.127	6.710	3.882	3.100

***, **, and * represent significance at the 1%, 5%, and 10% levels, respectively with 5% being the commonly used significance level. *p*-values and critical values were obtained by simulating 5000 iterations using the Bootstrap method [56].

When taking Pre as the threshold variable and Urb as the threshold-dependent variable, single threshold effects, double threshold effects, and triple threshold effects of Pre were all significant at the 5% confidence level. However, further assessment was needed in conjunction with a test of the authenticity of the threshold values under these three types of threshold models [27,30]. On the other hand, when taking Temp as the threshold variable and Urb as the threshold-dependent variable, the single threshold and double threshold models were significant at the 1% and 5% significance levels, respectively, while the triple threshold model was not significant. Therefore, the triple threshold effect of Temp was preliminarily ruled out.

3.2.2. Test of Threshold Value Authenticity

After detecting the existence of threshold effects, further estimation and validation of the threshold values were conducted. The estimated threshold values and their corresponding confidence intervals under the different threshold models are presented in Table 5. When "Pre" was used as the threshold variable, the estimated value for a single threshold was 532.342 mm, with a 95% confidence interval of [413.342, 559.629]. The estimated values for the second threshold (694.178 mm) and the third threshold (285.953 mm) were both outside of this interval, indicating significant differences between the second and third threshold values compared to the first single threshold value [29]. However, by examining the results of the LR test in Figure 6, the LR values were consistently below the critical threshold (Figure 6d), suggesting that the confidence interval for the third threshold is excessively wide and cannot effectively contribute to the threshold convergence. Therefore, it was eventually decided to adopt the double threshold model, that is, to take 532.342 mm and 694.178 mm as the threshold values of Pre.

Table 5. Results of authenticity test of threshold values.

Threshold Variable	Threshold-Dependent Variable	Threshold Model Selection	The Threshold Estimate	95% Confidence Interval
Pre	Urb	Single	532.342	[413.342, 559.629]
		Double	694.178	[184.199, 811.225]
			532.342	[413.293, 564.808]
		Triple	285.953	[184.199, 748.854]
Temp	Urb	Single	10.105	[0.238, 13.873]
		Double	9.191	[−2.012, 15.092]
			10.105	[9.335, 10.357]
		Triple	8.942	[−2.012, 14.570]

The threshold estimate is the value at which the likelihood ratio test statistic LR = 0.

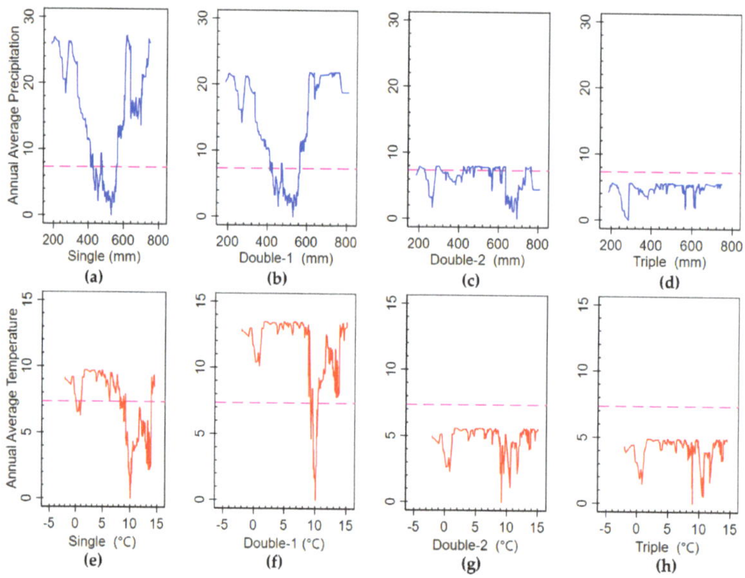

Figure 6. Confidence interval construction with each threshold model. (**a**) Likelihood ratio (LR) map corresponding to the single threshold estimate (532.342) for annual average precipitation (Pre). (**b**) LR map corresponding to the first threshold estimate (532.342) under the double threshold model for Pre. (**c**) LR map corresponding to the second threshold estimate (694.178) under the double threshold model for Pre. (**d**) LR map corresponding to the threshold estimate (285.953) under the triple threshold model for Pre. (**e**) LR map corresponding to the single threshold estimate (10.105) for annual average temperature (Temp). (**f**) LR map corresponding to the first threshold estimate (10.105) under the double threshold model for Temp. (**g**) LR map corresponding to the second threshold estimate (9.191) under the double threshold model for Temp. (**h**) LR map corresponding to the threshold estimate (8.942) under the triple threshold model for Temp. Note: in the figure, the dashed horizontal line (LR = 7.352) corresponds to a significance level of 0.05 [27].

Similarly, when "Temp" was considered as the threshold variable, the confidence interval for a single threshold was relatively narrower compared to the confidence intervals for the second and third thresholds, indicating more precise estimation results [28] (Figure 6). Although the double threshold model was significant at the 5% confidence level, the estimated threshold values were considered unreasonable for two main reasons: Firstly, both estimated threshold values of the double threshold fall within the confidence interval of the single threshold. Secondly, the confidence interval for the first threshold value (9.191 °C) of the double threshold model encompasses almost the entire sample space and exhibits mixed signs at the confidence interval boundary [27]. Since there is no evidence of a triple threshold effect, no further discussion is conducted regarding the third threshold estimation value.

3.2.3. Threshold Model Results

The estimation results of the double-threshold model for Pre and the single-threshold model for Temp are presented in Table 6. For the subset with precipitation as the threshold variable (Column 1), using two threshold values, 532.342 mm and 694.178 mm, we divided the samples into three subgroups: P_1, P_2, and P_3. The coefficient of "Urb" on "Eco" changed with an increase in Pre. Initially, it progressively reduced its negative impact, and then it shifted from negative to positive, indicating the ecological promotion effect of urbanization. Specifically, in P_1, which mainly includes the county-level samples from Qinghai, Inner Mongolia, and Ningxia, where there are high proportions of the primary industries, which has displaced the forest cover and ecological functions. Additionally, these areas experience the plateau continental climate, temperate continental climate, and temperate continental monsoon climate with distinct seasonal characteristics, including hot and rainy summers and cold, dry winters. Limited precipitation and ecological fragility restrict the growth and recovery of forest vegetation in P_1. Consequently, urbanization exerts the most pronounced stress on forest ecological function, with the largest absolute coefficient value of 0.197.

Table 6. Regression results of threshold model.

Variable	(1)	(2)
Slope	0.00469 *** (11.00)	0.00503 *** (11.72)
Altitude	−0.00718 ** (−2.03)	−0.00721 ** (−1.97)
GDP	−0.00644 ** (−2.02)	−0.00706 ** (−2.19)
Economic density	0.00169 (0.81)	0.00164 (0.77)
Population density	0.0936 *** (4.10)	0.0992 *** (4.24)
Disposable income of urban residents	−0.0155 *** (−2.95)	−0.0133 ** (−2.47)
Disposable income of rural residents	0.0142 * (1.73)	0.0121 (1.44)
Contiguous area level	0.00472 * (1.96)	0.00489 ** (1.98)
Reforestation intensity	0.00509 (1.31)	0.000283 (0.08)
Grassland restoration intensity	−0.0188 *** (−3.14)	−0.0206 *** (−3.38)
Wetland restoration intensity	0.00713 ** (2.06)	0.00891 ** (2.53)
Urb_1	P_1: −0.197 *** (−3.50)	T_1: −0.141 ** (−2.38)
Urb_2	P_2: 0.0134 (0.34)	T_2: 0.012 (0.30)
Urb_3	P_3: 0.306 ** (2.48)	—
Constant	0.453 *** (14.33)	0.463 *** (14.42)
R^2	0.435	0.405
Ajusted R^2	0.416	0.387
F	23.15	22.12
N	436	436

Standard deviations are provided in parentheses. ***, **, and * represent significance at the 1%, 5%, and 10% levels, with the 5% level being the commonly used significance level.

In P_2, which mainly covers counties in Shanxi, Shaanxi, Gansu, and Sichuan, the seasons are well-defined, with cold winters, hot summers, and often a rainy season in the Huang–Huai–Hai warm temperate zone, where temperatures exceed 30 °C in July and August. Here, the expansion of urbanization has led to a reversal in its impact on forest ecological function, resulting in a promotional effect with a coefficient of 0.0134 (not statistically significant).

In P_3, mainly comprising counties along the southern YRB, the expansion of urbanization enhances the promotion effect on forest ecological function, with a significant coefficient of 0.306. These areas experience a significant Northern Hemisphere subtropical monsoon climate, with high-pressure systems forming in the interior of northwest Asia during the winter, leading to prolonged cold north winds. The favorable climatic conditions endow these areas with the capacity for forest growth and self-repair. In these regions, the progress of urbanization contributes to the improvement of forest ecological function, indicating a higher degree of coordinated development between the two.

Similarly, when taking "Temp" as the threshold variable (Table 6, Column 2), the coefficients of "Urb" on "Eco" shifted from negative to positive as temperature increased. However, the absolute value of these coefficients did not exhibit a clear trend. With a single threshold value of 10.105 °C, within the resulting subintervals, T_1 and T_2, the coefficients were significant at −0.141 and not significant at 0.012, respectively. This suggests that in counties with higher annual average temperatures, factors other than urbanization may play a more significant role in influencing forest ecological function. The direction of the control variable coefficients aligns with the results in Table 6 (Column 1).

The coordinated development of urbanization and forest ecological function depended not only on the climate conditions of a county, but also on various natural factors, such as slope, elevation, and contiguous forest area, as well as human factors like economic development, population density, household income, and ecological engineering. Grassland restoration has squeezed the space for forest growth and inhibited the enhancement of forest ecological function. Conversely, increased afforestation and wetland restoration efforts have had a positive impact on forest ecological function.

4. Discussion

4.1. The Role of Climate in Shaping Urb-Eco Relationship

Coupling coordination is the result of the comprehensive scores of subsystems and the comprehensive effects of coupling degree [57]. Despite the urbanization development in the YRB promoting economic growth and improving people's living standards, it has also caused substantial harm to the forest ecosystem, leading to the relatively poor overall coordination between the two. The positive correlation between Urb and Eco coordination and both categories of climate factors suggests that a warm and humid climate is conducive to the harmonious development of urbanization and forest ecological function.

On a deeper dive, the explanatory role of climate in the relationship between urbanization and forest ecological function was examined. There are two structural change points for the threshold variable "Pre" in the context of Urb, located at 532.342 mm and 694.178 mm, respectively, which are reasonable, because, according to China's precipitation classification, these thresholds fall within the semi-humid range, leading to the division into 3 distinct zones: P_1, P_2, and P_3. Additionally, when considering "Temp" as the threshold variable for Urb, there exists only a single threshold value at 10.105 °C, resulting in categorization into two zones: T_1 and T_2. To be specific, firstly, the direction of Urb's impact changes from negative to positive, suggesting that the detrimental effect of urbanization on forest ecological function tends to manifest in colder and drier counties, while its beneficial effect tends to manifest in warmer and wetter counties. Secondly, for counties with lower (P_1) and higher (P_3) humidity levels in the basin, urbanization has a greater impact on the forest ecological function. This pattern is not easily discernible when taking annual average temperature as the threshold variable since there is only a single threshold. Thirdly, a comparison of the threshold effects of the two climate factors, Temp and Pre, shows that Urb's impact is much more sensitive on Pre. This suggests that urbanization development is driving the transformation of agricultural land into urban land, the migration of surplus rural labor to the urban areas, industrial optimization and upgrading, as well as the expansion of green spaces. The release of ecological space signals a positive contribution to the enhancement of forest ecological function. The results indicate that climate explains the spatial differences in the impact of urbanization on forest ecological function in the YRB.

4.2. Mechanism

This study explored the intricate interplays between climate and nonlinear impacts of urbanization on the forest ecological function. The results show that climate plays a moderating role in the effects of urbanization on the forest ecological function, manifesting a trajectory that transitioned from constraint to facilitation. Specifically, in the process of urbanization affecting forest ecological function, a dual threshold effect emerged concerning annual average precipitation, while a single threshold effect was observed for annual average temperature. A climate threshold framework was subsequently established based on these results.

Favorable water and thermal conditions serve to mitigate the adverse consequences of urbanization on the forest ecological function. In regions with a more humid and warm climate, urbanization's influence on these functions shifts from constraining to facilitating, in stark contrast to the arid and cold climate regions. This transformation can be attributed to two positive influences on the forest ecological function: the regulatory impact of superior thermal and hydrological conditions, and the ameliorating effect of heightened urbanization. Specifically, warmer and wetter climate effects promote forest growth and ecosystem resilience, as substantiated by this study. It underscores the role of hydrothermal conditions in shaping the extent to which urbanization impacts forest ecological function. Areas with favorable precipitation conditions generally exhibit superior vegetation growth dynamics and restoration capacity compared to areas with insufficient precipitation [58]. Over shorter timeframes, seasonally arid forests exhibit lower resilience than their humid tropical counterparts [59]. Yao et al. (2021) explored the agro-pastoral zones experiencing significant land use changes, whereby ecosystem resilience was significantly enhanced in wetter environments but weakened in drier ones [60]. Moreover, climate influences the forest ecological function by affecting forest species distribution [47], forest structure [49], forest productivity [48], plant functional traits [50], and forest ecological services [51,52]. As forests progressively assume the role of ecological "reservoirs" under climate regulation, their capacity to self-repair and accommodate human-generated emissions and pollutants increases [61]. This aids in mitigating the adverse impacts of urbanization on forests.

On the flip side, urbanization contributes to the enhancement of forest ecological function. Following the classic Environmental Kuznets Curve theory, there exists an "inverted U-shaped" relationship between economic development and environmental quality. Once a certain threshold is surpassed, environmental quality improves as economic development levels rise due to the predominance of positive effects over negative ones [62]. This is exemplified by the "promotion effect" of economic development on the ecological environment [38]. Wen et al.'s (2018) research aligns with the outcomes of this study, demonstrating that precipitation plays a significant positive moderating role in the urbanization process affecting vegetation cover [16]. Urbanization accompanies the influx of numerous production factors, leading to increased production efficiency and the transformation of various forest resource types [4], thereby augmenting centralized environmental management [7]. Furthermore, as urbanization progresses, both the government and public awareness of environmental protection intensify, benefiting the enhancement of forest ecological environments and functions. Li et al. (2021) reported that the implementation of reforestation projects could diminish forest fragmentation [17]. Ehrhardt-Martines et al. (2002) also noted a decreasing rate of natural resource consumption, including forest land, due to urbanization [12].

5. Strengths and Limitations

5.1. Strength of the Study

This study offers a proactive roadmap for addressing the challenges posed by global climate change and achieving a harmonious balance between urban development and environmental conservation. It serves as an innovative reference for the development of sustainable and climate-resilient urbanization policies. The main differences between our study and previous studies are as follows.

Firstly, our study introduces a climate perspective to investigate how climate affects the relationship between urbanization and forest ecological function. Previous studies often used urbanization itself as a threshold variable to explore the nonlinear impact of urbanization on the ecological environment under the different urbanization gradients [63–66]. In our study, climate factors serve as threshold variables, with urbanization as the threshold-dependent variable, allowing for an examination of the potential moderating effect of climate on the nonlinear impact of urbanization on the forest ecological function.

Secondly, previous studies predominantly relied on the urban population proportion (Urb) as a representation of urbanization indicators, often overlooking a comprehensive assessment of urbanization levels [43]. In contrast, our study not only considered the transition from agricultural to urban populations, but also incorporated various aspects of urban development, including changes in land use, adjustments in industrial structure, and modifications in urban green coverage. Consequently, it introduced variables, such as land urbanization, social urbanization, spatial urbanization, and ecological urbanization to construct a comprehensive urbanization index. On the other hand, compared to using forest coverage as a single proxy indicator of forest quality, the forest ecological index can more accurately and comprehensively reflect the structure, ecological functions, and overall efficiency of the forest [67,68]. Therefore, our study adopts a more comprehensive forest ecological functional indicator.

Thirdly, our study focuses on the county-level scale, as this is a crucial governance unit that bridges urbanization and ecological civilization construction. However, research at the county level and even smaller scales has been limited due to a lack of ground-based experiments [14]. Through county-level research, we have established a climate threshold system that can characterize how urbanization affects forest ecological function, and our results will be instrumental in offering decision support to the management authorities.

5.2. *Limitation of the Study*

Firstly, this study employs cross-sectional data and does not delve into the dynamic changes of variables. Forest ecological function is influenced by both natural and human factors. Natural factors exhibit long-term stability, whereas human activities, such as socio-economic factors, are more dynamic and constitute essential components of current research on the vegetation cover impacts [69]. Thus, this study considers it feasible to investigate climate distribution differences within the same year. Secondly, this study compensates for missing data by incorporating data from other counties within the same city, which might have introduced the disparities between the estimated values and actual values. In future research, the fuzzy numbers can be employed to represent the uncertain data [70], facilitating further exploration of the impacts of different types of urbanization on the forest ecological function. Thirdly, we have investigated only two primary climatic factors, temperature, and precipitation. Thus, future research needs to include other climatic factors. Fourthly, in exploring the impact of urbanization on the forest ecological function under the climate context, we chose to utilize data at an annual scale. In future studies, the consideration of data at a finer scale, such as quarterly or monthly intervals, can allow for a more precise and comprehensive examination of the effects of climate-shaped urbanization on the forest ecological function. Lastly, the threshold values of temperature and precipitation were not considered as fixed as they are subject to variations depending on the study year, study area, and selection of indicators. Within each climate gradient, the impact coefficient of urbanization on forest ecological function will also change.

6. Conclusions and Recommendations

This study, based on the county-level data acquired from the Yellow River Basin in China, examined the shaping role of climate factors in the relationship between urbanization and forest ecological function. With this motivation, the research unfolds along two lines. The first line of inquiry employs a coupled coordination modeling to qualitatively assess the coupling and coordination between urbanization and forest ecological function, as well as their connection to climate factors, temperature, and precipitation. The second line of inquiry utilizes a threshold regression model that estimates how the impact coefficient of urbanization on forest ecological function varies under the different precipitation and temperature gradients. This approach aims at constructing a climate threshold system that elucidates the impact of urbanization development on the forest ecological function.

The major conclusions are: (1) Both the systematic coupling degree and the coupling coordination degree between urbanization and forest ecological function exhibit positive correlations with two climatic factors: annual precipitation and annual temperature. This suggests that a warm and humid climate facilitates the coordinated development of urbanization and forest ecological function. (2) The spatial differentiation in the impact of urbanization on the forest ecological function can be explained by climate. Urbanization tends to exert a stress effect on the forest ecological function in colder and drier counties. Conversely, in warmer and more humid counties, the dominant role played by the "reservoir" effect of forest ecological regulation, which mitigates the stress effects of urbanization. Consequently, the promotional effect of urbanization becomes apparent. In counties with the lowest and highest humidity levels, urbanization has a greater impact on the forest ecological function. Since there is only a single threshold, this pattern is difficult to capture when annual temperature is used as the threshold variable. The impact coefficient of urbanization on the forest ecology is more sensitive to precipitation than temperature.

This research provides a fresh perspective and theoretical framework for tailoring region-specific strategies in managing the pace of urbanization, bolstering early warning mechanisms for forest ecological function, and crafting conservation strategies. These findings yield two notable policy recommendations. Firstly, by categorizing urban planning based on the precipitation and temperature gradients, it becomes feasible to effectively oversee the urbanization development, ensuring the harmonious coexistence of urbanization and forest ecological function. In regions marked by conspicuous stress effects induced by urbanization, it becomes imperative to rigorously control the pace of development, enforce stringent regulations concerning land use transitions, mitigate the rapid expansion of urban populations, carefully reconfigure the industrial structure, and direct capital and technological investments towards urban core areas to enhance land use efficiency [71]. This approach effectively mitigates resource misallocation. In areas where the beneficial impacts of urbanization become discernible, urbanization can be incrementally advanced, but it should refrain from hastily expanding towns and launching construction initiatives [16]. Moreover, it is vital to fortify ecological safety measures and nurture ecosystem development [5]. Secondly, it is crucial to continue promoting urban forest development while maintaining a balance between socio-economic growth and forest ecosystem construction. This entails the establishment of an integrated urban–rural forest ecosystem [72]. Implementing afforestation and wetland restoration projects plays a pivotal role in enhancing forest ecological function. Consideration should be given to creating peri-urban ecological buffer zones, primarily composed of forests and wetlands, in suburban areas. These zones should serve as the central framework for a network of forests and water bodies, fully realizing their ecological potential.

Supplementary Materials: The following supporting information can be downloaded at: https://www.mdpi.com/article/10.3390/land12112047/s1, Table S1: cover area by each climate zone and its proportion in the Yellow River Basin; Table S2: forest ecological function subsystem indicators. References [37,67] are cited in the supplementary materials.

Author Contributions: Conceptualization, X.Z.; funding acquisition, X.Z.; writing—original draft, X.G.; writing—review and editing, G.W., R.P.S. and S.Z.; formal analysis, L.F. (Liyong Fu), X.G. and X.Z.; data curation, L.F. (Linyan Feng), Q.W., S.Z., Y.D. and X.G.; visualization, H.Z. All authors have read and agreed to the published version of the manuscript.

Funding: This research was funded by National Natural Science Foundation of China: "Evaluation of Site Quality and Productivity Estimation of Natural Forests Based on Forest Biomass" (No. 31971653); the China Scholarship Council: "Chinese Government Scholarship" (No. 202203270009); the Fundamental Research Funds of IFRIT: "Key Technologies for Accurate Extraction of Tree Growth Phenotypes" (No. CAFYBB2022ZB002); and Land Greening and Ecological Restoration Management Projects of National Forestry and Grassland Administration under Grant: "Vegetation Suitability Evaluation in the Yellow River Basin" (No. 213205).

Data Availability Statement: The data presented in this study are openly available at https://data.cnki.net; https://www.resdc.cn/Data.Search.aspx, accessed on 24 September 2023.

Acknowledgments: We are indebted to the anonymous reviewers and editor.

Conflicts of Interest: The authors declare no conflict of interest.

References

1. Luan, Z. Climate change and forest restoration and management in Northern China. *Inn. Mong. For. Investig. Des.* **2007**, *30*, 47–49+74.
2. IPCC. *Climate Change 2007: Summary for Policymakers*; Cambridge University Press: Cambridge, UK, 2007.
3. Yu, W.; Ren, T.; Ren, W.; Li, W. Forest fragmentation and its relationship with urban expansion in Guangdong-Hong Kong-Macao Great Bay Area, China. *Acta Ecol. Sin.* **2020**, *40*, 8474–8481.
4. Wang, L.; Duan, Y.; Zhou, J. Analysis on changes of forest carbon sequestration and oxygen production in Pearl River Delta based on remote sensing. *J. Guizhou Norm. Univ.* **2010**, *28*, 36–39.
5. Feng, Y.; Dong, C.; Bao, Q. Impact of urbanization on forest ecological security in China. *Acta Ecol. Sin.* **2022**, *42*, 2984–2994.
6. Qian, C.; Zhao, Y.; Li, M.; Li, T. Dynamic analyses of forest ecological functions based on fixed plot data of continuous forest resources inventories from 1979 to 2012 in Guangzhou. *J. Nanjing For. Univ. Nat. Sci. Ed.* **2022**, *46*, 205–212.
7. McGranahan, G.; Jacobi, P.; Songsore, J.; Surjadi, C.; Earthscan, M.K. The citizens at risk: From urban sanitation to sustainable cities. *Int. J. Sustain. High. Educ.* **2004**, *5*, 455–456.
8. Liu, S.; Wen, Y.; Cai, D.; Zhu, H.; Huang, X.; Li, X. Impacts of climate change on forests and adaptive multi-scales management: A review. *Guangxi Sci.* **2014**, *21*, 419–435.
9. Yang, Q.; Duan, X.; Jin, Z.; Wang, L. Spatial scenario simulation and landscape ecological effect based on construction land expansion of Nantong city. *Sci. Geogr. Sin.* **2017**, *37*, 528–536.
10. Deng, J.; Yang, Y.; Lou, J.; Liang, C.; Wen, H.; Jing, L.; Wang, W.; He, X. Physicochemical properties and water holding capacity of forest soil in Liaohe upstream. *Soils Crops* **2021**, *10*, 449–459.
11. Chen, D. The impact of urbanization on the evolution of industrial structure. *Financ. Econ.* **2017**, *11*, 65–77.
12. Ehrhardt-Martinez, K.; Crenshaw, E.M.; Jenkins, J.C. Deforestation and the environmental Kuznets Curve: A cross-national investigation of intervening mechanisms. *Soc. Sci. Q.* **2002**, *83*, 226–243. [CrossRef]
13. Song, Y.; Zhang, M. Public cognition and environmental governance: An analysis of the path to realize green development in China. *China Popul. Resour. Environ.* **2018**, *28*, 161–168.
14. Zhang, L.; Yang, L.; Zohner, C.M.; Crowther, T.W.; Li, M.; Shen, F.; Guo, M.; Qin, J.; Yao, L.; Zhou, C. Direct and indirect impacts of urbanization on vegetation growth across the world's cities. *Sci. Adv.* **2022**, *8*, eabo0095. [CrossRef] [PubMed]
15. Su, H. Natural endowment and regional development:an interpretation of the widening gap of north-south economy. *China Soft Sci.* **2022**, *10*, 61–71.
16. Wen, X.; Yao, S.; Zhao, M. Coordinating the development of urbanization and vegetation coverage based on precipitation. *Prog. Geogr.* **2018**, *37*, 1352–1361.
17. Li, L.; Deng, D.; Zhang, D.; Yang, W. Analysis on socio-economic determinants of forest fragmentation in Beiing-Tianjin-Hebei region. *For. Econ.* **2021**, *43*, 5–16.
18. Chen, W.; Liang, J.; Dian, J.; Zeng, J.; Pan, S. Impact of landscape fragmentation on soil conservation services in the Yellow River Basin. *Sci. Geogr. Sin.* **2022**, *42*, 589–601.
19. Ren, T.; Chen, J.; Liu, N.; Fan, Q. Temporal and spatial variation characteristic and driving factors of seasonal vegetation coverage in the Yellow River Basin in recent 40 years. *China Rural. Water Hydropower* **2023**, *7*, 94–104.
20. Jian, S.; Mao, Z.; Wen, Y.; Yan, D.; Wang, J. Optimization of climate model for extreme climate change in the Yellow River Basin. *Yellow River* **2022**, *44*, 83–88.
21. Cao, Y.; Sun, Y.; Chen, Z.; Yan, H.; Qian, S. Dynamic changes of vegetation ecological quality in the Yellow River Basin and its response to extreme climate during 2000–2020. *Acta Ecol. Sin.* **2022**, *42*, 4524–4535.
22. Zhang, W.; Wang, L.; Xiang, F.; Qin, W.; Jiang, W. Vegetation dynamics and the relations with climate change at multiple time scales in the Yangtze River and Yellow River Basin, China. *Ecol. Indic.* **2020**, *110*, 105892. [CrossRef]
23. Wang, J.; Wang, S.; Li, S.; Feng, K. Coupling analysis of urbanization and energy-environment efficiency: Evidence from Guangdong province. *Appl. Energy* **2019**, *254*, 113650. [CrossRef]
24. Zhao, W.; Liu, L. Analysis on the coordinated development and obstacle factors of agricultural eco economic system in Shanxi province. *Chin. J. Agric. Resour. Reg. Plan.* **2023**, 1–14. Available online: http://kns.cnki.net/kcms/detail/11.3513.S.20230221.1828.002.html (accessed on 24 September 2023).
25. Yan, X.; Chen, C.; Yi, G.; Bai, J. Economic threshold effect of urbanization on energy consumption: Take the Yangtze River Economic Zone as an example. *Econ. Geogr.* **2019**, *39*, 73–81.
26. Li, Y.; Wang, K.; Huang, C.; Zhang, D. Threshold effect of government forestry investment on forest ecology: Taking the Yangtze River Economic Zone as an example. *Ecol. Econ.* **2021**, *37*, 119–127.
27. Yi, S.; Xiaoli, A. Application of threshold regression analysis to study the impact of regional technological innovation level on sustainable development. *Renew. Sustain. Energy Rev.* **2018**, *89*, 27–32. [CrossRef]

28. Li, J.; Luo, N. The double threshold effects of the urbanization process on urban land use efficiency. *Econ. Geogr.* **2015**, *35*, 156–162.
29. Lu, J.; Ren, E.; Guo, X.; Zhou, Z.; Wang, Y.; Zhang, N. The role of pet attachment in alleviating the negative effects of loneliness on a health-promoting lifestyle: An empirical study based on threshold effects for pet owners. *Int. J. Older People Nurs.* **2023**, *18*, e12554. [CrossRef]
30. Hansen, B.E. Threshold effects in non-dynamic panels: Estimation, testing, and inference. *J. Econ.* **1999**, *93*, 345–368. [CrossRef]
31. Yuan, Y.; Liu, Z.; Dong, L. Evaluation and analysis of forest ecological function level in Greater Khingan Mountains Pangu forest farm based on GIS. *J. Cent. South Univ. For. Technol.* **2016**, *36*, 108–114.
32. Ye, X. GIS spatial analysis of forest ecological function evaluation and its changes. *For. Prospect. Des.* **2022**, *42*, 21–25.
33. Wu, T.; Deng, X.; Huang, W.; Li, C.; Kang, W.; Cui, F.; Fang, X. Evaluation of forest ecosystem service value in Nan County. *J. Cent. South Univ. For. Technol.* **2015**, *35*, 109–115.
34. Zhang, Z.; Zhang, Z.; Yu, X.; Li, G. Spatial patterns and drivers of forest ecological functions in a typical soil-rock mountain area of northern China along an elevational gradient. *J. Clean. Prod.* **2022**, *361*, 132073. [CrossRef]
35. Fang, X. Evaluation of forest ecological function in Aojiang Watershed. *For. Prospect. Des.* **2022**, *1*, 36–39+61.
36. Ning, F.; Linyan, Y.; Dasheng, W.; Xinyu, Z.; Shimei, L. Assessment of forest ecological function levels based on multi-source data and machine learning. *Forests* **2023**, *14*, 1630.
37. National Public Service Platform for Standards Information. Technical regulations for continuous forest inventory (GB/T 38590-2020). Available online: https://std.samr.gov.cn/ (accessed on 24 September 2023).
38. Sun, S. *Research on the Impact of Urbanization on China's Ecological Efficiency*; Southeast University: Nanjing, China, 2020.
39. Liu, S.; Wei, C.; Guo, Y.; Zeng, X.; Liu, S. Exploring harmonious development between urbanization and eco-environment based on climate analysis-A study in Changsha, China. *J. Cent. South Univ. Technol.* **2011**, *18*, 101–107. [CrossRef]
40. Chen, M.; Lu, D.; Zhang, H. Comprehensive evaluation and the driving factors of China's urbanization. *Acta Geogr. Sin.* **2009**, *64*, 387–398.
41. Pan, A.; Liu, Y. The degree of imbalance between population urbanization and land urbanization of Xiangjiang River Basin. *Econ. Geogr.* **2014**, *34*, 63–68.
42. Lv, T.; Wu, C.; Li, H.; You, H.; Cai, X. The coordination and its optimization about population and land of urbanization: A case study of Nanchang City. *Sci. Geogr. Sin.* **2016**, *36*, 239–246.
43. Hou, M.; Deng, Y.; Yao, S. Coordinated relationship between urbanization and grain production in China: Degree measurement, spatial differentiation and its factors detection. *J. Clean. Prod.* **2022**, *331*, 129957. [CrossRef]
44. Xiao, Q.; Tao, J.; Xiao, Y. Dynamic vegetation cover change over the past 10 years on the Loess Plateau, China. *Acta Ecol. Sin.* **2016**, *36*, 7594–7602.
45. Yi, L.; Ren, Z.; Zhang, C.; Liu, W. Vegetation cover, climate and human activities on the Loess Plateau. *Resour. Sci.* **2014**, *36*, 166–174.
46. Zhang, C.; Ren, Z.; Li, X. Research on vegetation response to temperature and precipitation in Loess Plateau. *Sci. Agric. Sin.* **2012**, *45*, 4205–4215.
47. Davis, M.B.; Shaw, R.G. Range shifts and adaptive responses to Quaternary climate change. *Sci. Adv.* **2001**, *292*, 673–679. [CrossRef] [PubMed]
48. Boisvenue, C.; Running, S.W. Impacts of climate change on natural forest productivity–evidence since the middle of the 20th century. *Glob. Change Biol.* **2006**, *12*, 862–882. [CrossRef]
49. Sevegnani, L.; Uhlmann, A.; Gasper, A.L.d.; Meyer, L.; Vibrans, A.C. Climate affects the structure of mixed rain forest in southern sector of Atlantic domain in Brazil. *Acta Oecologica* **2016**, *77*, 109–117. [CrossRef]
50. Meng, T.; Ni, J.; Wang, G. Plant functional traits, environments and ecosystem functioning. *Chin. J. Plant Ecol.* **2007**, *31*, 150–165.
51. Ray, D.; Bathgate, S.; Moseley, D.; Taylor, P.; Nicoll, B.; Pizzirani, S.; Gardiner, B. Comparing the provision of ecosystem services in plantation forests under alternative climate change adaptation management options in Wales. *Reg. Environ. Change* **2015**, *15*, 1501–1513. [CrossRef]
52. Ding, H.; Chiabai, A.; Silvestri, S.; Nunes, P.A.L.D. Valuing climate change impacts on European forest ecosystems. *Ecosyst. Serv.* **2016**, *18*, 141–153. [CrossRef]
53. Wei, R. Thoughts on the promotion of green development in the process of urbanization. *Co-Oper. Econ. Sci.* **2020**, *24*, 16–18.
54. Wu, Q.; Zhu, J.; Zhao, X. Effects of human social-economic activities on vegetation suitability in the Yellow River Basin, China. *Forests* **2023**, *14*, 234. [CrossRef]
55. Zhang, S.; Gu, X.; Zhao, X.; Zhu, J.; Zhao, Y. Influences of climatic factors and human activities on Forest–Shrub–Grass suitability in the Yellow River Basin, China. *Forests* **2023**, *14*, 1198. [CrossRef]
56. Hansen, B.E. Inference when anuisance parameter is not identified under the null hypothesis. *Econometrica* **1996**, *1*, 413–430. [CrossRef]
57. Fan, Y.; Yuan, X. Synergy among "Five Constructions" in China from the perspective of ecological civilization. *J. Xi'an Jiaotong Univ.* **2017**, *37*, 19–27.
58. Liu, S.; Yan, D.; Shi, X.; Jie, Y. Inter-annual variability of vegetation NDVI, accumulated temperature and precipitation and their correlations in China. *Arid. Land Geogr.* **2014**, *37*, 480–489.
59. Poorter, L.; Bongers, F.; Aide, T.M.; Zambrano, A.M.A.; Balvanera, P.; Becknell, J.M.; Boukili, V.; Brancalion, P.H.S.; Broadbent, E.N.; Chazdon, R.L.; et al. Biomass resilience of Neotropical secondary forests. *Nature* **2016**, *530*, 211–214. [CrossRef] [PubMed]

60. Ying, Y.; Yanxu, L.; Yijia, W.; Bojie, F. Greater increases in China's dryland ecosystem vulnerability in drier conditions than in wetter conditions. *J. Environ. Manag.* **2021**, *291*, 112689.
61. Luo, W.; Wang, F.; Rong, M. Dynamic evolution of ecological-economic-social system coupling coordination in national key ecological function areas: Take the Altay region of Xinjiang as an example. *Acta Ecol. Sin.* **2022**, *42*, 4729–4741.
62. Grossman, G.K.A. Economic growth and the environment. *Q. J. Econ.* **1995**, *110*, 353–377. [CrossRef]
63. Saboori, B.; Sulaiman, J. Environmental degradation, economic growth and energy consumption: Evidence of the environmental Kuznets curve in Malaysia. *Energy Policy* **2013**, *60*, 892–905. [CrossRef]
64. Fan, P.; Ouyang, Z.; Dinh, N.D.; Nguyen, H.; Thuy, P.H.T.; Chen, J. Urbanization, economic development, environmental and social changes in transitional economies: Vietnam after Doimoi. *Landsc. Urban Plan.* **2019**, *187*, 145–155. [CrossRef]
65. Xia, Q.; Li, L.; Zhang, B.; Dong, J. Nonlinear influence of land-use transition on carbon emission transfer: A threshold regression analysis of the middle reaches of the Yangtze River in China. *Landsc. Urban Plan.* **2022**, *11*, 1531. [CrossRef]
66. Liu, M. *The Impact of Urbanization on Urban Eco-Efficiency in Yangtze River Economic Belt*; Jiangxi University of Finance and Economics: Nanchang, China, 2021.
67. Wang, B.; Zhang, W.; Yang, Z. Application of forest ecological indexes in forest resource planning, designing, and surveys. *East China For. Manag.* **2007**, 52–53+56.
68. Zhao, G. Application of forest ecological indexes in forest resource quality assessment. *For. Prospect. Des.* **2007**, *2*, 23–26.
69. Bai, W.; Zhao, S. A comprehensive description of the models of land use and land cover change study. *J. Nat. Resour.* **1997**, *12*, 169–175.
70. Yu, Q.; Sun, Z.; Shen, J.; Xu, X.; Han, Q.; Zhu, M. The nonlinear effect of new urbanization on water pollutant emissions: Empirical analysis based on the panel threshold model. *J. Environ. Manag.* **2023**, *345*, 118564. [CrossRef]
71. Geng, N. *Spatial Effect Analysis of the Influence of China's Urbanization on the Optimization of Forestry Industry Structure*; Northeast Forestry University: Harbin, China, 2019.
72. Peng, Z. Urban forest construction in china: A keynote report at the international seminar on urbanization and urban forest construction in China and Europe. *J. Chin. Urban For.* **2013**, *11*, 5–7.

land

Article

Spatio-Temporal Dynamics of Economic Density and Vegetation Cover in the Yellow River Basin: Unraveling Interconnections

Benxu Wang [1,†], Xuanqin Yang [1,†], Yaquan Dou [2], Qingjun Wu [3], Guangyu Wang [4], Ya Li [1] and Xiaodi Zhao [2,4,*]

[1] College of Economics and Management, Southwest Forestry University, Kunming 650224, China; wangbenxu@swfu.edu.cn (B.W.); xuanqin23@swfu.edu.cn (X.Y.); liy@swfu.edu.cn (Y.L.)
[2] Research Institute of Forestry Policy and Information, Chinese Academy of Forestry, Beijing 100091, China; douyq@caf.ac.cn
[3] College of Economics and Management, China Agricultural University, Beijing 100083, China; wqj2020@cau.edu.cn
[4] Faculty of Forestry, The University of British Columbia, Vancouver, BC V6T 1Z4, Canada; guangyu.wang@ubc.ca
* Correspondence: zhaoxiaodi@caf.ac.cn; Tel.: +86-136-7105-3260
† These authors contributed equally to this work.

Abstract: Vegetation, serving as the primary constituent of terrestrial ecosystems, plays a crucial role in regulating energy flow and material cycles and providing vital resources for human socio-economic activities. This study analyzes the spatio-temporal patterns of economic density and vegetation coverage in the Yellow River Basin (YRB) based on forest resource inventory and socio-economic data from 448 counties in 2008, 2013, and 2018. A three-tiered criterion layer is constructed using economic density as the core explanatory variable, encompassing social development factors, land use factors, and natural factors. A two-way fixed effects model is then utilized to analyze the impact of economic density on vegetation coverage. Results reveal that: (1) Spatially, economic density demonstrates a "low in the west and high in the east" pattern, with an overall upward trend in the YRB. Conversely, vegetation cover exhibits a "high in the west and low in the east" pattern, displaying a downward trend. (2) Over the 2008–2018 period, a significant negative correlation between economic density and vegetation cover is observed in each county of the YRB, with vegetation cover decreasing by 1.108% for every 1 unit increase in economic density. Notably, the upstream areas of the YRB experience a significant increase in vegetation coverage, while the middle and lower reaches witness a decrease. (3) Considering control variables, the proportion of the primary industry, urbanization rate, forest protection level, and cultivated land area exert a significant influence on vegetation coverage across the entire basin. Policymakers should formulate relevant policies to achieve sustainable development in the YRB, as discussed in the proposed countermeasures. This study delineates a practical pathway for high-quality economic development and high-level ecological protection in the YRB, offering a valuable reference for analogous research in other regions.

Keywords: economic density; vegetation cover; spatio-temporal pattern; two-way fixed effects model; the Yellow River Basin

1. Introduction

In the context of global warming and extreme climatic events, the loss of biodiversity is increasingly evident [1], making the study of vegetation cover changes a crucial topic in discussions about global ecosystem shifts. Forest ecosystems play a vital role in mitigating climate change, preserving biodiversity, and safeguarding water resources [2]. However, climate change can alter forest vegetation distribution, impact forest growth and development, and potentially lead to ecosystem instability and species loss [3]. Vegetation dynamics, heavily influenced by climate change, are crucial for forest ecosystems to effectively respond to human disturbances and the significant impacts of climate change. In recent decades, intensified human activities combined with ongoing climate change have led to rapid changes in vegetation. This is particularly evident in the Yellow River Basin (YRB), where vegetation restoration has become a critical aspect of ecological protection and socioeconomic development. The degradation and dysfunction of ecosystems, driven by human activities and climate change, pose significant threats to biodiversity, productivity, and the basin's ecological services, impacting regional socio-economic development and environmental protection [4].

As an important ecological shield and economic zone in China, the YRB plays an important role in China's economic and social development and ecological security. The YRB plays an important role in preventing wind, stabilizing sandy soil and maintaining ecological balance [5,6]. However, the fluctuations in vegetation cover within the YRB not only impact regional ecological equilibrium but also exert profound effects on climate and the environment in the basin and beyond. In addition to climate factors, human activities have become a major contributing factor to vegetation cover change [7]. The study revealed that climatic factors enhance vegetation suitability in the YRB, whereas the influence of human activities on vegetation suitability is dependent on specific circumstances [8]. Excessive human activities will interfere with vegetation suitability, such as over-irrigation and over-fertilization [9], economic density [10], and gross domestic product [11]. Scientifically driven vegetation restoration is crucial for promoting socioeconomic development in the basin to mitigate the YRB's ecological fragility.

The potential contributions of this paper are multifaceted. Firstly, historically, scholars have often concentrated on segmenting research areas, focusing primarily on specific river basins or sections. This research tended to explore the impact of natural factors or human activities on vegetation while overlooking the similarities and differences in economic density. Therefore, it is imperative to examine vegetation coverage while considering regional heterogeneity from an economic density perspective. This involves comparing and analyzing the economic development and resource endowment of different regions. Secondly, previous studies have predominantly taken a macro perspective when discussing the influence of socio-economic factors on vegetation coverage. Despite a general consensus within the academic community, further research is needed to delve deeper into the depth and specific mechanisms of this influence. Thirdly, in terms of index selection, many scholars have focused on factors such as economic development level, industrial structure, and economic development speed. However, there is limited research on the relationship between economic density and vegetation coverage. In this study, we integrate three primary indicators—social development factors, land use factors, and climate factors—along with nine secondary indicators, comprehensively considering various aspects. This approach minimizes the impact of external factors on experimental results.

The YRB encompasses diverse landscapes, ranging from densely populated urban areas to rural agricultural regions, spanning multiple provinces. Understanding the relationship between economic density and vegetation cover in this vital watershed is crucial for informing sustainable development strategies, environmental conservation efforts, and land use planning initiatives. This study utilizes data from the 7th to 9th National Forest Resource Inventory and ArcGIS (version 10.8) software to analyze the spatio-temporal patterns of economic density and vegetation cover changes across 448 counties in the YRB. We employ a two-way fixed effects model to investigate the impact of economic density on vegetation cover, explore underlying mechanisms, and identify spatial variations in these impacts. Based on the findings, actionable recommendations are proposed to enhance ecological protection and support the socioeconomic development of the YRB.

The structure of the paper is as follows: The second part comprises the literature review, followed by the third part detailing the materials and methods, including an introduction to the study area, data sources, the Fraction of Vegetation Cover (FVC) index system, ArcGIS spatial analysis, two-way fixed effects model, and variable selection. The fourth part presents the empirical results, followed by the discussion in the fifth part, and concluding remarks in the sixth part.

2. Literature Review

As the paramount socio-economic center in northern China and a traditionally delicate ecological zone, changes in vegetation coverage within the YRB have consistently captured the attention of scholars. However, previous studies have predominantly focused on specific river basins or sections. For instance, in the Gansu section of the YRB, vegetation cover has exhibited a continuous decline from the southwest to the northeast between 2000 and 2018 [12]. In the Henan section, areas with high and medium-high vegetation cover are predominantly concentrated in the western Funiu Mountain, northern Taihang Mountain, and eastern plain areas. Conversely, regions with medium, medium-low, and low vegetation cover are mainly found in the central Luoyang to Zhengzhou section, with bare land primarily distributed along the water system and both sides of the YRB [13]. The Inner Mongolia section, situated in an arid-semiarid region, relies on groundwater for vegetation, and increased rainfall stands as the primary driver for heightened vegetation cover [14]. In the Shanxi section, the most notable increase in vegetation cover occurs in the reforestation (grass) areas in the north of Shanxi [15]. Evidently, substantial variations in vegetation cover exist across different sections of the YRB.

In terms of influencing factors, climate change emerges as a crucial element affecting vegetation growth, while human activities also wield influence on changes in vegetation cover, often with predominant negative effects. However, there are periods where positive effects become more pronounced [16,17]. Specifically, in the middle section of the YRB, vegetation cover demonstrates a positive correlation with precipitation and temperature. The impact of topographic factors on vegetation cover is primarily manifested through elevation and slope. As elevation and slope increase, vegetation cover experiences an initial rise followed by a subsequent decline. Under diverse topographic conditions, vegetation cover appears to be more influenced by temperature than rainfall. From 2000 to 2002, the FVC of the urban agglomeration in central Yunnan showed an overall improvement trend, with an average value of 0.496 and an annual growth rate of 0.0024, showing a temporal and spatial distribution of high in the west and low in the east, while human activities had a positive impact on its FVC, accounting for 55.1% of the regional proportion [18]. Vegetation cover stands as a primary indicator of ecosystem alterations, and understanding vegetation dynamics and their responses to both natural and anthropogenic factors play a pivotal role in enhancing ecosystem structure and function amid the backdrop of global warming [19].

Regarding research methodologies, previous research has utilized various indicators to measure vegetation cover, encompassing FVC [20], Ecological Vulnerability Index (EVI) [21], MODIS NDVI vegetation index [22,23], Moran's index and coefficient of variation [24], and Ratio Vegetation Index (RVI) [25], among others. These indices provide a comprehensive framework for monitoring vegetation dynamics and offer critical insights for environmental management and conservation strategies. Various statistical approaches, such as the Dimidiate Pixel Model, linear regression, mean method, partial correlation analysis, and regression residual method, have been employed for spatio-temporal analyses of vegetation cover in the YRB [26–28]. Geographic detectors are commonly utilized to analyze the driving factors behind spatio-temporal changes in vegetation cover [29]. Existing studies have predominantly concentrated on vegetation cover changes within specific sections of the YRB, with limited research encompassing the entire basin. Therefore, the comparability of research conclusions necessitates further examination. Additionally, current studies on the influencing factors of vegetation cover tend to adopt a more macroscopic approach. While scholars generally agree on the impact of socio-economic factors on vegetation cover, further exploration is required to understand the extent and mechanisms of this impact.

Economic density refers to the concentration of labor, population, and tangible capital in a specific area, reflecting the efficient operation of economic activities and the intensive utilization of land resources. Various factors such as population, land, production, basic environmental conditions, transportation infrastructure, socio-economic infrastructure, industrialization level, structure, and regional economic development strategy and policies contribute to measuring the overall economic density level [30]. In the Yangtze River Delta, the quality of human capital has notably improved, attributed mainly to advancements in human capital levels, urbanization promotion, foreign direct investment influx, and economic scale expansion [31]. This has led to enhanced economic prosperity and ecological health in the YRB, signifying a high degree of coordination and balance between economic development and environmental protection [32]. The harmonious coexistence of economic development and ecological environment bodes well for the region's socioeconomic development. The ecological status of the YRB has shown a trend of initial decline followed by improvement, indicating a correlation between ecological environment enhancement and economic development. However, significant disparities exist in ecological coordination among different regions, with light intensity being linked to urban economic development levels, showing a consistent upward trajectory [33]. Furthermore, scholars have evaluated economic development and ecological situations, as well as the coupling degree, in the YRB based on economy, energy consumption, ecology, and water resources [34]. Given the critical ecological protection and socioeconomic development role of areas such as the YRB in China, economic development, albeit progressing towards in a good direction, advances at a relatively slow pace. Notably, the reduction in resource and environmental pressures suggests a relaxation of the coupling relationship between economic growth and resource utilization, with the environmental decoupling index significantly influencing the overall decoupling scenario [35]. In China, achieving the coordinated development of the industrial economy, resource conservation, and environmental protection has become imperative for promoting ecological civilization construction and socioeconomic development [36]. The eco-economic coordination model serves as an effective tool for evaluating the consistency between ecosystem value and economic growth, often measured by gross domestic product. From 2001 to 2015, the ecological carrying capacity of the YRB exhibited a unique changing trend, with the total ecological carrying capacity and ecological carrying capacity per unit area showing a "U" pattern. While about 55.6% of the region experienced an increase in ecosystem service value (ESV), ESV decline was mainly concentrated in the western and northern regions of the basin, illustrating the complex interplay between economic development and ecological protection in the YRB [37]. As an indicator of efficient economic activity and intensive land resource utilization, economic density comprehensively reflects various economic aspects through indicators such as labor concentration, population density, capital investment, land utilization rate, gross domestic product, and industrial structure. This provides a comprehensive and in-depth perspective for evaluating regional economic development. Given the YRB's significance as an ecological protection and socioeconomic development experimental area, special attention to the coordinated relationship between economic development and the ecological environment is necessary to achieve sustainable development.

In terms of economic density and vegetation coverage, the Vegetation Coverage Rate (VFC) stands as a critical global indicator used to assess land use changes, ecosystem health, and landscape characteristics. It quantifies the ratio of vegetation to land surface, offering insights into land use alterations, vegetation distribution shifts, and ecosystem functionality enhancements [38]. Economic density is intertwined with factors such as wages, innovation, rent, accessibility to amenities, public service provision costs, transportation, and environmental conditions [39]. Furthermore, economic density indirectly influences vegetation coverage trends through urbanization, vegetation types, gross domestic product, and population density. For instance, urban economic density serves as a robust measure of a city's economic scale and elucidates the drivers behind rapid urbanization efforts in the region. Against the backdrop of global urbanization, vegetation coverage in urban areas is undergoing transformations. While vegetation coverage is generally declining in the southern hemisphere, it is increasing in eastern North America and certain European urban locales. Moreover, numerous urban areas with high vegetation coverage are witnessing expansion [40]. Changes in vegetation coverage in the urban core and peripheral areas (buffer zones 5 km, 10 km, and 15 km away from the urban core) correlate with variations in urbanization rates, gross domestic product, population density, and urban development stages among cities [41]. Despite their advantageous geographical locations, urban expansions result in an overall loss of vegetation. However, San Diego's urban dynamic index has risen, albeit with only a slight increase in vegetation coverage in its urban core area [42]. In the Qaidam Basin, vegetation coverage varies across different vegetation types. The Vegetation Classification Inversion (VCI) can estimate the Fraction of Vegetation Cover (FVC). In areas with relatively high vegetation coverage density (FVC > 60%), the empirical model (EM) for mixed vegetation FVC estimation differs from that of mixed vegetation inversion (VMI). Pixel Decomposition Model (PDM)-derived VCI yields lower inversion values for very low-density vegetation cover (FVC $\leq$ 10%) and higher inversion values for high-density vegetation cover (FVC > 80%) [43]. Nevertheless, research on whether economic density directly impacts vegetation coverage remains nascent and necessitates further development. Through a thorough investigation of this relationship, we can better strike a balance between economic development and the ecological environment, ultimately achieving sustainable social and economic development.

In conclusion, previous studies predominantly focused on segmented research areas, particularly specific river basins or sections, often discussing the impact of natural factors or human activities on vegetation while neglecting differences in economic density. To comprehensively understand vegetation coverage, it is crucial to fully consider regional heterogeneity from the perspective of economic density and conduct a detailed analysis of the region's economic development and resource endowment. Moreover, although past studies have analyzed the impact of socio-economic factors on vegetation coverage from a macro level, further exploration is needed to delve into the deep-seated and specific mechanisms of this impact. Investigating the internal logic of this relationship can reveal its essence more comprehensively. Additionally, while many scholars focus on indicators such as economic development level, industrial structure, and economic development speed, research on economic density and vegetation coverage is relatively scarce. Hence, this paper integrates social development, land use, and climate and selects nine secondary indicators—including the proportion of primary and secondary industries—to enhance the comprehensiveness of the study and reduce the influence of external factors on experimental results. Furthermore, previous research data have been relatively macroscopic and rough, necessitating improved data accuracy. In this paper, is employed to ensure data accuracy based on forest resource inventory and socio-economic data from 448 counties in 2008, 2013, and 2018. Lastly, regarding the research method, the two-way fixed effects model adopted herein can simultaneously model causal and reverse relationships, providing a comprehensive perspective on the influence between variables. This model not only addresses endogeneity issues but also fully considers individual and time heterogeneity, thereby enhancing the accuracy and reliability ofl research results.

The effects of economic density on vegetation coverage are analyzed in this paper based on forest resource inventory and socio-economic data from the YRB in 2008, 2013, and 2018. ArcGIS software is employed, utilizing the existing literature to analyze the spatio-temporal patterns of economic density and vegetation cover changes in the YRB, with a two-way fixed effects model utilized to analyze the relationship between economic density and vegetation cover. Through result analysis and comparison with relevant scholars' research conclusions, suggestions for economic development and ecological protection in the YRB and other areas are proposed. The research hypotheses in this paper are as follows:

Hypothesis 1: *economic density in the YRB exhibits a negative correlation with vegetation cover.*

Hypothesis 2: *economic density varies in its impact on vegetation cover across different watersheds.*

The YRB spans a total length of 5464 km, with a water surface difference of 4480 m and a total basin area of 795,000 km^2. The geographical landform, climate, and ecological environment of the upper, middle, and lower reaches significantly differ, manifesting varying degrees of conflict between human activities and ecological preservation across different sections. In the upper reaches, characterized by steep terrain, soil erosion, and desertification pose significant challenges. Prolonged periods of excessive human activities, irrational agricultural practices, and overgrazing have resulted in extensive soil erosion and desert encroachment. Conversely, the downstream areas exhibit high levels of urbanization, with industrial and mining activities surrounding urban centers leading to significant river pollution. Rapid urbanization exacerbates environmental degradation, marked by land overexploitation and indiscriminate logging.

The YRB spans nine provinces in China, with downstream regions experiencing higher population densities. Increased human activities, particularly urbanization, contribute to forest degradation. This study considers economic density as the core explanatory variable to test the aforementioned hypotheses and incorporates nine control variables, such as social development, land use, and natural factors. The study utilizes ArcGIS software spatial analysis and a two-way fixed effects model to analyze forest resource inventory data from 2008, 2013, and 2018, exploring the influence of economic density and other variables on vegetation cover across different areas of the YRB.

3. Materials and Methods

3.1. Study Area

The YRB is situated between 96° E and 119° E, 32° N and 42° N [44]. By the end of 2021, the total population of the YRB will be 420.63 million, accounting for 29.14% of China, while Gross Domestic Product is USD 396634 $\times$ 10^4, accounting for 25.08%; the forest area is 7.32 $\times$ 10^7 hm^2, accounting for 31.72%; the forest stock is 4.60 $\times$ 10^8 m^3, accounting for 26.22%. The topography of the region exhibits higher elevations in the west, with the average elevation in the western source area surpassing 4000 m, and lower elevations in the east, ranging from 1000–2000 m in the central area [45]. The YRB, spanning 19% of China's land area, the YRB is divided into three basins: upstream, middle and downstream (refer to Figure 1), traverses nine provincial-level administrative regions: Qinghai, Sichuan, Gansu, Ningxia, Inner Mongolia, Shaanxi, Shanxi, Henan, and Shandong. Within the basin, there are 168 counties in the upper reaches, 219 in the middle reaches, and 61 in the lower reaches (refer to Appendix A). The basin exhibits three distinct climate zones: the plateau climate zone, the temperate zone, and the warm temperate zone [46]. Water resources within the basin are inherently scarce, characterized by an average annual precipitation of less than 450 mm and notable regional differences in average temperatures [47]. The vegetation coverage in the YRB generally exhibits a distribution feature with high coverage in the southeast and west and lower coverage in the northeast and north [48]. The primary land use types contributing to vegetation coverage include forest land, grassland, and cultivated land [49]. Population growth, land reclamation, deforestation, and urbanization have hastened the exploitation of land and water resources in the YRB, presenting significant challenges to ecosystem restoration in recent decades.

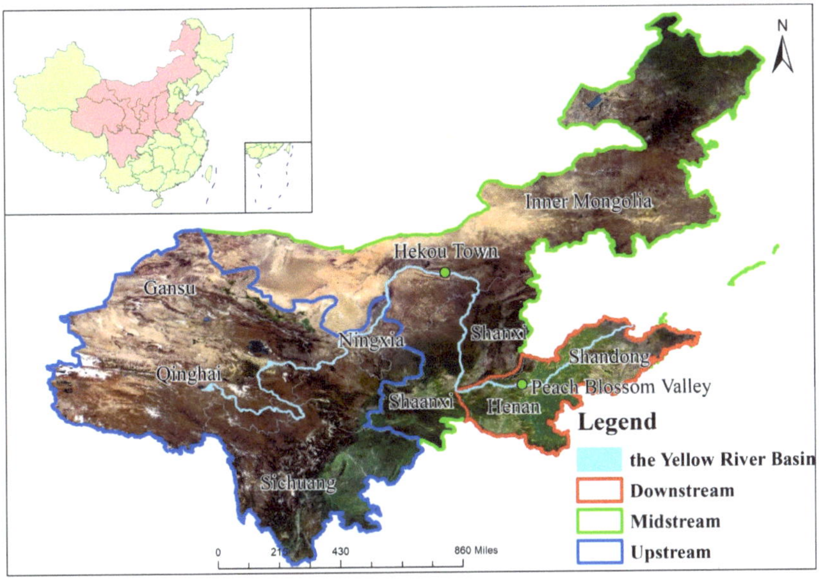

Figure 1. Map of the study area. In the regional map of China above, pink represents the YRB and green represents other regions. The boundary point between the middle and upper reaches of the YRB is Hekou Town, Toktuo County, Inner Mongolia Autonomous Region, and the boundary point between the middle and lower reaches of the YRB is Taohuayu, Xingyang County, Zhengzhou City, Henan Province.

3.2. Data Sources and Variable Selection

3.2.1. Data Sources

Vegetation cover: Vegetation cover data are sourced from the forest inventory data in periods 7–9 (2008, 2013, 2018). The vegetation cover level for each county is calculated utilizing the weighted average method based on the sample plot data.

Economic and social development: data from the China County Statistical Yearbook.

Land use: land use data is extracted from the datasets of 2008, 2013, and 2018, as released by the Resource and Environmental Science Data Center of the Chinese Academy of Sciences.

Climate: climate data are obtained from the World Climate Database (https://www.worldclim.org/, accessed on 12 January 2024).

3.2.2. Variable Selection

(1) Explained variable: vegetation cover. Vegetation cover, an essential indicator for gauging regional ecological changes, typically denotes the ratio of forest area to the total land area [50]. As a widely used indicator of vegetation response, it mirrors the growth status of surface vegetation and serves as a measure for regional ecological environment changes [51]. Thus, this study adopts the vegetation coverage of each county (district) as the dependent variable for the empirical model.

(2) Core explanatory variable: The correlation between socio-economic factors and vegetation cover or land use change in urban systems has garnered escalating interest [52]. Thus, delineating the relationship between socio-economic indicators and vegetation cover assumes pivotal significance in enhancing both socio-economic development and ecological environments. Economic density: This study investigates the influence of economic development levels on vegetation coverage, selecting economic density as the key measure based on previous research findings. "Economic density" refers to the measure of economic activity per unit area of land, often quantified as the output

value per square kilometer of land. It is typically calculated as the gross regional product divided by the administrative area (billion yuan/km^2). Beyond measuring the efficiency of economic activities and land use intensity in a city, economic density also represents the intensity of economic activities within a region [53]. Moreover, it serves as a crucial foundation for devising regional policies and evaluating implementation processes [54]. Consequently, this article designates the economic density of each county (district) as the core explanatory variable.

(3) Control variables: Previous studies indicate that, apart from economic density, numerous factors can influence the vegetation coverage of each county (district). This article establishes three primary indicators by building upon prior research conclusions [55]: social development factors, land use factors, and natural factors. Corresponding secondary indicators are constructed, encompassing the proportion of the primary industry, the proportion of the secondary industry, population density, urbanization rate, forest protection level, proportion of cultivated land area, proportion of construction land, annual precipitation, and monthly average temperature. Specific variable descriptions are outlined in Table 1.

Table 1. Description of variables.

Variable Types		Variable Names	Abbreviation	Description
Explained variable		Vegetation cover	vc	The ratio of forest area to total land area
Core explanatory variable		Economic density	eco	Gross regional product/area of the administrative region (billion yuan/km^2)
Control variables	Social development factors	The proportion of primary industry	agri	The ratio of the value added of primary industry to the gross regional domestic product of the county (%)
		The proportion of secondary industry	indus	The ratio of the value added of secondary industry to the gross regional domestic product of the county (%)
		Population density	ps	The ratio of permanent population to county area (thousands/km^2)
		Urbanization rate	urbi	Urbanization rate of permanent population (%)
		Forest protection level	protect	Based on the weighted average of "forest land protection level" in the sample land attribute table
	Land use factors	Proportion of cultivated land area	farmland	The ratio of cultivated land area to county area (%)
		Proportion of construction land	constru	The ratio of the construction area to county area (%)
	Natural factors	Annual precipitation	rainfall	Annual precipitation in each county and district (mm)
		Monthly average temperature	tempe	Annual average temperature in each county and district (°C)

3.3. *Methods*

3.3.1. ArcGIS Spatial Analysis

Spatial analysis involves the manipulation of geospatial information using geographic technology and associated mathematical models. It encompasses a set of functions that describe spatial correlations within and beyond a single dataset. In recent years, GIS technology has assumed an increasingly pivotal role in surveying and data processing. ArcGIS software, being a well-established GIS software, holds the highest market share and popularity in the industry [56]. It boasts robust tools and functionalities for map production and analysis. This study employs ArcGIS software to process forest resource inventory data from the 7th to 9th periods, extracting vector data. Subsequently, it conducts an analysis of the vector data pertaining to vegetation coverage and economic density in the YRB.

3.3.2. Two-Way Fixed Effects Model

This study employs a two-way fixed effects model for fundamental regression analysis to assess the extent of the influence of socio-economic activities on vegetation cover in the YRB [57]. The two-way fixed effects model is a statistical technique primarily utilized for panel data analysis. It excels in managing inter-regional heterogeneity that remains constant over time, effectively mitigating biased coefficient estimation arising from omitted variables.

The equation of the model is as follows:

$$Y_{it} = \beta_0 + \beta_1 eco_{it} + \beta_2 X_{it} + \mu_i + \sigma_t + \varepsilon_{it}$$

where Y_{it} denotes the vegetation cover of the 448 counties in the YRB, eco_{it} is the economic density of each county (district), and β_1 is the main coefficient under observation in this study, representing the coefficients reflecting the influence of socio-economic development on vegetation cover. X_{it} denotes the control variables, encompassing variables that may impact vegetation cover, excluding the level of socio-economic development. μ_i represents the individual effect, which remains constant over time, σ_t represents the time-fixed effect, unchanging based on individual differences between cities, and ε_{it} denotes the random interference term. This paper's data analysis is grounded in panel data, and hence, the two-way fixed effects model is utilized for analysis. This approach offers advantages such as improved accuracy in observations.

4. Results

4.1. Spatial-Temporal Evolution of Vegetation Cover and Economic Density

This study delves into the spatio-temporal evolution patterns of vegetation coverage and economic density in 2008, 2013, and 2018, as illustrated in Figures 2–7. Figures 2–4 depict the spatiotemporal evolution patterns of vegetation coverage in 2008, 2013, and 2018. Figures 5–7 depict the spatiotemporal evolution patterns of economic density in 2008, 2013, and 2018.

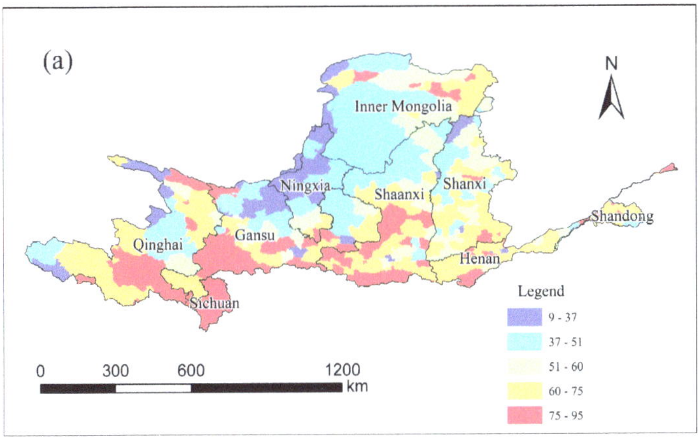

Figure 2. Spatio-temporal pattern of vegetation cover from 2008.

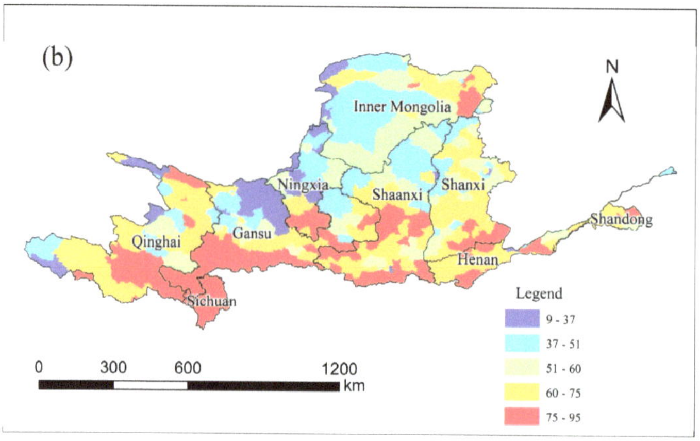

Figure 3. Spatio-temporal pattern of vegetation cover from 2013.

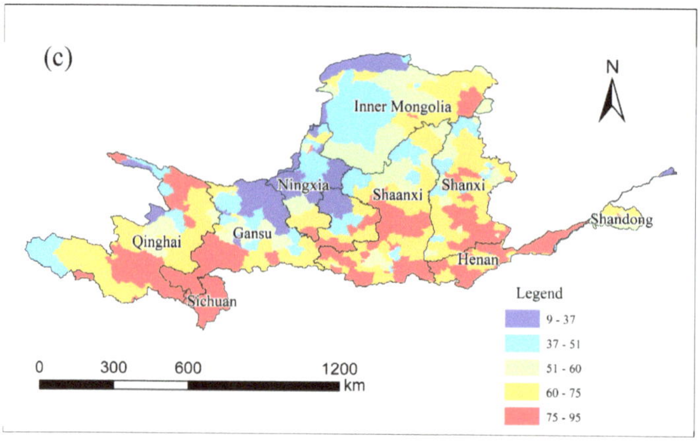

Figure 4. Spatio-temporal pattern of vegetation cover from 2018.

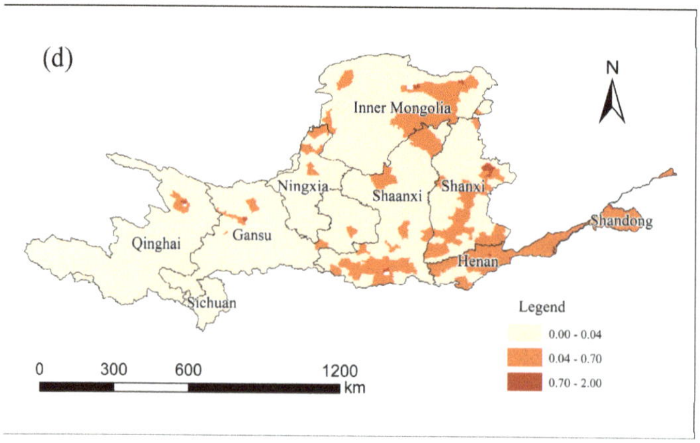

Figure 5. Spatio-temporal pattern of economic density from 2008.

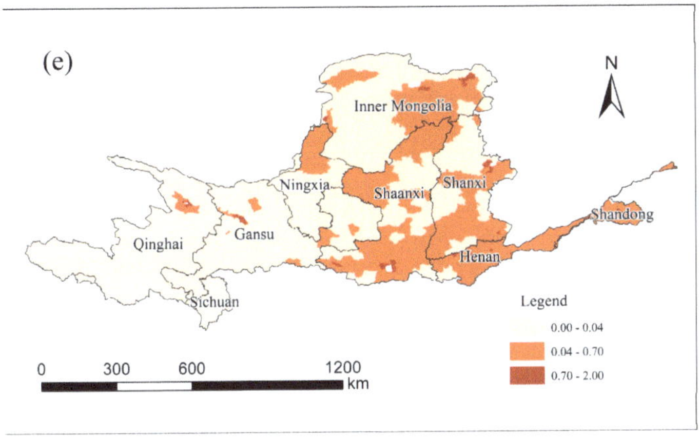

Figure 6. Spatio-temporal pattern of economic density from 2013.

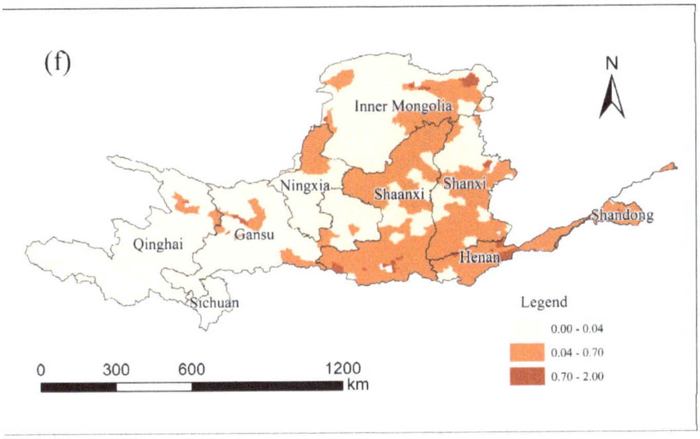

Figure 7. Spatio-temporal pattern of economic density from 2018.

Vegetation cover serves as a reliable measure of an area's ecological environment [58]. The analysis results reveal that in 2008, certain regions in the middle and upper reaches of the YRB exhibited low vegetation cover. However, as the government increasingly prioritizes ecological environment protection and various ecological projects are gradually implemented, there has been a significant improvement in vegetation cover in the middle and upper reaches of the YRB over the 10-year period from 2008 to 2018. Particularly noteworthy is the sustained high vegetation cover in certain counties and cities in Sichuan and Qinghai, with a more concentrated distribution of areas with elevated vegetation cover. In 2008, the vegetation cover in the lower reaches of the YRB was relatively robust, with some counties and cities in Henan Province exceeding 75%. However, over the past decade, the vegetation cover in the lower reaches of the YRB has experienced a noteworthy decline compared with the previous period. The number of areas with higher vegetation cover has become scarce in this region.

In terms of spatial distribution, the economic development in the middle and lower reaches of the YRB is more pronounced, particularly in Shandong and Henan, where economic density exhibits a substantial difference compared to the other seven provinces. Over the period of 2008 to 2018, the regions within the YRB witnessed rapid economic development, marked by a significant increase in economic density compared with the preceding period. In the middle reaches of the YRB, the economic level has experienced a substantial and notable increase in the past decade. The economic development of the lower reaches continues to maintain a high growth rate, with a more concentrated distribution of areas characterized by high economic density. However, the upper reaches of the YRB are developing at a slower pace, and the distribution of areas with high economic density is more dispersed.

As illustrated in Figures 2–7, there exists a correlation between the temporal and spatial changes in economic density and vegetation coverage. As economic density continues to rise across various regions of the YRB, vegetation coverage has, to some extent, experienced a decrease. The subsequent discussion employs a two-way fixed effects model to provide a more intuitive description of the relationship between economic density and vegetation coverage in diverse regions of the YRB.

4.2. Regression Results

In this section, the fixed effects model is applied to examine the forest resource inventory data in the YRB. Following a comprehensive analysis of the YRB, the study further divides it into three segments: upstream, midstream, and downstream. The outcomes of this analysis are presented in Table 2.

Table 2. Regression results.

Variable	The Entire Basin		Upper		Middle		Lower	
Variable	Coefficient	t	Coefficient	t	Coefficient	t	Coefficient	t
eco	−1.108 ***	−3.920	−0.579	−0.460	−1.103 ***	−3.580	−0.706	−0.340
agri	0.114 **	2.010	0.197 ***	3.600	0.019	0.150	−0.152	−0.440
indus	−0.019	−0.350	0.065	1.270	−0.092	−1.370	−0.004	−0.020
ps	−0.109	−1.120	0.023	0.040	−0.102	−1.020	0.081	0.020
urbi	0.095 *	1.710	−0.049	−0.610	0.145 **	2.080	−0.017	−0.050
protect	−2.165 *	−1.850	−0.053	−0.030	−3.364 *	−1.890	−0.340	−0.110
farmland	0.120 *	1.800	−0.133	−1.020	0.224 **	2.510	0.015	0.080
constru	0.171	1.600	−0.167	−1.090	0.133	0.740	0.689	1.570
rainfall	0.001	0.170	0.024 **	2.540	0.008	0.840	−0.015	−0.570
tempe	0.959	1.490	1.196	1.590	1.800 *	1.660	−1.270	−0.180

Note: ***, **, and * represent significant levels at 10%, 5%, and 1%, respectively.

Simultaneous with socio-economic development, it exerts a certain impact on vegetation coverage in various regions [59]. Owing to substantial differences in economic development levels and quality among counties in the YRB, the influence of economic density on vegetation coverage also varies. According to the results of the regression analysis covering the entire basin, the economic density of each county in the YRB exhibited a negative impact on vegetation coverage from 2008 to 2018. The correlation coefficient of economic density is −1.108, indicating a particularly significant impact on the dependent variable, vegetation coverage. Every increase of 1 km^2 in land output value corresponds to a 1.108% decrease in vegetation coverage. There was no significant correlation between economic density and vegetation coverage in the upper and lower reaches of the YRB from 2008 to 2018 after analyzing the watershed-specific regressions. However, a significant negative correlation was observed between economic density and vegetation coverage in the middle reaches of the YRB. For every 1 km^2 increase in land output value, vegetation coverage experiences a decrease of 1.103%. This suggests that the economic density in the middle reaches of the YRB has the most substantial impact on vegetation coverage, resulting in a negative correlation between economic density in the YRB and vegetation coverage.

Considering the control variables, it is evident that the proportion of the primary industry, urbanization rate, forest protection level, and cultivated land area in the entire YRB significantly impact vegetation coverage. Specifically, in the following:

Every 1% increase in the proportion of the primary industry results in a 0.114% increase in vegetation coverage; every 1% increase in the urbanization rate leads to a 0.095% increase in vegetation coverage; every 1 unit increase in the forest land protection level corresponds to a 2.165% decrease in vegetation coverage; every 1% increase in the proportion of arable land leads to a 0.120% increase in vegetation coverage.

Moreover, there is a notable correlation between the proportion of the primary industry, annual precipitation in the upper reaches of the YRB, and vegetation coverage. Specifically, in the following:

Every 1% increase in the proportion of the primary industry results in a 0.197% increase in vegetation coverage; every 1 mm increase in annual precipitation leads to a 0.024% increase in vegetation coverage.

In the middle reaches of the YRB, economic density, urbanization rate, forest protection level, and the proportion of cultivated land area also exhibit significant correlations with vegetation coverage. Specifically, in the following:

Every 1 km^2 increase in land output value leads to a 1.103% decrease in vegetation coverage; every 1% increase in the urbanization rate results in a 0.145% increase in vegetation coverage; every 1 unit increase in the forest land protection level corresponds to a decrease of 3.364% in vegetation coverage; every 1% increase in the proportion of arable land leads to a 0.224% increase in vegetation coverage.

4.3. Robust Test

This paper employs a variable replacement method to conduct a robustness test on the original basic regression model to enhance the robustness and credibility of the research findings. The core variable, economic density, is replaced by a comparable variable, GDP (the final outcome of production activities in the region over a certain period of time), for a two-way fixed effects model test. The regression results obtained are presented in Table 3.

Table 3. Robustness test results.

Variable	The Entire Basin		Midstream	
Variable	Coefficient	t	Coefficient	t
GDP	−0.009 ***	−2.690	−0.010 **	−2.140
agri	0.098 *	1.720	−0.003	−0.020
indus	−0.018	−0.420	−0.093	−1.370
ps	−0.128	−1.310	−0.113	−1.120
urbi	0.110 **	1.990	0.160 **	2.280
protect	−2.370 **	−2.020	−3.594 **	−2.000
farmland	0.103	1.520	0.197 **	2.190
constru	0.139	1.290	0.060	0.340
rainfall	0.000	0.000	0.008	0.820
tempe	0.982	1.510	2.004 **	1.810

Note: ***, **, and * represent significant levels at 10%, 5%, and 1%, respectively.

The results indicate that the correlation coefficient of the replaced core variable GDP is negative, and it has a significant impact on the explained variable, vegetation coverage. Therefore, the economic development level of the entire basin and the middle reaches of the region align with the basic regression results in terms of direction and significance.

5. Discussion

5.1. Analysis of Spatio-Temporal Pattern

5.1.1. Vegetation Coverage Trend

From 2008 to 2018, the upper reaches of the YRB experienced a significant increase in vegetation coverage, with notable improvements in regions like Ningxia Hui Autonomous Region and Gansu. However, some areas, such as Sichuan, Gansu, and Qinghai, maintain high vegetation coverage. The middle reaches of the YRB witnessed a significant decrease in vegetation coverage, particularly in areas like Shaanxi and Shanxi, which experienced a substantial decline compared with 2008. Although there was an increase from 2013 to 2018, it remained lower than 2008 levels. The lower reaches of the YRB saw a significant decrease in vegetation coverage, especially from 2008 to 2013. However, there was a certain degree of improvement in vegetation coverage by 2018. Some areas in Henan Province that had high vegetation coverage in 2008 experienced a significant decrease by 2018.

5.1.2. Economic Density Trend

From 2008 to 2018, economic development in the upper reaches of the YRB was relatively slow, significant changes occurred in the overall economic density pattern in the middle reaches, and rapid economic development was observed in the lower reaches, particularly in Shandong and Henan. The YRB has been an agricultural area for a long time in history, and the industrial development started slowly, which has a certain impact on the development of the social economy. The geographical location of this region is relatively harsh, natural disasters such as soil erosion are serious, and the ecological environment is fragile, which brings no small challenge to the development of an urban economy.

5.2. Analysis of the Relationship between Vegetation Coverage and Economic Density

Economic density growth has had a certain impact on vegetation coverage, but the decline in vegetation coverage is attributed to multifaceted factors such as climate change, human activities, and natural disasters [60]. Rapid social and economic development, coupled with human construction activities and the acceleration of industrialization, have led to vegetation destruction. Initiatives like the Grain Greening Project and the Natural Forest Protection Project in the middle and upper reaches of the YRB have promoted vegetation growth [61]. However, despite these efforts, overall vegetation coverage continues to exhibit a downward trend.

5.2.1. The Whole Basin of the YRB

Upon a comprehensive analysis of the YRB, it becomes evident that economic density exhibits a negative correlation with vegetation coverage. The intricate interplay of various factors contributes to the fluctuation in vegetation, with key influences being the proportion of primary industry, urbanization rate, and cultivated land area.

A notable insight arises from the regression findings, suggesting a seemingly negative association between forest protection levels and vegetation coverage. However, a more nuanced interpretation emerges when considering the classification criteria for forest resource inventory's continuous forest land protection levels. Under the classification standards of forest protection levels in forest resource inventory, there are four levels ranging from I to IV. Interestingly, lower forest protection levels correspond to higher vegetation coverage. The forest resource inventory process entails continuously addressing structural changes in demand for forest products and services, as well as ecological environment protection, in response to social, economic, and environmental developments. Starting from the growth, distribution patterns, and characteristics of forest resources themselves, combined with China's national conditions, forest conditions, and forest resource management characteristics, it comprehensively prices the current situation and changes in forest resources and provides corresponding technical drawings. Therefore, the higher the protection level, the more positively it influences vegetation coverage, aligning with on-ground protection practices. The positive correlation between forest land protection and vegetation coverage highlights the beneficial impact of conservation efforts. This study underscores a crucial point—economic development and ecological protection in the YRB can coexist harmoniously. The path forward involves prioritizing the improvement of industrial quality and ensuring that economic growth aligns with environmental preservation. The complex economic system, encompassing factors like economic growth, urbanization, agricultural activities, and commercialization, collectively shapes and influences vegetation coverage.

5.2.2. The Sub-Basin of the YRB

Considering the significant disparity in economic development levels among the upper, middle, and lower reaches of the YRB, industrial development priorities vary accordingly. Upon integrating the analysis of economic density's role, it becomes evident that the upper reaches of the YRB, characterized by relatively lagging economic development, seek to enhance their economic status by leveraging geographical advantages. Initiatives such as the establishment of the Qingtongxia Hydropower Station and the Laxiwa Hydropower Station have not only safeguarded the ecological integrity of the upper reaches but also spurred economic growth. Meanwhile, the middle reaches serve as a crucial energy hub and industrial center in China. However, the absence of incentives for traditional industrial transformation and upgrading has led to the proliferation of resource-intensive industries, resulting in environmental degradation and diminished vegetation cover. As the economy continues to evolve, downstream counties and cities witness escalating levels of economic prosperity. The Huanghuaihai Plain, occupying a substantial portion of arable land in the downstream region, offers opportunities for enhancing vegetation coverage through the development of a modernized agricultural economy.

The ecological sensitivity and strategic geographical location designate the upper reaches of the YRB as a key restricted development zone at the national level. Despite a relatively gradual pace of economic development, this region holds immense potential for large-scale forest nurturing [62]. Notably, the fluctuations in economic density have minimal repercussions on vegetation coverage in this area. Instead, the paramount influence emerges from natural environmental factors, with annual precipitation significantly shaping the vegetative landscape. While emphasizing the preservation of vegetation coverage, it remains imperative for the upper reaches of the YRB to address economic development. The resolute implementation of the Western Development Strategy [63] stands as a crucial directive. This strategy encompasses a comprehensive approach, aiming not only at ecological conservation but also at enhancing the economic landscape. It calls for concerted efforts to improve the income and welfare of residents in the upper reaches, addressing the challenge of economic backwardness. By striking a harmonious balance between ecological well-being and economic progress, the upper reaches can chart a sustainable and prosperous future.

The middle reaches of the YRB witnessed robust economic development during the period from 2008 to 2018, accompanied by significant environmental challenges, particularly evident in the Loess Plateau, where soil erosion became a prominent issue. The region, endowed with abundant coal resources, underwent extensive coal exploitation as part of its economic expansion. Unfortunately, this rapid development, coupled with an outdated technological infrastructure, contributed to a range of ecological problems, hastening the decline of the local ecological environment [64]. A notable consequence of this economic surge was the concurrent increase in urbanization rates. As urbanization progressed, there was a positive correlation observed with vegetation coverage—a phenomenon often driven by public awareness of the importance of ecological sustainability. The rise in wealth acted as a catalyst, intensifying the demand for both high-quality and substantial vegetation coverage [55]. Recognizing the ecological toll of economic activities, various regions in the middle reaches of the YRB have taken proactive measures to address environmental concerns. These initiatives include the implementation of ecological and environmental protection measures and a steadfast commitment to policies encouraging the conversion of farmland back to forests and grasslands. Through such concerted efforts, the aim is to mitigate the adverse impact of economic activities on vegetation coverage and restore the regional ecological equilibrium.

The lower reaches of the YRB, boasting an advantageous geographical location with proximity to the capital, Beijing, and encompassing coastal cities, laid a strong foundation for robust economic development [65]. In comparison to the upper and middle reaches of the YRB, the lower reaches exhibited a higher level of economic development, a superior ecological environment, and a more pronounced degree of ecological and economic coordination, albeit still below the national average [66]. The economic density in the lower reaches of the YRB demonstrated a minimal impact on vegetation coverage. However, the continuous march of industrialization underscores the need to guard against potential declines in vegetation coverage arising from ecological and economic imbalances. The heightened population density in the lower reaches amplifies the importance of judiciously utilizing limited land resources, presenting a pivotal factor in enhancing vegetation coverage. As the lower reaches navigate the intricate interplay between economic prosperity and environmental preservation, it becomes imperative to implement strategic measures that foster a harmonious coexistence. Addressing ecological and economic imbalances, especially in the face of increasing industrialization, stands as a key consideration for sustaining and enhancing vegetation coverage in this vital region.

5.3. Limitation

This study delineates a pragmatic trajectory for the high-level preservation of forest resources and the high-quality advancement of the social economy across the upper, middle, and lower reaches of the YRB, offering valuable insights for other developing nations and regions. Nonetheless, it is subject to certain limitations. Primarily, in terms of study scope, our analysis is grounded in socio-economic and vegetation coverage data from the YRB spanning 2008 to 2018, thus limiting its generalizability to other countries and regions. The YRB, characterized by ecological fragility, exhibits a weak resistance to interference, with human activities exerting a profound impact on its ecological landscape. The basin's unique geographical features, predominantly sloping hills and its passage through the Loess Plateau, render it susceptible to soil erosion and other natural calamities. In contrast, regions like the Yangtze River Basin boast high vegetation suitability owing to their favorable geographic positioning, with forest reserves accounting for 25% of the national total [67]. Consequently, the findings of our study are specific to the YRB. Future research endeavors will expand the scope to explore the nexus between China's national economic development and forest resource conservation. Secondly, our analysis may overlook variables, potentially omitting other factors influencing vegetation cover in the YRB. Subsequent research will comprehensively consider the interplay of climate change and economic development, which jointly impact forest vegetation cover.

6. Conclusions

6.1. Conclusions

Based on vector data, the spatial-temporal pattern of vegetation coverage and economic density in the YRB from 2008 to 2018 was analyzed using ArcGIS software. The results revealed that vegetation coverage displayed a pattern of high levels in the west and low levels in the east, with an overall declining trend across the YRB. Conversely, economic density exhibited a pattern of low levels in the west and high levels in the east, displaying an overall increasing trend. The outcomes of the two-way fixed effects model indicated a negative correlation between economic density and FVC in the YRB, particularly evident in the middle reaches of the YRB. Specifically, across the entire YRB, vegetation coverage decreased by 1.108% for every 1 unit increase in economic density, while in the middle reaches of the Yellow River, it decreased by 1.103% with a similar increase in economic density. The encroachment of large-scale industrial activities on forest land has adversely impacted vegetation coverage [68], suggesting that the development and utilization of natural resources may undermine vegetation coverage in the Yangtze River Delta amid economic growth. Notably, no significant correlation between economic density and vegetation coverage was observed in the upper and lower Yellow River regions during 2008–2018. Furthermore, factors such as forest protection level, urbanization rate, and proportion of primary industry influence vegetation coverage through various mechanisms, exacerbating spatial disparities in vegetation coverage within the YRB. Results indicate a positive impact of the primary industry on vegetation coverage in the YRB, attributed to advancements in agricultural modernization, adoption of water-saving irrigation technologies, and improved production conditions, all contributing to agricultural improvements and increased vegetation coverage. Additionally, higher forest protection levels in the YRB correspond to more pronounced improvements in vegetation coverage. Ecological restoration projects initiated by the Chinese government since the early 21st century have augmented vegetation coverage across parts or even the entire YRB [69]. Previous research on the dynamic nature of China's vegetation area underscores the significant influence of socio-economic factors on vegetation change, offering valuable insights for sustainable development strategies in the YRB and beyond [70].

6.2. Countermeasures

In conclusion, the YRB, endowed with abundant resources but facing ecological fragility, necessitates a transformative approach to economic development, prioritizing green practices and emphasizing responsible resource exploitation and utilization. Several key recommendations emerge.

Upstream: Environmental Conservation

Implement afforestation policies by closing mountain areas to human activities, allowing natural ecology to recover. Strengthen ecological protection in mountainous regions, focusing on safeguarding vegetation and water sources and preventing overgrazing and deforestation. Enhance the sustainable development of the primary industry, encouraging green agricultural practices to minimize damage to vegetation.

Midstream: Economic and Ecological Harmony

Recognize the negative correlation between vegetation coverage and economic density in the middle reaches. Reduce human activities based on the environmental carrying capacity of vegetation, continuing initiatives like returning farmland to forest and grassland for promoting vegetation restoration. Control the growth rate of economic density, implement rational land use planning, and mitigate industrial impacts on vegetation. Plan urban construction thoughtfully, prioritizing the protection of natural landscapes and ecological environments, strengthening forestland protection, and improving forestry management systems.

Downstream: Urban Development with Ecological Focus

Focus on ecological restoration projects in downstream areas, including wetland protection, riverbank greening and urban forest and green space development. Implement measures to control pollution and promote regional ecological protection and socio-economic development in human settlement areas. By incorporating these recommendations, the YRB can strike a balance between economic development and ecological preservation, fostering sustainability and resilience in the face of environmental challenges.

Themes such as "global forest coverage and changes", "global land degradation situation", and "global major natural disasters and their impacts" have become focal points in the ecological discourse, with the coordinated development of green ecology and the economy emerging as a shared objective for nations worldwide. Since the onset of the 21st century, there has been a slight decline in the global forest coverage area, prompting other countries to consider emulating China's green ecological policies. Firstly, countries should prioritize the protection and restoration of the ecological environment, striving to enhance and sustain forest vegetation coverage to contribute significantly to maintaining the global forest coverage equilibrium. Secondly, adopting China's ethos of "Lucid waters and lush mountains are Invaluable Asset" is essential, wherein economic development is pursued alongside the preservation of green ecology. Improving global forest vegetation coverage not only mitigates land degradation but also helps mitigate the impacts of climate change.

Author Contributions: Conceptualization, X.Z. and Y.D.; methodology, Q.W. and B.W.; validation, G.W.; formal analysis, X.Y. and B.W.; writing—original draft preparation, B.W. and X.Y.; writing—review and editing, Y.D., Q.W. and X.Z.; visualization, Y.L.; supervision, G.W.; funding acquisition, X.Z. All authors have read and agreed to the published version of the manuscript.

Funding: This research was funded by the Fundamental Research Funds of CAF, Theoretical Framework and Method for Accurate Estimation of the Full Life-Cycle Carbon Sink Potential in Masson Pine Artificial Monoculture Forests (CAFYBB2023MB019), China Scholarship Council Chinese Government Scholarship (202203270009), and the Fundamental Research Funds of CAF, Precision Estimation Study of Carbon Sink Potential in Deciduous Forests in the Yellow River Basin (CAFYBB2023PA006-02).

Informed Consent Statement: This article does not contain any studies with human participants or animals performed by any of the authors. Informed consent was obtained from all the individual participants included in the study.

Data Availability Statement: The data presented in this study are available upon request from the corresponding author.

Acknowledgments: Thank you to the anonymous reviewer and editor.

Conflicts of Interest: The authors declare no conflicts of interest.

Appendix A

Basin	Province	City	Country
Upper reaches	Qinghai	Xining	Chengdong, Central, Chengxi, Chengbei, Huangzhong, Datong Hui and Tu Autonomous, Huangyuan
		Haidong	Ledu, Ping An, Minhe Hui and Tu Autonomous, Huzhu Tu Autonomous, Hualong Hui Autonomous, Xunhua Salar Autonomous
		Haibei Tibetan Autonomous Prefecture	Menyuan Hui Autonomous, Qilian, Haiyan, Gangcha, Tongren, Jainca, Zeku, Henan Mongol Autonomous, Gonghe, Tongde, Guide, Xinghai, Guinan, Maqin, Banma, Gande, Dari, Jiuzhi, Maduo, Chengduo, Qumalai, Dulan, Tianjun
	Sichuan	Aba Tibetan and Qiang Autonomous Prefecture	Malcolm, Songpan, Aba, Ruoergai, Hongyuan
		Ganzi Tibetan Autonomous Prefecture	Shiqu
	Gansu	Lanzhou	Chengguan, Qilihe, Xigu, Anning, Honggu, Yongdeng, Gaolan, Yuzhong
		Baiyin	Baiyin, Pingchuan, Jingyuan, Huining, Jingtai
		Tianshui	Qinzhou, Maiji, Qingshui, Qin'an, Gangu, Wushan, Zhangjiachuan Hui Autonomous
		Wuwei	Gulang, Tianzhu Tibetan Autonomous
		ZhangYe	Shandan
		Pingliang	Kongtong, Jingchuan, Lingtai, Chongxin, Huating, Zhuanglang, Jingning
		Qingyang	Xifeng, Qingcheng, Huan, Huachi, Heshui, Zhengning, Ning, Zhenyuan
		Dingxi	Anding, Tongwei, Longxi, Weiyuan, Lintao, Zhang, Min
		Longnan	Dangchang, Li
		Linxia Hui Autonomous Prefecture	Linxia (a), Linxia (b), Kangle, Yongjing, Guanghe, Hezheng, Dongxiang Autonomous Condado, Jishishan Bao'an Dongxiang Salar Autonomous
		Gannan Tibetan Autonomous Prefecture	Hezuo, Lintan, Zhuoni, Diebu, Maqu, Luqu, Xiahe
	Ningxia Hui Autonomous Region	Yinchuan	Xingqing, Xixia, Jinfeng, Yongning, Helan, Lingwu
		Shizuishan	Dawukou, Huinong, Pingluo
		Wuzhong	Litong, Hongsibao, Yanchi, Tongxin, Qingtongxia
		Guyuan	Yuanzhou, Xiji, Longde, Jingyuan, Pengyang
		Zhongwei	Shapotou, Zhongning, Haiyuan

Basin	Province	City	Country
Upper reaches	Inner Mongolia Autonomous Region	Hohhot	Xincheng, Huimin, Yuquan, Saihan, Tumote Left Banne, Tokto, Helingeer, Qingshuihe, Wuchuan
		Baotou	Donghe, Kundulun, Qingshan, Shiguai, Bayan Obo, Jiuyuan, Tumed Right Banner, Guyang, Darhamo Ming'an United Banner
		Wuhai	Haibowan, Hainan, Wuda
		Ordos	Dongsheng, Kangbashi, Dalad Banner, Jungar Banner, Otog Front Banner, Etuoke Banner, Hangjin Banner, Wushen Banner, Ejin Horo Banner
		Bayan Nur	Linhe, Wuyuan, Dengkou, Urat Front Banner, Urad Middle Banner, Urat Rear Banner, Hanggin Rear Banner
		Ulanqab	Zhuozi, Liangcheng, Chahar Right Middle Banner, Siziwang Banner
		Alxa	Alxa Left Banner
Middle reaches	Shaanxi	Xi'an	Xincheng, Beilin, Lianhu, Baqiao, Weiyang, Yanta, Yanliang, Lintong, Chang'an, Gaoling, Huayi, Lantian, Zhouzhi
		Tongchuan	Wangyi, Yintai, Yaozhou, Yijun
		Baoji	Weibin, Jintai, Chencang, Fengxiang, Qishan, Fufeng, Mei, Long, Qianyang, Linyou, Feng, Taibai
		Xianyang	Qindu, Yangling, Weicheng, Sanyuan, Jingyang, Qian, Liquan, Yongshou, Changwu, Xunyi, Chunhua, Wugong, Xingping, Binzhou
		Weinan	Linwei, Huazhou, Tongguan, Dali, Heyang, Chengcheng, Pucheng, Baishui, Fuping, Hancheng, Huayin
		Yan'an	Baota, Ansai, Yanchang, Yanchuan, Zhidan, Wuqi, Ganquan, Fu, Luochuan, Yichuan, Huanglong, HuangLing, Zichang
		Hanzhong	Foping
		Yulin	Yuyang, Hengshan, Fugu, Jingbian, Dingbian, Suide, Mizhi, Jia, Wubao, Qingjian, Zizhou, Shenmu
		Ankang	Ningshan
		Shangluo	Shangzhou, Luonan, Danfeng, Zhashui
	Shanxi	Taiyuan	Xiaodian, Yingze, Xinghualing, Jiancaoping, Wanbailin, Jinyuan, Qingxu, Yangqu, Loufan, Gujiao
		Datong	Zuoyun
		Yangquan	urban area, mining area, suburbs, Pingding, Yu
		Changzhi	Shangdang, Tunliu, Zhangzi, Wuxiang, Qin, Qinyuan
		Jincheng	urban area, Qinshui, Yangcheng, Lingchuan, Zezhou, Gaoping
		Shuozhou	Shuocheng, Pinglu, Youyu
		Jinzhong	Yuci, Taigu, Yushe, Heshun, Xiyang, Shouyang, Qi, Pingyao, Lingshi, Jiexiu
		Yuncheng	Yanhu, Linyi, Wanrong, Wenxi, Jishan, Xinjiang, Jiangxian, Yuanqu, Xia, Pinglu, Ruicheng, Yongji, Hejin
		Xinzhou	Xinfu, Ningwu, Jingle, Shenchi, Wuzhai, Kelan, Hequ, Baode, Pianguan, Yuanping
		Linfen	Yaodu, Quwo, Yicheng, Xiangfen, Hongdong, Gu, Anze, Fushan, Ji, Xiangning, Daning, Xi, Yonghe, Pu, Fenxi, Houma, Huozhou
		lvliang	Lishi, Wenshui, Jiaocheng, Xing, Lin, Liulin, Shilou, Lan, Fangshan, Zhongyang, Jiaokou, Xiaoyi, Fenyang
	Henan	Zhengzhou	Jinshui, Shangjie, Huiji, Zhongmou, Gongyi, Xingyang, Xinmi, Dengfeng
		Luoyang	Old Town, Xigong, Chanhe Hui Autonomous Region, Jianxi, Jili, Luolong, Mengjin, Xin'an, Luanchuan, Song, Ruyang, Yiyang, Luoning, Yichuan, Yanshi
		Pingdingshan	Ruzhou
		Anyang	Hua, Neihuang
		Hebi	Xun
		Jiaozuo	Bo'ai, Wuzhi, Wen, Qinyang, Mengzhou
		Sanmenxia	Hubin, Shaanzhou, Mianchi, Lushi, Yima, Lingbao
		Nanyang	Xixia
		Jiyuan	Jiyuan

Basin	Province	City	Country
Lower reaches	Henan	Kaifeng	Longting, Shunhe Hui, Gulou, Yuwangtai, Xiangfu, Lankao
		Xinxiang	Hongqi, Xinxiang, Huojia, Yuanyang, Yanjin, Fengqiu, Weihui, Changyuan
		Puyang	Hualong, Fan, Taiqian, Puyang
	Shandong	Jinan	Lixia, Shizhong, Huaiyin, Tianqiao, Licheng, Changqing, Zhangqiu, Jiyang, Laiwu, Gangcheng, Pingyin
		Zibo	Zichuan, Boshan, Gaoqing, Yiyuan
		Dongying	Dongying, Hekou, Kenli, Lijin
		Jining	Wenshang, Sishui, Liangshan, Qufu
		Tai'an	Taishan, Daiyue, Ningyang, Dongping, Xintai, Feicheng
		Linyi	Pingyi, Mengyin
		Dezhou	Qihe
		Binzhou	Bincheng, Huimin, Boxing, Zouping
		Heze	Mudan, Yuncheng, Juancheng, Dongming

References

1. Johnson, C.N.; Balmford, A.; Brook, B.W.; Buettel, J.C.; Galetti, M.; Guangchun, L.; Wilmshurst, J.M. Biodiversity losses and conservation responses in the Anthropocene. *Science* **2017**, *356*, 270–275. [CrossRef] [PubMed]
2. Forzieri, G.; Dakos, V.; McDowell, N.G.; Ramdane, A.; Cescatti, A. Emerging signals of declining forest resilience under climate change. *Nature* **2022**, *608*, 534–539. [CrossRef] [PubMed]
3. Kijowska-Oberc, J.; Staszak, A.M.; Kamiński, J.; Ratajczak, E. Adaptation of Forest Trees to Rapidly Changing Climate. *Forests* **2020**, *11*, 123. [CrossRef]
4. Wang, H.; Zhou, S.; Li, X.; Liu, H.; Chi, D.; Xu, K. The influence of climate change and human activities on ecosystem service value. *Ecol. Eng.* **2016**, *87*, 224–239. [CrossRef]
5. Zhang, W.; Wang, Y.; Li, J.; Hao, Z. Coupling Coordination Network Analysis of Ecological Protection and High-quality Economic Development in the Yellow River Basin. *Ecol. Econ.* **2022**, *38*, 179–189.
6. Jin, F.; Ma, L.; Xu, D. Environmental Stress and Optimized Path of Industrial Development in the Yellow River Basin. *Resour.* **2020**, *42*, 127–136. [CrossRef]
7. Deng, C.; Bai, H.; Gao, S.; Liu, R.; Ma, X.; Huang, X.; Meng, Q. Spatial-temporal variation of the vegetation coverage in Qinling Mountains and its dual response to climate change and human activities. *J. Nat. Resour.* **2018**, *33*, 425–438.
8. Zhang, S.L.; Gu, X.; Zhao, X.B.; Zhu, J.F.; Zhao, Y.R. Influences of climatic factors and human activities on Forest–Shrub–Grass suitability in the Yellow River Basin, China. *Forests* **2023**, *14*, 1198. [CrossRef]
9. Wu, Q.J.; Zhu, J.F.; Zhao, X.D. Effects of human social-economic activities on vegetation suitability in the Yellow River Basin, China. *Forests* **2023**, *14*, 234. [CrossRef]
10. Zhang, Y.; He, Y.; Li, Y.; Jia, L. Spatiotemporal variation and driving forces of NDVI from 1982 to 2015 in the Qinba Mountains, China. *Environ. Sci. Pollut. Res.* **2022**, *29*, 52277–52288. [CrossRef]
11. Zhang, M.; Du, H.; Zhou, G.; Mao, F.; Li, X.; Zhou, L.; Zhu, D.; Xu, Y.; Huang, Z. Spatiotemporal Patterns and Driving Force of Urbanization and Its Impact on Urban Ecology. *Remote Sens.* **2022**, *14*, 1160. [CrossRef]
12. Lu, Y.; Zhang, Q.; Li, X.H.; Yang, S.X.; Yang, Q. Temporal and spatial variation of vegetation coverage and its response to climate factors in Gansu section of Yellow River Basin. *Bull. Soil. Water Conserv.* **2020**, *40*, 232–238.
13. Wang, L.; Li, N.; Wen, G.C.; Yang, Y.H. Vegetation coverage changes and driving forces in Henan section of the Yellow River basin. *Bull. Soil. Water Conserv.* **2022**, *42*, 393–399.
14. Zhang, H.Y.; Zhan, C.S.; Xia, J.; Hu, S.; Ning, L.K.; Deng, X.J. Spatio-temporal variations and strip pattrns of vegetation in the Inner Mongolia of Yellow River Basin. *Acta Ecol. Sin.* **2022**, *42*, 8818–8829.
15. He, H.J.; Wang, Z.; Dong, J.F.; Wang, J.; Zou, J.Y. Synergy and trade-off between vegetation change and urbanization development in the Yellow River Basin of Shanxi Province based on satellite remote sensing data. *Acta Ecol. Sin.* **2022**, *42*, 3536–3545.
16. Sheng, Y.; Liu, L.F.; Yuan, Y.; Wang, S.; Li, J.; An, L. Effects of climate change and human activities on vegetation coverage in arsenic sandstone area of Yellow River basin. *Bull. Soil. Water Conserv.* **2023**, *43*, 412–420.
17. Li, Y.J.; Chen, Q.C.; Fang, H.; Li, J. Spatial heterogeneity of vegetation evolution and its influencing factors in the Yangtze River Basin based on multi-scale geographical weighted regression. *China Environ. Sci.* **2024**, *44*, 352–362.
18. Wang, R.; Ding, X.; Yi, B.J.; Wang, J.L. Spatiotemporal characteristics of vegetation cover change in the Central Yunnan urban agglomeration from 2000 to 2020 based on Landsat data and its driving factors. *Geocarto. Int.* **2024**, *39*, 2316643. [CrossRef]
19. Zhang, Z.Q.; Liu, H.; Zuo, Q.T.; Yu, J.T.; Li, Y. Spatiotemporal change of fractional vegetation cover in the Yellow River Basin during 2000–2019. *Resour. Sci.* **2021**, *43*, 849–858.
20. Zhang, W.; Wang, L.; Xiang, F.; Qin, W.; Jing, W. Vegetation dynamics and the relations with climate change at multiple time scales in the Yangtze River and Yellow River Basin, China. *Ecol. Indic.* **2020**, *110*, 105892. [CrossRef]

21. Zhang, Q.; Wang, G.; Yuan, R.; Singh, V.P.; Wu, W.; Wang, D.Z. Dynamic responses of ecological vulnerability to land cover shifts over the Yellow River Basin, China. *Ecol. Indic.* **2022**, *144*, 109554. [CrossRef]
22. Sun, G.P.; Liu, X.F.; Wang, X.H.; Li, S.S. Changes in vegetation coverage and its influencing factors across the Yellow River Basin during 2001-2020. *J. Desert Res.* **2021**, *41*, 205–212.
23. Zhang, X.J.; Wang, G.Q.; Xue, B.L.; Ying, L.A. Changes in vegetation cover and its influencing factors in the inner Mongolia reach of the Yellow River basin from 2001 to 2018. *Environ. Res.* **2022**, *215*, 114253. [CrossRef] [PubMed]
24. Liu, C.X.; Zhang, X.D.; Wang, T.; Chen, G.Z.; Zhu, K.; Wang, Q.; Wang, J. Detection of vegetation coverage changes in the Yellow River Basin from 2003 to 2020. *Ecol. Indic.* **2022**, *138*, 108818. [CrossRef]
25. Chen, S.; He, L.; Yan, F. Spatio-temporal evolution of vegetation coverage in Beijing-Tianjin-Hebei and its response to natural anthropogenic changes. *China Environ. Sci.* **2024**, 1–18.
26. Zhang, J.; Du, J.Q.; Sheng, Z.L.; Zhang, Y.C.; Wu, J.H.; Liu, B. Spatio-temporal changes of vegetation cover and their influencing factors in the Yellow River Basin from 1982 to 2015. *Ecol. Environ.* **2021**, *30*, 929.
27. Wang, X.L.; Shi, S.H.; Chen, Z.X. Change and driving factors of vegetation coverage in the Yellow River Basin. *China Environ. Sci.* **2022**, *42*, 5358–5368.
28. Ciccone, A.; Hall, R.E. Productivity and density of economic activity. *The American Economic Review.* **1996**, *86*, 54–70.
29. Jie, L.I. A simulation approach to optimizing the vegetation covers under the water constraint in the Yellow River Basin. *Forest Policy Econ.* **2021**, *123*, 102377.
30. Zhang, P.; Pang, B.; Li, Y.; He, J.; Hong, X.; Qin, C.; Zheng, H. Analyzing spatial disparities of economic development in Yellow River Basin, China. *GeoJournal* **2019**, *84*, 303–320. [CrossRef]
31. Wang, Y.; Yang, N. Differences in High-Quality Development and Its Influencing Factors between Yellow River Basin and Yangtze River Economic Belt. *Land* **2023**, *12*, 1461. [CrossRef]
32. Ke Liu, K.; Qiao, Y.R.; Shi, T.; Zhou, Q. Study on coupling coordination and spatiotemporal heterogeneity between economic development and ecological environment of cities along the Yellow River Basin. *Environ. Sci. Pollut. R.* **2021**, *28*, 6898–6912. [CrossRef] [PubMed]
33. Zhang, Z.; Li, H.; Cao, Y. Research on the Coordinated Development of Economic Development and Ecological Environment of Nine Provinces (Regions) in the Yellow River Basin. *Sustainability* **2022**, *14*, 13102. [CrossRef]
34. Zhao, Y.; Hou, P.; Jiang, J.; Zhai, J.; Chen, Y.; Wang, Y.; Bai, J.; Zhang, B.; Xu, H. Coordination Study on Ecological and Economic Coupling of the Yellow River Basin. *Int. J. Environ. Res. Public Health* **2021**, *18*, 10664. [CrossRef] [PubMed]
35. Zhao, X.; Li, X.; Deng, G.; Xi, Y. Decoupling Relationship between Resource Environment and High-Quality Economic Development in the Yellow River Basin. *Sustainability* **2023**, *15*, 9385. [CrossRef]
36. Song, C.; Yin, G.; Lu, Z.; Chen, Y. Industrial ecological efficiency of cities in the Yellow River Basin in the background of China's economic transformation: Spatial-temporal characteristics and influencing factors. *Environ. Sci. Pollut. R.* **2022**, *29*, 4334–4349. [CrossRef] [PubMed]
37. Guo, A.; Zhang, Y.; Zhong, F.; Jiang, D. Spatiotemporal Patterns of Ecosystem Service Value Changes and Their Coordination with Economic Development: A Case Study of the Yellow River Basin, China. *Int. J. Environ. Res. Public. Health* **2020**, *17*, 8474. [CrossRef] [PubMed]
38. Hill, M.J.; Guerschman, J.P. The MODIS Global Vegetation Fractional Cover Product 2001–2018: Characteristics of Vegetation Fractional Cover in Grasslands and Savanna Woodlands. *Remote Sens.* **2020**, *12*, 406. [CrossRef]
39. Ahlfeldt, G.M.; Pietrostefani, E. The economic effects of density: A synthesis. *J. Urban Econ.* **2019**, *111*, 93–107. [CrossRef]
40. Henderson, J.V.; Nigmatulina, D.; Kriticos, S. Measuring Urban Economic Densit. *J. Urban Econ.* **2019**, *125*, 103188. [CrossRef]
41. Richards, D.R.; Belcher, R.N. Global Changes in Urban Vegetation Cover. *Remote Sens.* **2020**, *12*, 23. [CrossRef]
42. Zhang, Y.; Wang, P.; Wang, T.; Li, J.; Li, Z.; Teng, M.; Gao, Y. Using Vegetation Indices to Characterize Vegetation Cover Change in the Urban Areas of Southern China. *Sustainability* **2020**, *12*, 9403. [CrossRef]
43. Zhang, S.; Chen, H.; Fu, Y.; Niu, H.; Yang, Y.; Zhang, B. Fractional Vegetation Cover Estimation of Different Vegetation Types in the Qaidam Basin. *Sustainability* **2019**, *11*, 864. [CrossRef]
44. Li, C.H.; Yang, Z.F. Spatio-temporal changes of NDVI and their relations with precipitation and runoff in the Yellow River Basin. *Geogr. Res.-Aust.* **2004**, *23*, 753–759.
45. McVicar, T.R.; Van Niel, T.G.; Li, L.; Hutchinson, M.F.; Mu, X.; Liu, Z. Spatially distributing monthly reference evapotranspiration and pan evaporation considering topographic influences. *J. Hydrol.* **2007**, *338*, 196–220. [CrossRef]
46. Zhou, K.; Wang, Y.M.; Chang, J.X.; Zhou, S.; Guo, A.J. Spatial and temporal evolution of drought characteristics across the Yellow River basin. *Ecol. Indic.* **2021**, *131*, 108–207. [CrossRef]
47. Liu, H.; Liu, F.; Zheng, L. Effects of climate change and human activites on vegetation cover change in the Yellow Rive Basin. *J. Soil. Water Conserv.* **2021**, *35*, 143–151.
48. Li, Q.Q.; Cao, Y.P.; Miao, S.L. Spatio-temporal variation in vegetation coverage and its response to climate factors in the Yellow River Basin, China. *Acta Ecol. Sin.* **2022**, *42*, 4041–4054.
49. Yuan, L.H.; Jiang, W.G.; Shen, W.M.; Liu, Y.H.; Wang, W.J.; Tao, L.L.; Zheng, H.; Liu, X.F. The spatio-temporal variations of vegetation cover in the Yellow River Basin from 2000 to 2010. *Acta Ecol. Sin.* **2013**, *33*, 7798–7806.
50. Wang, K.; Zhou, J.; Tan, M.L.; Lu, P.; Xue, Z.; Liu, M.; Wang, X. Impacts of vegetation restoration on soil erosion in the Yellow River Basin, China. *Catena* **2024**, *234*, 107547. [CrossRef]

51. Piao, S.L.; Fang, J.Y. Dynamic vegetation cover change over the last 18 years in China. *Quat. Sci. Rev.* **2001**, *21*, 294–302.
52. Lopez, E.; Bocco, G.; Mendoza, M.; Duhau, E. Predicting land cover and land-use change in the urban fringe, A case in Morelia city, Mexico. *Landsc. Urban Plan.* **2001**, *55*, 271–285. [CrossRef]
53. Percoco, M. Path dependence, institutions and the density of economic activities: Evidence from Italian cities. *Pap. Reg. Sci.* **2014**, *93*, 53–76. [CrossRef]
54. Wheeler, D.; Hammer, D.; Kraft, R.; Dasgupta, S.; Blankespoor, B. Economic dynamics and forest clearing: A spatial econometric analysis for Indonesia. *Ecol. Econ.* **2013**, *85*, 85–96. [CrossRef]
55. Chen, W.Y.; Wang, D.T. Economic development and natural amenity: An econometric analysis of urban green spaces in China. *Urban. For. Urban Green.* **2013**, *12*, 435–442. [CrossRef]
56. Ding, Y. Spatial distribution characteristics and impact analysis of national wetland parks in the Yellow River basin from the perspective of arcgis. *SPIE* **2023**, *12797*, 105–110.
57. Cicia, G.; Giudice, T.; Scarpa, R. Consumers' perception of quality in organic food: A random utility model under preference heterogeneity and choice correlation from rank-orderings. *Brit. Food J.* **2002**, *104*, 200–213. [CrossRef]
58. Liu, X.; Pan, Y.; Zhu, X.; Li, S. Spatio-temporal variation of vegetation coverage in Qinling-Daba Mountains in relation to environmental factors. *Acta Geogr. Sin.* **2015**, *70*, 705–716.
59. Luck, G.W.; Smallbone, L.T.; O'Brien, R. Socio-economics and vegetation change in urban ecosystems: Patterns in space and time. *Ecosystems* **2009**, *12*, 604–620. [CrossRef]
60. Ren, Z.; Tian, Z.; Wei, H.; Liu, Y.; Yu, Y. Spatio-temporal evolution and driving mechanisms of vegetation in the Yellow River Basin, China during 2000–2020. *Ecol. Indic.* **2022**, *138*, 108832. [CrossRef]
61. Fang, J.; Yu, G.; Liu, L.; Hu, S.; Chapin, F.S. Climate change, human impacts, and carbon sequestration in China. *Proc. Nat. Acad. Sci. USA* **2018**, *115*, 4015–4020. [CrossRef] [PubMed]
62. Liu, W.; Shi, C.; Zhou, Y. Trends and attribution of runoff changes in the upper and middle reaches of the Yellow River in China. *J. Hydro.-Environ. Res.* **2021**, *37*, 57–66. [CrossRef]
63. Li, P.; Tian, Y.; Wu, J.; Xu, W. The Great Western Development policy: How it affected grain crop production, land use and rural poverty in western China. *Agricultural Economic Review.* **2021**, *13*, 319–348. [CrossRef]
64. Zhang, Z.; Chang, T.; Qiao, X.; Yang, Y.; Guo, J.; Zhang, H. Eco-economic coordination analysis of the Yellow River Basin in China: Insights from major function-oriented zoning. *Sustainability* **2021**, *13*, 2715. [CrossRef]
65. Zhang, G.X.; Su, Z.X. Construction and measurement of high-quality development evaluation system for the central cities in the Yellow River Basin. *Ecol. Econ.* **2020**, *36*, 37–43.
66. Chen, Y.; Syvitski, J.P.; Gao, S.; Overeem, I.; Kettner, A.J. Socio-economic impacts on flooding: A 4000-year history of the Yellow River. *China Ambio* **2012**, *41*, 682–698. [CrossRef] [PubMed]
67. Qu, S.; Wang, L.; Lin, A.; Yu, D.; Yuan, M. Distinguishing the impacts of climate change and anthropogenic factors on vegetation dynamics in the Yangtze River Basin, China. *Ecol. Indic.* **2020**, *108*, 105724. [CrossRef]
68. Zhao, W.; Li, J.; Chu, L.; Wang, T.; Li, Z.; Cai, C. Analysis of spatial and temporal variations in vegetation index and its driving force in Hubei Province in the last 10 years. *Acta Ecol. Sin.* **2019**, *39*, 7722–7736.
69. Ma, F.; Zhuo, J.; He, H.J.; Han, S.S. Ecological evolution and driving mechanism of vegetation in Yulin City, Shaanxi Province. *Bull. Soil. Water Conserv.* **2020**, *40*, 257–261.
70. Lu, Y.; Zhang, L.; Feng, X.; Zeng, Y.; Fu, B.; Yao, X.; Li, J.; Wu, B. Recent ecological transitions in China: Greening, browning, and influential factors. *Sci. Rep.* **2015**, *5*, 8732. [CrossRef]

land

MDPI

Article

Bridging Gaps towards the 2030 Agenda: A Data-Driven Comparative Analysis of Government and Public Engagement in China towards Achieving Sustainable Development Goals

Hongpeng Fu [1], Lingbo Fu [2], Lóránt Dénes Dávid [3,4,5], Qikang Zhong [6,*] and Kai Zhu [7,*]

1 Khoury College of Computer Science, Northeastern University, Seattle, WA 98122, USA; fu.hongp@northeastern.edu
2 School of Cultural Industries Management, Communication University of China, Beijing 100024, China; fulingbo@cuc.edu.cn
3 Department of Tourism and Hospitality, Faculty of Economics and Business, John von Neumann University, 6000 Kecskemét, Hungary; david.lorant.denes@uni-mate.hu
4 Institute of Rural Development and Sustainable Economy, Hungarian University of Agriculture and Life Sciences, 2100 Godollo, Hungary
5 Savaria Department of Business Economics, Faculty of Social Sciences, Savaria University Centre, Eötvös Loránd University, 9700 Szombathely, Hungary
6 School of Architecture and Art, Central South University, Changsha 410083, China
7 Faculty of Resources and Environmental Science, Hubei University, Wuhan 430062, China
* Correspondence: 201311063@csu.edu.cn (Q.Z.); hizhukai@163.com (K.Z.)

Abstract: The United Nations' 17 Sustainable Development Goals (SDGs) provide a pivotal framework to guide government policies and mobilize public engagement for sustainability, although further exploration is still needed to analyze the attention given to the SDGs by these stakeholders using multi-source data. To comparatively analyze government and public engagement in the SDGs, based on a case study in China, this study employed a multi-source data and content analysis to assess the perceived importance and performance of the two stakeholders. Then, this study performed an importance–performance analysis (IPA) to assess the SDGs' priority for further improvement. The results highlight the government's emphasis on SDG 7, SDG 2, and SDG 9, contrasting with public attention predominantly on SDG 8, SDG 9, and SDG 4. Regarding the performance of the SDGs in China, the greatest achievements have been exhibited for SDG 4 and SDG 1, while SDG 10, SDG 15, and SDG 17 have shown the least progress. Concerning the individual indicators, 12 out of 98 have attained 100% completion, while 19 out of 98 remain below the 50% threshold. According to the IPA results, China should concentrate on SDG 10 and SDG 15 and improve SDGs 6, 7, 14, 16, and 17 as a relatively low priority. The stakeholder analysis indicated that these SDGs are neglected zones by both the public and government. This research innovatively assessed the priorities of SDGs and could strengthen the cooperation between the government and the public to drive the implementation of the Sustainable Development Goals more effectively. This study could contribute to guiding policy directions, inform strategies for public engagement, and enhance the comprehension of sustainable development in China. The framework could serve as a valuable reference for stakeholder engagement in the SDGs in other countries.

Keywords: Sustainable Development Goals; content analysis; multiple stakeholders; importance–performance analysis; multi-source data

1. Introduction

As the world increasingly becomes globalized and interconnected, the United Nations' Sustainable Development Goals (SDGs) provide a vital framework for guiding and steering global efforts toward a more equitable and sustainable future [1,2]. Adopted by the United Nations in 2015, the 17 goals address a wide array of challenges from poverty and inequality to climate change and peace and justice. In addressing the SDGs, a multi-stakeholder analysis is a widely used approach to analyze the importance of different stakeholders, including governments, international organizations, civil society, and the public [1]. Among these, governments and the public emerge as essential due to their direct influence on policy making and implementation, respectively, to achieve the SDGs [1,3]. As a crucial driver in policy making and implementation, the government plays an instrumental role in aligning resources and efforts toward these goals [4]. Simultaneously, the public wields significant influence through democratic engagement, activism, and consumer choices, effectively shaping the trajectory of sustainable development [5]. Assessing the prioritization of SDGs by integrating the subjective perceptions of the government and the public [5–7] could contribute to detecting the objective progress made toward these goals and help generate nuanced insights that guide effective and targeted interventions to reach the SDGs.

Since the SDGs' inception in 2015, certain policies have evolved, showing varied development focuses based on different economic and political backgrounds. Numerous international organizations and research institutions now provide guidelines to support the SDGs' implementation [6,8]. Despite the global progress, the unique societal and political context of China, where the government leads matters, especially regarding environmental policies, calls for an examination of the SDGs' implementation [6,7]. In China, the SDGs, as with environmental policies, are primarily government-led [8]. A performance assessment is crucial to encourage the government to engage in environmental governance [9,10]. The government and the public are important stakeholders in achieving the SDGs [11,12], and there is tension between the roles of government and the public based on their needs. However, a performance assessment may incentivize local governments to prioritize political achievements, potentially neglecting some of the public's most pressing concerns and other governance responsibilities, resulting in environmental challenges [13]. The untapped potential of the public remains a concern, emphasizing the immediate efforts needed to enhance public awareness, attention, and oversight of the SDGs to facilitate collaborative governance and achieve these goals with greater efficiency and better resource allocation. However, gaps often exist between governmental actions and public interest, presenting potential challenges to the effective achievement of the SDGs [14,15]. An approach to effectively coordinate the two stakeholders and achieve the collaborative governance of the SDGs more efficiently is urgently needed.

A comparative analysis of the attention given to SDGs by the government and the public would be insightful, revealing alignment or divergence, which would assist in refining policy making and public engagement strategies [5,16]. The effective implementation of the SDGs could be facilitated if governmental actions aligned with public interests [1,5,17,18]. For instance, Tiefenbeck demonstrated that households significantly reduced their energy consumption when energy conservation was treated as a shared societal goal [19]. Conversely, the government significantly influences the promotion and realization of SDGs through policy making and resource mobilization [4,20,21]. Therefore, identifying areas of potential disconnection or synergy by comparing governmental and public attention can effectively direct efforts and resources toward mutual sustainability objectives [22]. Although previous research has examined the roles of governmental policies and public engagement in achieving the SDGs [16], the specific degree of attention allocated to these goals by both sectors still needs to be explored. This research gap is particularly noteworthy considering the central role of the SDGs in guiding the global development agenda. To the best of our knowledge, the previous studies have yet to provide a comparative analysis of the attention given to the SDGs by the government and the public, indicating a significant gap in the existing research that our study aims to address.

The digital revolution and the widespread use of social media platforms have profoundly transformed how the public discourse is formed and disseminated [23,24]. In the context of the SDGs, these platforms serve as digital spaces where diverse stakeholders express their perceptions, opinions, and levels of engagement with these global goals [25]. The vast amount of user-generated data offer an invaluable resource, providing unfiltered insights into public attention and opinions toward the SDGs [26,27]. Multi-source data refer to the combination of two or more distinct data sources, aimed at integrating the advantages of different data sources. Previous research has recognized the importance of utilizing multi-source data, especially big data, in monitoring and understanding various societal trends and issues [24,28–30]. In the context of the SDGs, multi-source data can provide valuable insights into the awareness and priorities of both public and government stakeholders, thereby enabling policymakers to pinpoint areas in need of heightened attention and public engagement [4]. However, despite the potential benefits of utilizing multi-source data in SDG-related research, the full potential still needs to be explored due to certain technical and methodological challenges. Overcoming these challenges would enable researchers to leverage multi-source data to better understand the public's engagement with the SDGs, thereby contributing to a more effective and responsive policy-making process.

This research seeks to bridge the identified gaps by establishing a multi-source-data-based framework to assess the priority of improving the SDGs and analyze the attention devoted to the SDGs by the government and the public. Through this comparative analysis involving a case study of China, the research seeks to answer the following key questions:

(1) What are the consistencies and differences in the level of attention given to the SDGs by the two stakeholders in China?
(2) Who are the main leading stakeholders in the various SDGs in China?
(3) Which SDGs require increased attention as high-priority areas for improvement in China?

This research aims to provide a nuanced understanding of the online discourse around the SDGs and shed light on the differing perspectives between the public and government. The findings of this study have the potential to guide policy directions, inform strategies for public engagement, and enhance the comprehension of sustainable development in the digital era.

2. Material and Methods

2.1. Research Framework

For this paper, we proposed a research framework to utilize muti-source data and a content analysis to compare the attention given to the SDGs by the government and the public (Figure 1). This study leveraged multi-source data, including Chinese Voluntary National Reviews, SDG reports, and big data from Sina, China News, and Toutiao. We constructed an SDG keywords list, extracting and supplementing seed words from official SDG reports and China National Knowledge Infrastructure (CNKI) database searches and using the word2Vec model to expand the seed words. After data collection and preprocessing from muti-source data based on the SDG keywords list, we analyzed the relative frequency of references to the 17 SDGs and popular SDG topics made by the public and the government, discerning consistencies or disparities in the attention directed towards these goals. Then, following previous studies [31,32], we took the relative frequency of references to the 17 SDGs as the perceived importance and the SDG realization status in the annual report on the SDGs published by the United Nations as the performance score. On this basis, in this study, we used the IPA analysis to understand the attention towards and prioritization of the SDGs comprehensively.

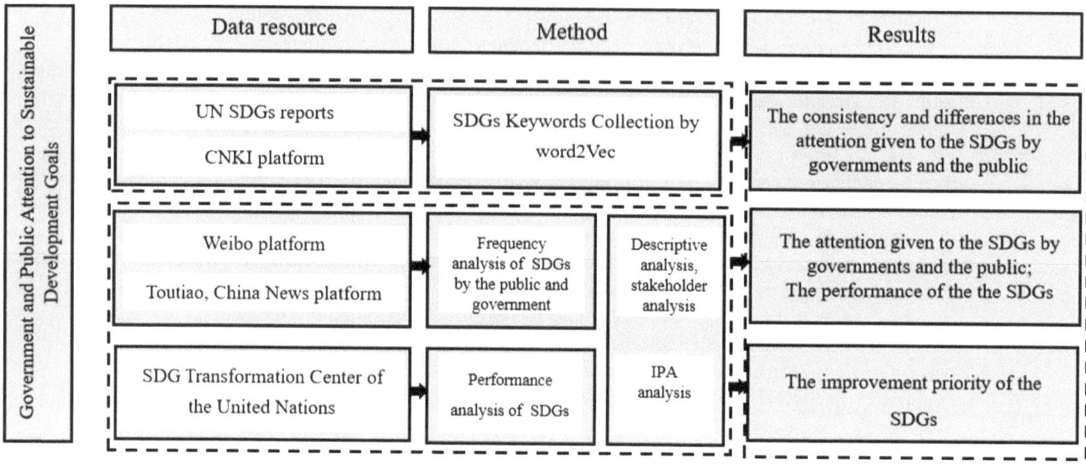

Figure 1. Research framework.

2.2. Data Collection, Preprocessing, and Analysis

This study first involved a comprehensive SDG keyword collection process, beginning by extracting seed words from the official report, expanding the list through database searches and synonyms, and refining the collection using the word2Vec model for semantic alignment, followed by a manual review for precision. The initial stage of our research involved creating a dedicated SDG keywords collection (Table 1) based on the official SDGs report. The additional keywords were identified and incorporated through a thorough search on CNKI in 2023—the preeminent Chinese research database. According to the official SDGs report and data from CNKI, word frequency calculations were incorporated to extract the top 500 most commonly used words associated with SDGs, resulting in a foundational SDG seed words list [33–35]. After constructing the seed words list, five experts in this field assigned these words to the corresponding SDG categories. The coding was done in Excel based on the consensus of two SDG experts each time. In cases of disagreement, a third expert was consulted for adjudication. Furthermore, we expanded the SDG keywords collection by referencing the Chinese Synonym Thesaurus from the Harbin Institute of Technology. The word2Vec model is beneficial for natural language processing tasks, as it can capture semantic relationships and word similarities in a vector space, enabling more accurate language understanding and word embedding processes [36,37]. Additionally, it has the potential to enhance various applications, such as in information retrieval and text classification, by providing context-aware word representations [38]. Following data preprocessing, we employed the word2Vec model to assess the similarity and filter keywords that aligned with the existing seed words, whereby we identified 1384 seed words within 17 SDG categories. All extension steps involved a manual review to ensure the inclusion of only the most pertinent keywords in the SDG keywords collection.

For the data collection, based on the SDGs keywords list, we utilized web crawling to gather data from Sina Weibo, representing the public's attention and focus on the SDGs, and from China News and Toutiao, representing the government's attention and focus on the SDGs. Sina Weibo, one of China's largest social media networks, allows users to share and engage with content, including short texts, images, videos, and links [39]. It serves as a platform for topic-based discussions, making it a vital source for assessing public discourse and opinion [40]. In parallel, we gathered data from China News and Toutiao to represent the government's attention and focus on the SDGs. China News is a prominent national-level news organization in China, providing timely, authoritative, and comprehensive news coverage across various fields such as domestic and international politics, economics, culture, and society [41]. Toutiao is a leading personalized news recommendation platform in China that covers various topics, including current affairs, technological innovations, and finance, both domestically and internationally [42,43]. The platform offers interactive features for users to participate in discussions and share their viewpoints, such as commenting, liking, and sharing [44]. However, Toutiao has special sections such as "society" and "new era" for state media coverage of government statements. At the same time, its moderation and filtering systems can manage user-generated content to ensure alignment with government priorities. All of the platforms, including Weibo, Toutiao, and China News, employ sophisticated search engines and recommendation algorithms that are capable of identifying, matching, and prioritizing content based on SDG-related keywords and synonyms.

Table 1. SDG keywords collection.

SDGs	Context	Example
SDG 1	No Poverty	Poverty, Impoverished, Impoverishment, Poverty Alleviation, Poverty-Stricken, Destitution, Indigence, etc.
SDG 2	Zero Hunger	Food, Land, Agriculture, Grain, Food Security, Farmland, Staple Food, Hunger, Arable Land, etc.
SDG 3	Good Health and Wellbeing	Happiness, Health, Wellbeing, Physical and Mental Health, Disease Prevention, Fitness, Mental Health, Strength, Physical Activities, Robust, etc.
SDG 4	Quality Education	Go to School, Study, Read Books, Education, Online Education, Cultural Education, Education Quality, Education Resources, Pursue Education, School Environment, Attend School, Education Opportunities, Education Policy, Further Education, etc.
SDG 5	Gender Equality	Gender Discrimination, Feminism, Women's Rights, Gender Diversity, Women's Status, Women's Safety, Gender Equality, Gender Roles, etc.
SDG 6	Clean Water and Sanitation	Clean, Hygiene, Cleanliness, Purification, Clean Water, Rainwater, Drinking Water, Environmental Sanitation Policy, Sustainable Hygiene, etc.
SDG 7	Affordable and Clean Energy	Energy, Clean Technology, Renewable Energy, Low-Carbon Economy, Solar Energy, Electric Vehicles, Renewable Fuel, Bioenergy, Hydropower, Wind Energy, etc.
SDG 8	Decent Work and Economic Growth	Business, Task, Work, Economy, Trade, Matter, Occupation, Job, Employment, Affair, Industry, Labor Protection, etc.
SDG 9	Industry, Innovation, and Infrastructure	Internet, Ports, Railways, Aviation, Bridges, Innovation, Roads, Technology, Research and Development (R&D), Telecommunications, etc.
SDG 10	Reduced Inequalities	Equal Opportunities, Fairness, Justice, Equity, Equality, Impartiality, Unequal Economic Development, etc.
SDG 11	Sustainable Cities and Communities	Cities, Urban Areas, Sustainable Transportation, Ecological Cities, Housing, Residential, Sustainable Buildings, Disaster Resilience, etc.
SDG 12	Responsible Consumption and Production	Consumption, Resource Recycling, Production, Expenditure, Output, Spending, Green Technology, Output, Energy Conservation and Emissions Reduction, etc.
SDG 13	Climate Action	Climate Models, Ecosystems, Climate Justice, Glacier Melting, Fossil Fuels, Global Warming, Extreme Weather, Climate Policies, Sea Level Rise, Carbon Emissions, Greenhouse Gases, etc.
SDG 14	Life Below Water	Marine Conservation, Fishing Regulations, Aquatic Habitats, Lake Ecology, Water Conservation Areas, Marine Biodiversity, Water Quality Management, Marine Parks, Wetland Ecology, Fishing Restrictions, etc.
SDG 15	Life on Land	Law, Rule of Law, Rights Protection, Legislation, Criminal Justice, Legal Procedures, Policy Making, Regulations, Legal Framework, etc.
SDG 16	Peace, Justice, and Strong Institutions	Law, Rule of Law, Rights Protection, Legislation, Criminal Justice, Legal Procedures, Policy Making, Regulations, Legal Framework, etc.
SDG 17	Partnerships for the Goals	Partnership, International Cooperation, Agreement, Global Trade, etc.

Along with the annual SDG report published by the United Nations in June 2023 [45], our data collection period spanned from 1 January 2022 to 19 June 2023. This period reflects China's annual efforts toward sustainable development and aligns with the previous SDG evaluation timeline. The data collection period coincided with phases of stable macroeconomic growth, enhanced social wellbeing, intensified ecological initiatives aimed at achieving a "beautiful China", and active participation in global development initiatives. The collected data were then filtered using the SDG keywords collection as the keywords, allowing us to extract relevant texts for each specific SDG. For China News and Toutiao, which primarily represent the government's focus on and initiatives regarding the SDGs, we collected information on publication times and posted content. The data from these platforms help track the alignment of government communications with the Sustainable Development Goals. When collecting Weibo user information, we collected each individual's user name, region, number of followers, number of Weibo posts, certification conditions, post content, number of collections, number of comments, number of likes, post times, and other information. To ensure the quality and consistency of the data, a thorough cleaning process was undertaken, including removing irrelevant, meaningless texts, as well as texts with confusing formats or missing content. To accurately reflect the public's attention using the Weibo data, we filtered out messages posted by official or government accounts on Sina Weibo. Finally, we obtained over 160,000 data points from Sina Weibo and over 60,000 data points from China News and Toutiao.

Due to the original dataset being intermixed with a substantial amount of text irrelevant to the event, even after removing invalid results during data collection and preprocessing, there remained a significant presence of duplicate and irrelevant texts. XGBoost models based on decision trees, which can process various types of data and feature selection capabilities, as well as being less sensitive to ranges of variable values, are suitable for the classification problems encountered in this study [46–48]. Consequently, following precedents in previous studies, the XGBoost model was chosen as the classifier to filter out irrelevant microblogs, yielding content that was highly pertinent to each event [49]. We manually set up the training and sample sets to construct a machine learning model for efficient classification. The training and validation sets were allocated at a ratio of 8:2, with 500 microblogs randomly selected from each SDG theme in the preprocessed dataset, totaling 8500 topics for manual annotation. According to our calculations, this classification model achieved an accuracy rate of 90.36% and an F1 score of 88.74% for the microblog dataset. These results underscore the model's effective classification capabilities, significantly enhancing the filtration of texts for relevance to an event, thereby validating their effectiveness in capturing relevant content.

2.3. Evaluation of the Prioritization of SDGs Based on the Importance–Performance Analysis (IPA) Method

The IPA method allows for a dual-dimensional assessment framework that evaluates both the importance and performance at the same time to show the urgency and priority of the attributes directly. The IPA method, initially developed as a decision-making tool in the marketing sector, facilitates the evaluation of attributes and informs the creation of strategic management measures to enhance the user experience [50]. This technique has extensive applications across a range of sectors, proving its versatility and efficacy, including in tourism [51,52], transportation [53], and e-commerce [54]. The IPA method provides a straightforward way to analyze and display the prioritization of and urgency towards implementing the SDGs. The IPA framework enables the classification of the SDGs into four distinct quadrants (Figure 2) based on their perceived importance and performance levels. The "keep up the good work zone" (Q1) represents attributes or services that are perceived as of high importance and performance levels to stakeholders. These elements are considered strengths for achieving the SDGs in this country and should be maintained at their current high performance levels. In the "maintenance zone" (Q2) quadrant, the stakeholders perceive the attributes as being less important but showing high performance levels. These elements should be consistently maintained to ensure high performance levels to achieve the 2030 Agenda. The "neglected zone" (Q3) includes attributes perceived as less important by stakeholders and yielding low performance levels. Long-term strategies should address these areas to improve the stakeholders' perceptions and enhance the overall performance to achieve the 2030 Agenda. The "improvement zone" (Q4) consists of attributes or services that are perceived as highly important but delivering low performance. These elements required immediate attention and improvement to achieve the 2030 Agenda. However, using the IPA method to analyze the SDGs and achieve the 2030 Agenda remains an underexplored area.

To avoid reaching a ceiling in our efforts using the IPA, namely, that most attributes often fail in the "keep up the good work" quadrant, since the respondents are prone to give high performance and importance ratings, a data-scale-centered method was proposed in some studies, where average values of importance and performance were used as the crosshair [9]. Meanwhile, the IPA diagram revealed the improvement priorities clearly for decision making. We then classified the IPA results for easy decision making.

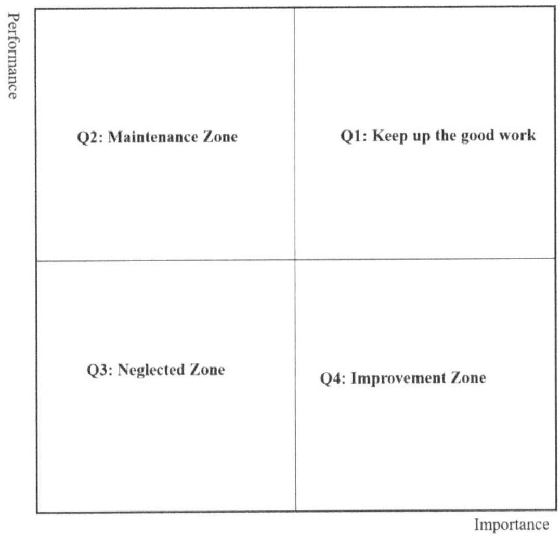

Figure 2. Quadrants of IPA results.

2.4. Calculating the Perceived Importance Scores Using a Relative Frequency Analysis

In this study, we employed the SDGs as a coding scheme to gauge the relative frequency of each SDG from the total comments, aiming to explore the attention devoted to these goals by both the public and government [31,32]. We assessed the frequency of SDG comments to identify any consistencies or differences in attention towards these SDGs. Finally, we examined the focus on popular SDG-related topics within both government and public discussions, entailing an assessment of the relative frequency rates of words associated with the SDGs to identify any consistencies or differences in attention towards these subjects. The relative frequency is calculated as follows:

$$I_i = \frac{(n_{i,p} + n_{i,g})}{(N_{i,p} + N_{i,g})} \times 100\% \quad (1)$$

where i represents the different SDGs; I_i is the absolute importance of the different SDGs; $n_{i,p} + n_{i,g}$ represent the number of times an SDG is mentioned in the public and government comments, respectively; and $N_{i,p} + N_{i,g}$ represent the total number of public and government comments, respectively. The theoretical value of I_i ranges from 0% to 100%; the closer to 100%, the more important the attribute, while the closer to 0%, the less important the attribute.

2.5. *Calculating the Performance Scores of Each SDG*

In this study, we calculated the performance scores of each SDG in China based on the SDG Report 2023 by The SDG Transformation Center for the United Nations [55]. Initiated in 2015 as part of the United Nations' 2030 Agenda for Sustainable Development, the 17 SDGs are complemented by 169 specific targets, aimed at guiding global efforts towards sustainable development. Due to the extensive system of indicators, many scholars have found it particularly challenging to collect comprehensive data to monitor SDG progress [56,57]. For instance, as of 2023, only 65 indicators were accessible for China in the United Nations SDG database (https://unstats.un.org/sdgs/dataportal/database, accessed on 6 June 2024). Therefore, by comparing the availability, recency, and previous studies of the data, we selected the Sustainable Development Report 2023 by The SDG Transformation Center, which is an initiative of the UN Sustainable Development Solutions Network (SDSN) for the United Nations (https://sdgtransformationcenter.org/, accessed on 6 June 2024). The Sustainable Development Report 2023 provides the most comprehensive assessment of the SDG performances of all 193 UN Member Nations. The methodology of the SDSN Index was audited by the EU's Joint Research Commission in July 2019 and has undergone peer review by Nature Geoscience and Cambridge University Press [58]. This annual report and regional editions have become world-leading references for monitoring progress on the SDGs. The Sustainable Development Report is accessed online over 300,000 times each year by government officials, researchers, investors, policymakers, and consulting firms [55]. For the 2023 SDG Index Edition, two-thirds of the data were sourced from official statistics, typically provided by UN-hosted agencies, while the remaining one-third came from non-traditional sources such as research centers, universities, and NGOs. The SDG Report 2023 by the SDG Transformation Center for the United Nations includes the 98 indicators of the 17 SDGs and their corresponding values.

Following the instructions proposed by the SDG Transformation Center, the procedure for calculating the SDG Index comprises three steps [58]: (i) establish performance thresholds and censor extreme values from the distribution of each indicator; (ii) rescale the data to ensure comparability across indicators (normalization); (iii) aggregate the indicators within and across SDGs. Following the methodology proposed by the SDG Transformation Center, the lower bound is specifically defined at the 2.5th percentile of the distribution for each indicator for all of the countries. The upper bounds for each indicator are determined by the following decision tree:

1. For SDGs with specific targets, such as zero poverty, universal education, and full gender equality, the use of absolute quantified thresholds is prescribed.
2. In the absence of explicit targets, the principle of "leaving no one behind" is applied, setting universal access or zero deprivation as the upper limits.
3. For goals with a future achievement date, such as zero carbon emissions from greenhouse gases by no later than 2050 or limiting global warming to 1.5 °C, relevant scientific targets are used to establish the 100% upper limits.
4. For all other indicators, the average of the top five performing indicators is used.

As indicated in the SDG Report 2023 by the SDG Transformation Center for the United Nations, several rounds of expert consultations on the early drafts of the SDG Index clearly indicated that there is almost no consensus among different academic communities regarding the weighting of some SDGs over others. Therefore, as a normative assumption, the SDG Report assigns a fixed and equal weight to each SDG, reflecting policymakers' commitment to treat all SDGs equally and as a cohesive and indivisible set of goals. This means that countries need to focus on all goals to improve their SDG score, although special attention should be paid to those goals that are farthest from achievement [55], being expected to show the quickest incremental progress. To calculate the SDG scores, the SDG Report first estimated the score for each goal using the arithmetic mean of the goal's indicators. These goal scores were then averaged across all 17 SDGs to obtain the final index score. Various sensitivity analyses were conducted for the SDG Report, the results of which are available online, including comparisons between arithmetic and geometric means and Monte Carlo simulations at the index and goal levels. The score for each SDG indicator is calculated as follows:

$$P_{i,j} = \frac{x - \min(x)}{\max(x) - \min(x)} \times 100\% \tag{2}$$

$$P_j = \frac{\sum_1^i P_{i,j}}{i} \times 100\% \tag{3}$$

where $P_{i,j}$ is the absolute performance of indicator i of SDG j; x is the raw data value; max and min denote the upper and lower bounds, respectively; x′ is the normalized value after rescaling; and P_j is the absolute performance of SDG j. The theoretical value of P_i ranges from 0% to 100%; the closer to 100%, the better the performance of the attribute, while the closer to 0%, the poorer the performance of the attribute. A score of 0% represents the poorest performance, indicating that the SDG falls below the 2.5th percentile for the distribution of all nations worldwide. A score of 50% denotes a median performance. This implies that the country has either met or surpassed half of the targeted goal values. A score of 100% signifies optimal or ideal performance, corresponding with fully achieving specific targets of the SDGs, such as achieving zero poverty, complete universal education, or full access to water and sanitation facilities.

3. Results

3.1. Analysis of the Government and Public Attention on SDGs

3.1.1. Frequency Analysis of the Government and Public Attention on SDGs

The results showed that the government has strongly focused on SDG 7, SDG 2, and SDG 9, while allocating less attention to SDG 16, SDG 12, and SDG 4 (Figure 3). Notably, Affordable and Clean Energy (SDG 7) received the highest level of attention (23.75%). Correspondingly, Industry, Innovation, and Infrastructure (SDG 9) and Zero Hunger (SDG 2) also received relatively high levels of attention, accounting for 12.15% and 12.61%, respectively. However, in comparison to other goals, the government's attention given to Peace, Justice, and Strong Institutions (SDG 16) and Responsible Consumption and Production (SDG 12) was relatively low, accounting for only 0.78% and 1.13%, respectively.

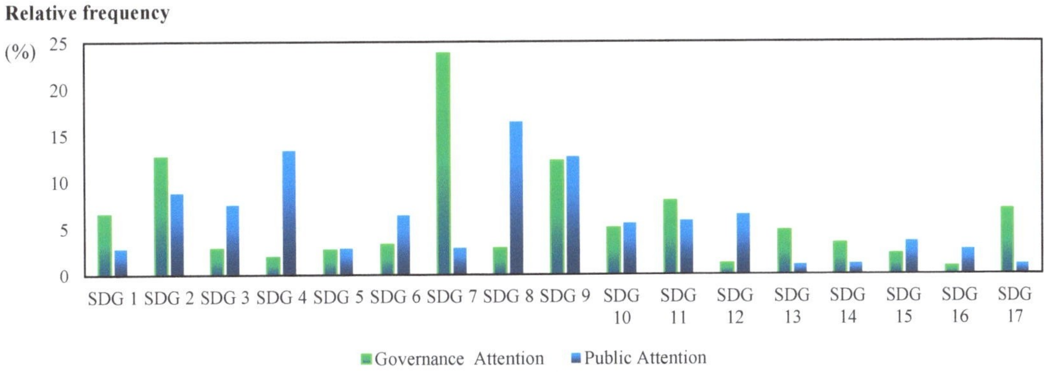

Figure 3. Government and public attention given to the SDGS.

The public's attention is mainly concentrated on SDG 8, SDG 9, and SDG 4, with the least attention given to SDG 13, SDG 14, and SDG 17 (Figure 3). Among the SDGs,

Decent Work and Economic Growth (SDG 8) leads with 16.41%, indicating the public's high level of attention given to employment and economic development. This is followed by Quality Education (SDG 4) and Industry, Innovation, and Infrastructure (SDG 9), accounting for 13.29% and 12.58%, respectively, possibly indicating that the public attaches great importance to the role of education and innovation in socio-economic development. However, the attention given to Climate Action (SDG 13), Life Below Water (SDG 14), and Partnerships for the Goals (SDG 17) is relatively low, accounting for 1.00%, 1.09%, and 1.04%, respectively.

When comparing the government and public's focus on each Sustainable Development Goal (SDG), we can identify areas of consensus and divergence in China.

Consensus is evident in the attention both stakeholders devote to Industry, Innovation, and Infrastructure (SDG 9), with the two stakeholders allocating 12.15% and 12.58% of their focus, respectively. Similarly, Zero Hunger (SDG 2) accounts for 12.61% of the government's focus and 8.74% of the public's focus, reflecting the importance of food security and agricultural development in society. However, divergence can be seen in several areas. Affordable and Clean Energy (SDG 7) receives 23.75% of the government's focus but only 2.81% from the public. Conversely, Decent Work and Economic Growth (SDG 8) accounts for 16.41% of the public's attention compared to a modest 2.81% from the government. At the same time, it is worth noting that although Climate Action (SDG 13) receives low attention from both stakeholders, other goals related to environmental and climate change, such as SDG 7 and Responsible Consumption and Production (SDG 12), receive a certain level of attention from both stakeholders in China.

3.1.2. Relative Frequency Analysis of the Government and Public Attention Given to the SDGs Topics

The government's popular topics mainly involve SDG 8 (Decent Work and Economic Growth), SDG 9 (Industry, Innovation, and Infrastructure), SDG 13 (Climate Action), and SDG 3 (Good Health and Wellbeing) (Figure 4a). For SDG 8 (Decent Work and Economic Growth), keywords such as economy (3.98%), enterprises (3.41%), market (2.27%), and industry (2.27%) frequently appear, which show the government's focus on economic development, enterprise development, market prosperity, and industrial innovation. This indicates a robust governmental effort to bolster economic growth and improve industrial competitiveness. For SDG 9 (Industry, Innovation, and Infrastructure), the keywords science and technology (2.10%), information (0.84%), and infrastructure (0.58%) frequently appear, indicating the government's high level of attention given to innovative technology, information technology, and their roles in promoting industrial progress. Regarding SDG 11 (Sustainable Cities and Communities), the frequent appearance of keywords such as city (1.91%), village (1.21%), and housing (0.46%) highlights the government's focus on the construction and improvement of cities and villages, including citizen's living conditions, to create more sustainable and livable communities. In terms of SDG 13 (Climate Action) and other environmental issues, the focus on keywords such as ecology (2.31%), carbon (1.97%), and environment (1.62%) shows the government's proactive stance on environmental protection and climate action. In particular, the focus here is on tackling climate change and promoting carbon neutrality. The appearance of keywords such as epidemic (1.21%), COVID-19 (0.44%), medical care (0.43%), and hospital (0.42%) relating to from SDG 3 (Good Health and Wellbeing) indicates that the government is paying attention and responding to public health crises, especially the COVID-19 epidemic.

(**a**)

(**b**)

Figure 4. (**a**) Government attention given to SDG topics. (**b**) Public attention given to SDG topics. Note: The sizes of the words reflect the results of the relative frequency analysis.

The analysis of the public attention around SDG topics revealed a concentrated focus on SDG 4 (Quality Education), SDG 8 (Decent Work and Economic Growth), SDG 3 (Good Health and Wellbeing), and SDG 5 (Gender Equality) (Figure 4b). Regarding SDG 4 (Quality Education), the public focuses mainly on children's education and the school environment. The high relative frequency of keywords such as children (2.50%), schools (1.25%), going to school (1.24%), students (1.10%), and teachers (1.10%) show the public's strong attention given to improving the quality of education, ensuring educational fairness, and providing education opportunities for all. Regarding SDG 8 (Decent Work and Economic Growth), the high relative frequency of the keywords enterprises (2.38%) and companies (2.28%) suggest public concerns about corporate social responsibility, the impact of corporate operations on the economy and society, and the state of the job market. Meanwhile, the appearance of the keywords market (1.38%) and economy (1.60%) reflects the public's attention given to economic trends, industry development trends, and market competition. Regarding SDG 3 (Good Health and Wellbeing), the public attention's is centered on epidemic prevention and control and public health services. The appearance of the keywords epidemic (2.33%) and COVID-19 (0.63%) reflects the public's concerns about epidemic prevention and control, the distribution of medical resources, and public health services. The appearance of the keywords hospitals (1.15%), doctors (0.63%), and cases (0.85%) shows the public's concerns about the quality of medical services, accessibility of medical resources, and disease treatment and prevention. For SDG 5 (Gender Equality), the public's attention is mainly focused on protecting women's rights and gender equality issues. The frequent appearance of the keywords women (1.44%), gender (0.50%), woman (0.05%), female students (0.40%), and women's rights (0.38%) underlines the public's concern for gender equality, the protection of women's rights, and gender discrimination.

The commonalities in the topics of interest related to the SDGs indicate shared concerns and priorities regarding SDG 3 (Good Health and Wellbeing) and SDG 8 (Decent Work and Economic Growth) (Table 2). SDG 3 (Good Health and Wellbeing) primarily focuses on pandemic prevention, public health services, and medical resources. This result shows that public health is a topic of common concern to both the public and the government. As for SDG 8 (Decent Work and Economic Growth), the focus on economic development is mainly reflected in business development, market conditions, and economic situations.

Table 2. Topics frequently discussed by the government and the public related to each SDG.

SDGs	Context	Government Top Topics	Public Top Topics
SDG 1	No Poverty	villagers (0.45%), people's livelihood (0.43%), common prosperity (0.43%)	money (1.23%), price (0.62%), poverty (0.31%)
SDG 2	Zero Hunger	agriculture (1.25%), food (0.75%), grain (0.57%), food products (0.57%)	food (1.89%), diet (0.88%), farming (0.31%)
SDG 3	Good Health and Wellbeing	epidemic (1.21%), COVID-19 (0.44%), medical care (0.43%), hospital (0.42%)	epidemic (2.33%), COVID-19 (0.63%), hospital (1.15%), doctor (0.63%), case (0.85%)
SDG 4	Quality Education	children (0.85%), students (0.74%), schools (0.74%)	children (2.50%), schools (1.25%), going to school (1.24%), students (1.10%), teachers (1.10%)
SDG 5	Gender Equality	female (0.74%), women (0.45%), gender (0.28%)	women (1.44%), gender (0.50%), woman (0.05%), female student (0.40%), women's rights (0.38%)
SDG 6	Clean Water and Sanitation	water (0.80%), water resources (0.23%), lake (0.13%)	lake (0.9%), water (0.75%), freshwater (0.25%), rainwater (0.24%)
SDG 7	Affordable and Clean Energy	energy (1.52%), electricity (1.01%), solar power (0.45%)	renewable energy (0.73%), energy (0.43%)
SDG 8	Decent Work and Economic Growth	economy (3.98%), enterprises (3.41%), market (2.27%), industry (2.27%)	enterprises (2.38%), companies (2.28%), market (1.38%), economy (1.60%)
SDG 9	Industry, Innovation, and Infrastructure	science and technology (2.10%), information (0.84%), infrastructure (0.58%)	technology (1.88%), project (1.60%), industry (1.50%), science and technology (1.23%), information (1.12%)
SDG 10	Reduced Inequalities	ethnicity (0.57%), equity (0.26%), justice (0.21%)	women (1.45%), gender (0.53%), woman (0.50%), female student (0.40%), women's rights (0.38%)
SDG 11	Sustainable Cities and Communities	city (1.91%), village (1.21%), housing (0.46%)	city (1.73%), countryside (0.98%), community (0.95%)
SDG 12	Responsible Consumption and Production	corporation (1.53%), product (1.36%), resources (1.46%)	product (1.15%), brand (0.75%), responsibility (0.58%)
SDG 13	Climate Action	ecology (2.31%), carbon (1.97%), environment (1.62%),	environment (1.28%), ecology (0.75%), carbon (0.35%)
SDG 14	Life Below Water	ocean (0.82%), Yangtze River (0.42%), Yellow River (0.26%)	ocean (0.25%)
SDG 15	Life on Land	ecology (2.27%), biology (1.01%)	panda (0.45%), wildlife (0.45%)
SDG 16	Peace, Justice, and Strong Institutions	system (1.11%), peace (1.05%), institution (0.94%)	crime (0.75%)
SDG 17	Partnerships for the Goals	cooperation (1.64%), alliance (0.77%), international (0.69%)	international (0.08%)

Regarding the differences in the topics related to the SDGs, the public is more concerned about education, health, and employment, while the government is more concerned about economic development, social equity, and environmental protection. To be specific, for SDG 2 (Zero Hunger), based on the frequent occurrence of the keywords food and diet, it can be seen that the public has a high degree of concern for food safety, healthy eating, and nutritional intake. For SDG 16 (Peace, Justice, and Strong Institutions), the public has a relatively high level of concern for social safety, justice, and various forms of violence, as indicated by the frequent occurrence of the keywords crime, law, and violence. However, compared to other issues, the government is less concerned about these topics. Conversely, SDG 9 (Industry, Innovation, and Infrastructure), SDG 11 (Sustainable Cities and Communities), and SDG 10 (Reduced Inequalities) are topics of high importance to the government. However, these topics are not considered as significant in the public's opinion.

3.2. Performance Analysis of the Government and Public Focus on SDGs

Regarding the performance indicators for the Sustainable Development Goals (SDGs), the greatest achievements were found for SDG 4 (Quality Education) and SDG 1 (No Poverty), while SDG 10 (Reduced Inequalities), SDG 15 (Life on Land), and SDG 17 (Partnerships for the Goals) demonstrated the least progress (Figure 5). Specifically, the targets set in the 2030 Agenda for SDG 4 and SDG 1 have nearly been achieved, with accomplishment rates of 97.99% and 99.24%, respectively. In contrast, approximately 50% of the 2030 Agenda goals have been reached for SDG 10, SDG 15, and SDG 17, with progress indicators of 34.93%, 49.24%, and 41.37%, respectively. These goals require increased focus to fulfill the objectives of the 2030 Agenda.

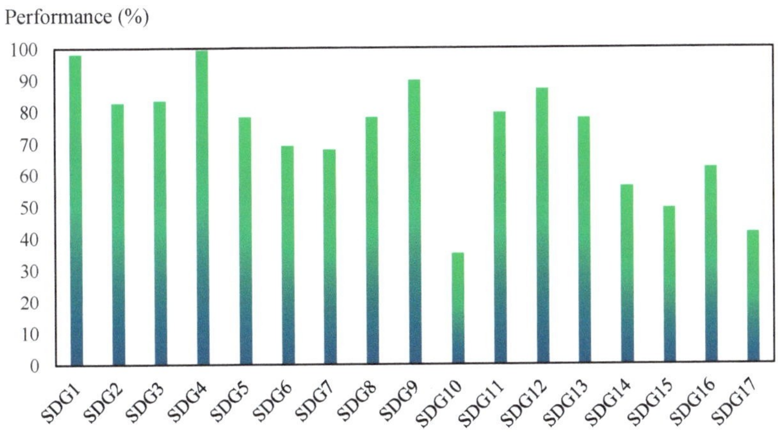

Figure 5. Performance scores for the SDGs.

Concerning the individual indicators, 100% completion has been attained for 12 out of 98, while 19 out of 98 remain below the 50% threshold (Table 3). Of these, 12 indicators across nine SDGs have been completed, with SDG 9 (Industry, Innovation, and Infrastructure) performing exceptionally well, fulfilling four indicators. However, there appears to be a discrepancy in the report, as it also states that 19 out of 98 indicators for 12 SDGs have reached 100%, with SDG 9 (Industry, Innovation, and Infrastructure) notably achieving all four of its indicators. This inconsistency requires clarification. For SDGs 14 (Life Below Water), 15 (Life on Land), 16 (Peace, Justice, and Strong Institutions), and 17 (Partnerships for the Goals), the situation is particularly severe, as they encompass the most underperforming indicators, each with three indicators significantly lagging. Notably, there are five indicators, including those in SDG 6 (Clean Water and Sanitation), 8 (Decent

Work and Economic Growth), 10 (Reduced Inequalities), and 16 (Peace, Justice, and Strong Institutions), which have not shown any progress, marked by a score of 0%.

Table 3. Performance results for each SDG. Note: the SDG values are sourced from the SDG Report 2023 by the SDG Transformation Center [55].

SDG Types	SDG Indicators	SDG Values	Performance Scores (%)
SDG 1—No Poverty	Poverty headcount ratio at USD 2.15/day (%)	0.80	98.90
	Poverty headcount ratio at USD 3.65/day (%)	1.50	97.10
SDG 2—Zero Hunger	Prevalence of undernourishment (%)	2.50	100.00
	Prevalence of stunting in children under 5 years of age (%)	4.60	90.80
	Prevalence of wasting in children under 5 years of age (%)	1.90	88.30
	Prevalence of obesity, BMI $\geq$ 30 (% of adult population)	6.20	89.50
	Human trophic level (best 2–3 worst)	2.20	61.00
	Cereal yield (tons per hectare of harvested land)	6.30	89.70
	Sustainable Nitrogen Management Index (best 0–1.41 worst)	0.70	41.70
	Exports of hazardous pesticides (tons per million population)	0.30	99.90
SDG 3—Good Health and Wellbeing	Maternal mortality rate (per 100,000 live births)	23.10	97.60
	Neonatal mortality rate (per 1000 live births)	3.20	94.60
	Mortality rate, under-5 (per 1000 live births)	6.90	96.60
	Incidence of tuberculosis (per 100,000 population)	55.00	90.20
	Age-standardized death rate due to cardiovascular disease, cancer, diabetes, or chronic respiratory disease in adults aged 30–70 years (%)	15.90	69.60
	Age-standardized death rate attributable to household air pollution and ambient air pollution (per 100,000 population)	95.30	74.20
	Traffic deaths (per 100,000 population)	17.40	53.40
	Life expectancy at birth (years)	77.40	80.70
	Adolescent fertility rate (births per 1000 females aged 15 to 19)	6.10	97.40
	Births attended by skilled health personnel (%)	99.90	99.90
	Surviving infants who received 2 WHO-recommended vaccines (%)	99.00	98.30
	Universal Health Coverage (UHC) Index of service coverage (worst 0–100 best)	82.00	70.90
	Subjective wellbeing (average ladder score, worst 0–10 best)	5.90	60.50
SDG 4—Quality Education	Lower secondary completion rate (%)	99.50	99.40
	Literacy rate (% of the population aged 15 to 24)	99.50	99.10
SDG 5—Gender Equality	Demand for family planning satisfied by modern methods (% of females aged 15 to 49)	96.60	95.90
	The female-to-male ratio of mean years of education received (%)	93.30	88.50
	The female-to-male ratio of labor force participation (%)	84.10	79.70
	Seats held by women in national parliament (%)	24.90	48.60
SDG 6—Clean Water and Sanitation	Population using at least basic drinking water services (%)	94.30	90.50
	Population using at least basic sanitation services (%)	92.40	91.60
	Freshwater withdrawal (% of available freshwater resources)	43.20	64.90
	Anthropogenic wastewater that receives treatment (%)	9.40	0.00
	Scarce water consumption embodied in imports (m^3 H_2O eq/capita)	305.70	98.10
SDG 7—Affordable and Clean Energy	Population with access to electricity (%)	100.00	100.00
	Population with access to clean fuels and technology for cooking (%)	79.40	79.00
	CO_2 emissions from fuel combustion per total electricity output ($MtCO_2$/TWh)	1.30	78.00
	Renewable energy share in total final energy consumption (%)	10.60	14.60
SDG 8—Decent Work and Economic Growth	Adjusted GDP growth (%)	3.10	90.40
	Victims of modern slavery (per 1000 population)	2.80	87.30
	Adults with an account at a bank or other financial institution or with a mobile money service provider (% of the population aged 15 or over)	88.70	87.70
	Unemployment rate (% of total labor force, ages 15+)	4.60	83.90
	Fundamental labor rights effectively guaranteed (worst 0–1 best)	0.30	0.00
	Fatal work-related accidents embodied in imports (per 100,000 population)	0.00	100.00
	Victims of modern slavery embodied in imports (per 100,000 population)	8.50	97.20

Table 3. *Cont.*

SDG Types	SDG Indicators	SDG Values	Performance Scores (%)
SDG 9—Industry, Innovation and Infrastructure	Rural population with access to all-season roads (%)	85.00	76.90
	Population using the internet (%)	73.10	72.50
	Mobile broadband subscriptions (per 100 population)	101.60	100.00
	Logistics Performance Index: Quality of trade and transport-related infrastructure (worst 1–5 best)	3.80	100.00
	The Times Higher Education Universities Ranking: Average score of top 3 universities (worst 0–100 best)	81.60	100.00
	Articles published in academic journals (per 1000 population)	0.60	50.00
	Expenditure on research and development (% of GDP)	2.40	64.90
SDG 10—Reduced Inequalities	Gini coefficient	38.20	69.90
	Palma ratio	3.90	0.00
SDG 11—Sustainable Cities and Communities	The annual mean concentration of particulate matter of less than 2.5 microns in diameter (PM2.5) (μg/m^3)	48.60	47.60
	Access to improved water source, piped (% of urban population)	91.40	90.80
	Satisfaction with public transport (%)	85.00	100.00
SDG 12—Responsible Consumption and Production	Municipal solid waste (kg/capita/day)	0.80	80.60
	Electronic waste (kg/capita)	7.20	70.00
	Production-based SO_2 emissions (kg/capita)	18.00	96.60
	SO_2 emissions embodied in imports (kg/capita)	0.50	98.30
	Production-based nitrogen emissions (kg/capita)	32.90	68.50
	Nitrogen emissions embodied in imports (kg/capita)	4.00	94.30
	Exports of plastic waste (kg/capita)	0.00	100.00
SDG 13—Climate Action	CO_2 emissions from fossil fuel combustion and cement production (tCO_2/capita)	8.00	60.00
	CO_2 emissions embodied in imports (tCO_2/capita)	0.20	95.60
			0.00
SDG 14—Life Below Water	The mean area that is protected in marine sites important to biodiversity (%)	7.10	7.10
	Ocean Health Index: Clean Waters score (worst 0–100 best)	45.90	24.20
	Fish caught from overexploited or collapsed stocks (% of total catch)	25.50	71.90
	Fish caught by trawling or dredging (%)	48.20	47.00
	Fish caught that are then discarded (%)	2.70	86.50
	Marine biodiversity threats embodied in imports (per million population)	0.00	100.00
SDG 15—Life on Land	The mean area that is protected in terrestrial sites important to biodiversity (%)	10.10	10.10
	The mean area that is protected in freshwater sites important to biodiversity (%)	9.60	9.60
	Red List Index of species survival (worst 0–1 best)	0.73	32.50
	Permanent deforestation (% of forest area, three-year average)	0.00	100.00
	Terrestrial and freshwater biodiversity threats embodied in imports (per million population)	0.60	94.00
SDG 16—Peace, Justice, and Strong Institutions	Homicides (per 100,000 population)	0.50	99.50
	Population who feel safe walking alone at night in the city/area where they live (%)	94.00	100.00
	Corruption Perceptions Index (worst 0–100 best)	45.00	42.30
	Exports of major conventional weapons (TIV constant million USD per 100,000 population)	0.10	97.10
	Press Freedom Index (worst 0–100 best)	23.00	0.00
	Access to and affordability of justice (worst 0–1 best)	0.70	92.30
	Timeliness of administrative proceedings (worst 0–1 best)	0.60	64.30
	Expropriations are lawful and adequately compensated (worst 0–1 best)	0.30	0.00
SDG 17—Partnerships for the Goals	Government spending on health and education (% of GDP)	6.60	44.00
	Other countries: Government revenue excluding grants (% of GDP)	14.10	13.70
	Corporate Tax Haven Score (best 0–100 worst)	63.00	61.70
	Statistical Performance Index (worst 0–100 best)	59.60	46.10

3.3. Analyzing the Leading Stakeholder for Each SDG

The 17 United Nations SDGs are divided into four quadrants according to the relative frequency of the topics mentioned by the two stakeholders, each corresponding to different states of focus, states of management, and drivers (Figure 6). Consistent with the IPA method used to avoid ceiling effects, the average public- and government-dominated values are used as the crosshair.

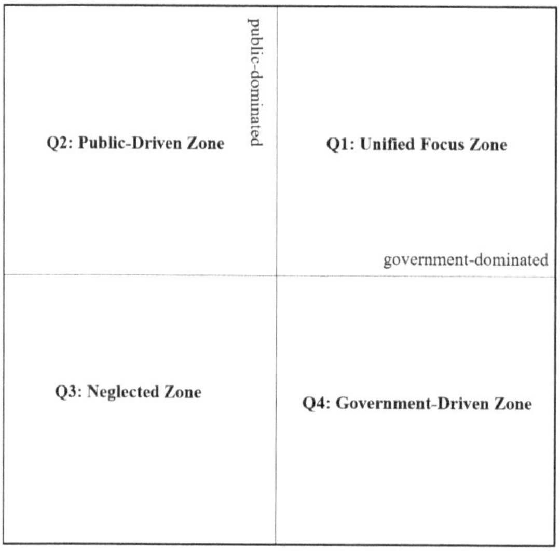

Figure 6. Stakeholder analysis for each SDG.

The results (Figure 7) show that the "unified focus zone" (quadrant 1) contains two SDGs, SDG 2 (Zero Hunger) and SDG 9 (Industry, Innovation, and Infrastructure). These goals demonstrate that the public and the government focus on the critical issues of hunger and infrastructure. The "public-driven zone" (quadrant 2) contains five SDGs, namely, SDG 3 (Good Health and Wellbeing), SDG 4 (Quality Education), SDG 6 (Clean Water and Sanitation), SDG 8 (Decent Work and Economic Growth), and SDG 12 (Responsible Consumption and Production). The high level of public attention to these goals indicates a strong desire for good health, educational opportunities, job security, a clean environment, and sustainable economic practices. The "neglected zone" (quadrant 3) contains six SDGs, namely, SDG 5 (Gender Equality), SDG 10 (Reduced Inequalities), SDG 13 (Climate Action), SDG 14 (Life Underwater), SDG 15 (Life on Land), and SDG 16 (Peace, Justice, and Strong Institutions). The goals in this quadrant are critically important, although they appear to be receiving insufficient attention from both the public and the government. Despite their significance, they might be overlooked due to perceived complexities or a lack of immediate visibility. The "government-driven zone" (quadrant 4) contains four SDGs, namely, SDG 1 (No Poverty), SDG 7 (Affordable and Clean Energy), SDG 11 (Sustainable Cities and Communities), and SDG 17 (Partnerships for the Goals). These goals reflect high levels of government attention, suggesting they are strategically important or a high priority at the policy level.

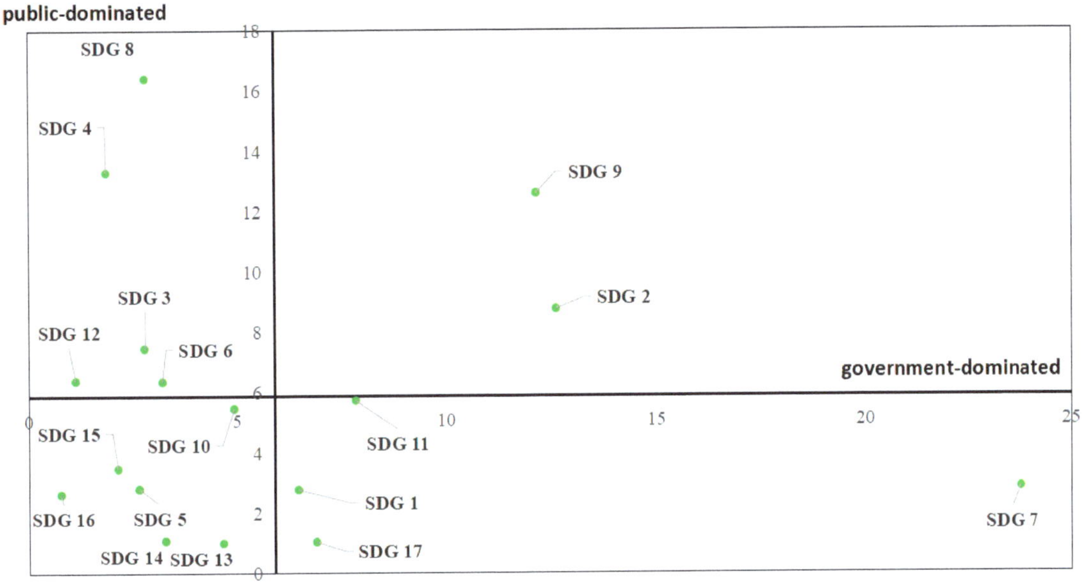

Figure 7. Stakeholder analysis results for each SDG.

3.4. SDG Assessment Using the Importance–Performance Analysis

According to the IPA results (Figure 8), China should concentrate on SDG 10, Reduced Inequalities, and SDG 15, Life on Land. These two SDGs are perceived as highly important but are associated with low performance, which is why they are in the "improvement zone". Additionally, these two SDGs are in the "neglected zone" according to the stakeholder analysis results (Table 4), so the public and government should pay more attention to the two SDGs. Five SDGs, namely, SDGs 6, 7, 14, 16, and 17, are perceived to hold low importance and show poor performances. Additionally, these SDGs are in the "neglected zone" according to the stakeholder analysis results, so the public and government should pay more attention to these SDGs. SDG 7 and SDG 17 are currently government-driven, while SDG 6 is public-driven. The corresponding stakeholders should assess the SDGs' status, integrate other stakeholders' strengths if necessary, and improve their performance accordingly to achieve the 2030 Agenda. SDGs 14 and 16 are in the neglected zone, so the public and government should pay more attention to these two SDGs. Five SDGs, namely, SDGs 1, 2, 3, 11, and 12, are in Q1, which shows the strength of China's progress towards achieving the 2030 Agenda goals. These SDGs are perceived as of high importance with high levels of performance. Additionally, these SDGs have received attention from both the public and the government, who should keep up their good work towards reaching the 2030 Agenda goals. Five SDGs, namely, SDGs 4, 5, 8, 9, and 13, are in Q2, which are perceived as having low importance but with high levels of performance. China should maintain these SDGs to achieve the 2030 Agenda goals.

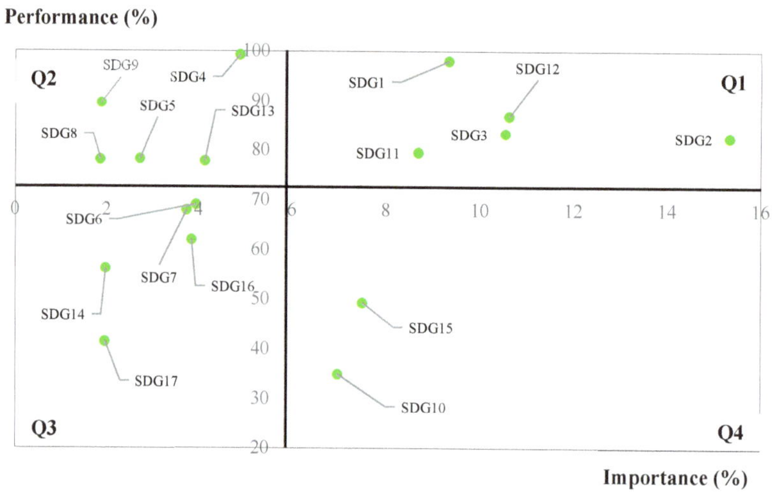

Figure 8. IPA results for each SDG.

Table 4. IPA results for each SDG and the associated stakeholder analysis results.

SDG Types	IPA Results	Stakeholder Analysis Results
SDG 1	Q1	Government-Driven
SDG 2	Q1	Unified Focus Zone
SDG 3	Q1	Public-Driven
SDG 4	Q2	Public-Driven
SDG 5	Q2	Neglected Zone
SDG 6	Q3	Public-Driven
SDG 7	Q3	Government-Driven
SDG 8	Q2	Public-Driven
SDG 9	Q2	Unified Focus Zone
SDG 10	Q4	Neglected Zone
SDG 11	Q1	Government-Driven
SDG 12	Q1	Public-Driven
SDG 13	Q2	Neglected Zone
SDG 14	Q3	Neglected Zone
SDG 15	Q4	Neglected Zone
SDG 16	Q3	Neglected Zone
SDG 17	Q3	Government-Driven

4. Discussion

4.1. Consistencies and Differences in Government and Public Attention Given to the SDGs

The United Nations' 17 SDGs provide a comprehensive framework that is crucial in guiding government policies and mobilizing public participation toward sustainability [3]. Comparative analyses have been employed to explore national adaptation pathways for the SDGs [59,60]. The previous research has predominantly focused on analyzing the attention dedicated to the SDGs by either the public, business stakeholders, or government stakeholders across various countries [61,62]. However, few studies have conducted a comparative analysis of the attention directed toward the SDGs by both the public and the government, who are the two most important stakeholders in implementing the SDGs. Our comparative analysis of these two entities' attention toward the SDGs is particularly insightful due to the potential disparities between governmental actions and public interests. Taking the example of China, in this research we employed multi-source data to comparatively analyze the attention dedicated to the SDGs by both stakeholders.

This study could help in achieving a more comprehensive understanding of the areas of attention of the two stakeholders in implementing the SDGs. These insights could inform the development of more effective collaborative strategies to achieve the common goals of sustainable development.

The results show that both stakeholders in China currently share common interests in certain SDGs and topics related to people's basic needs. They jointly prioritize SDG 2 (Zero Hunger) and SDG 9 (Industry, Innovation, and Infrastructure), SDG 3 (Good Health and Wellbeing), and some other themes in SDG 4 (Quality Education) and SDG 8 (Decent Work and Economic Growth). According to Maslow's hierarchy of needs, basic needs such as food, security, and health must be met before any society can focus on higher-order needs such as sustainability and justice [63]. The mutual focus is on SDGs related to basic welfare and economic stability, as fulfilling these needs is a prerequisite for achieving higher societal goals for both stakeholders. The importance of economic development is reflected in the attention given to business development, market conditions, and economic situations [64]. Additionally, both stakeholders prioritize certain aspects of SDG 5 (e.g., gender equality). The public focuses mainly on women's rights and gender equality, while the government's attention is reflected in the frequent use of the keyword "female", indicating that the government should critically reflect on whether their attention and policies can enhance gender equality [65].

The results reveal that the public tends to prioritize specific SDGs and topics that directly impact their welfare, while the SDGs associated with public goods require government intervention for their provision. According to the theoretical framework of Maslow's hierarchy of needs, individuals are motivated to fulfill basic needs such as health and economic security before addressing higher-level social issues [63]. The findings indicate that SDG 4 (Quality Education), SDG 3 (Good Health and Wellbeing), and SDG 8 (Decent Work and Economic Growth) are of particular interest as they relate closely to the public's wellbeing [64,65]. The SDGs related to the quality of children's education, family health, and employment status significantly affect their quality of life. SDG 16 (Peace, Justice, and Strong Institutions) garners public concern for social safety, justice, and violence. The public looks to the government to provide improved services and support in these areas. Moreover, the lack of public focus on SDGs related to macro-level concerns might stem from a gap in understanding the importance of these issues or the absence of an immediate personal connection. From the perspective of rational ignorance theory, the public might choose not to acquire knowledge about certain complex SDGs that require significant cognitive investment with comparatively little personal return [66–68]. Furthermore, according to the collective action theory [69,70], some SDGs, especially those involving collective goods, such as environmental sustainability or climate action, are more challenging for individuals to engage with due to the diffuse benefits and costs. Along with the public goods theory [71,72], these public goods, which are non-excludable and non-rivalrous, require government intervention for their provision, since the public is less motivated to invest in or provide goods and services that lead to sustainable development due to the lack of direct profitability. Hence, there is a significant need for the government to bridge this gap by enhancing public awareness and engagement through effective education and clear policy communication initiatives.

The research findings indicate the importance of the government in coordinating large-scale responses and policy measures for SDGs involving collective goods or public goods. The results suggest that the government is concerned about certain macro-level issues encapsulated in SDG 7 (Affordable and Clean Energy), SDG 10 (Reduced Inequalities), SDG 11 (Sustainable Cities and Communities), and SDG 17 (Partnerships for the Goals). However, these subjects do not command significant public attention. The government, operating with a broader administrative perspective, emphasizes macro-level issues and long-term planning encapsulated in goals such as SDG 11 and SDG 10 [73,74]. This divergence can be viewed through potential differences between government action and public interest. According to the collective action theory [69,70], the government may play a more prominent role in addressing these SDGs (especially those involving collective goods) because it can coordinate large-scale responses and policy measures beyond the scope of individual actions. The government prioritizes areas that require structured long-term strategies, such as SDG 17 and SDG 7, which involve extensive policy frameworks and international negotiations. Furthermore, due to their non-excludability and non-rivalrous nature, some SDGs require government intervention for their provision [75–78]. For instance, SDGs 11 and 13 often require government intervention and regulation. These areas require technological innovation, bolstering the national industrial competitiveness, and strategic urban and rural development planning [77–79], which may not yield immediately visible benefits, resulting in less public attention being attracted. These issues, although crucial for long-term sustainability, might not align closely with the immediate daily experiences or urgent needs of the general public.

4.2. Stakeholder Analysis for the Sustainable Development Goals

This study utilized a two-dimensional matrix to assess the attention given by the government and the public in China to different SDGs. By categorizing the SDGs into discrete quadrants based on their respective attention levels, this analysis unveiled the harmonies and disparities between the two stakeholders regarding the attention and prioritization of these goals. This approach provides a comprehensive framework for understanding the consistencies and differences between the two stakeholders in prioritizing and driving SDGs. It highlights the need for collaboration, targeted efforts, and increased public education to bridge gaps, align priorities, and enhance the progress towards sustainable development. The findings from this analysis can serve as a valuable guide for policymakers and stakeholders in making informed decisions about resource allocation, policy formulation, and advocacy efforts.

The application of the two-dimensional matrix framework effectively categorizes the SDGs into specific quadrants, providing a clear and structured means to assess their prominence and the need for focused action. The findings indicate that certain SDGs receive greater attention from both the government and the public in China, including SDG 2 (Zero Hunger) and SDG 9 (Industry, Innovation, and Infrastructure). This suggests that shared efforts and resources are dedicated to these goals, enhancing the likelihood of successful progress [79]. The "neglected zone" highlights SDGs that receive little attention from either stakeholder in China. The neglected zone encompasses SDG 5 (Gender Equality), SDG 10 (Reduced Inequalities), SDG 13 (Climate Action), SDG 14 (Life Under Water), SDG 15 (Life on Land), and SDG 16 (Peace, Justice, and Strong Institutions), which require increased attention and resources from both stakeholders in China. These goals risk being overlooked or stagnating without concerted efforts to raise awareness, allocate resources, and foster commitment [80,81]. Their multifaceted and long-term nature may present challenges that necessitate robust awareness-raising strategies and resource mobilization efforts to drive meaningful progress.

The approach used in this study could help enhance the understanding of consistencies and differences in attention given to SDGs by employing a two-dimensional matrix to distinctly categorize public-driven and government-driven SDGs. The public-driven zone comprises SDGs that garner high levels of public attention but comparatively lower levels of government attention. In the public-driven zone, SDG 3 (Good Health and Wellbeing), SDG 4 (Quality Education), and others are prioritized due to their direct impact on people's daily lives. However, the notably lower level of government attention in this quadrant underscores the necessity to align policy priorities with public expectations. While the public interest is conspicuous, driving these goals forward may require more substantial government support and policy interventions. Lastly, the government-driven zone encompasses SDGs that attract high levels of government attention but relatively lower levels of public interest. In the government-driven zone, SDGs such as SDG 1 (No Poverty), SDG 7 (Affordable and Clean Energy), and SDG 11 (Sustainable Cities and Communities) receive high levels of government attention, indicating their strategic importance. Promoting greater public understanding and engagement with these goals can help in leveraging the high level of trust that the Chinese public reportedly has in their government. This high level of trust could facilitate more effective SDG advocacy and execution. Previous research by Edelman Global Public Relations has shown that among the general populations from 26 countries and regions, the Chinese public has the highest level of trust in its government (88% of the informed public) [82,83]. Therefore, the proposed research method contributes to understanding the differences in attention between the government and the public, facilitating the better implementation of SDGs.

5. Conclusions

Based on our case study of China, this paper innovatively underscores the importance of bridging the gaps in attention given to SDGs between the government and the public using multi-source data. This study offers valuable insights into unveiling alignments or discrepancies in attention and commitment, aligning SDG priorities, and strengthening cooperation for more effective SDG implementation between the two stakeholders. This study provides a novel perspective and tool for policymakers and stakeholders to understand and address key issues in driving the implementation of SDGs. The findings of this study have the potential to guide policy direction processes, inform strategies for public engagement, and enhance the comprehension of sustainable development in the digital era.

The results reveal that the public tend to prioritize specific SDGs and topics that directly impact their welfare, while the SDGs involving collective goods or public goods require government intervention for their provision and in coordinating large-scale responses and policy measures. The findings indicate a pronounced emphasis by the government on SDG 7, SDG 2, and SDG 9, with relatively lower levels of attention allocated to SDG 16, SDG 12, and SDG 4. Conversely, the public attention is predominantly centered around SDG 8, SDG 9, and SDG 4, with minimal focus directed toward SDG 13, SDG 14, and SDG 17. Both governmental and public entities exhibited a shared priority for SDGs associated with hunger, industry, innovation, infrastructure, and economic growth.

In examining the progress of the SDGs, the findings highlight the uneven progress across the SDGs and underscores the need for targeted, equitable, and holistic strategies to bolster the lagging indicators towards the 2030 Agenda. SDG 4 and SDG 1 stand out for the considerable strides that have been made toward their targets, while SDG 10, SDG 15, and SDG 17 show a concerning lack of advancement. Delving into specific metrics, the objectives have been fully achieved for a subset of 12 indicators, although nearly one-fifth of the indicators hover below the halfway mark, signaling areas where intensified action is necessary.

Analyzing the stakeholder's focus zones provides valuable insights for policy direction recommendations to enhance the alignment of government and public efforts towards the SDGs. In the unified focus zone (quadrant 1), both stakeholders in China share high levels of attention towards SDG 2 and SDG 9. Policymakers should capitalize on this common ground and shift the policy direction by launching financial products and digital media platforms to promote agricultural products, providing preferential loans to start-up companies and traditional industries. In the public-driven zone (quadrant 2), the public places high importance on wellbeing, health, and education. Policymakers should focus on policies that directly impact the public's wellbeing, such as establishing a trustworthy health system, particularly focusing on primary healthcare, infectious disease control, and non-communicable disease management. Furthermore, given the public's strong emphasis on SDG 4, the government should consider a comprehensive overhaul of the education system. In the government-driven zone (quadrant 4), where the government places high levels of attention on certain SDGs with lower public interest, policymakers should initiate campaigns and policies to inform the public about the relevance of these goals from a macro-level perspective. This could include structured dialogues and communication campaigns that build trust, share knowledge, and clarify the importance of these SDGs for the public.

According to the IPA results, not all SDGs attracting the greatest levels of attention from the public and the government have shown the most progress towards achieving the 2030 Agenda goals. China should concentrate on SDG 10 and SDG 15 and improve SDGs 6, 7, 14, 16, and 17 as a relatively low priority. This study also reveals certain SDGs that receive low levels of attention from the two stakeholders, categorized as the neglected quadrant, necessitating increased attention and resource allocation from the two stakeholders.

This study still has some limitations that require further research in the future. First, it is likely that some content will be subject to censorship by social media platforms. Some content related to SDG barriers could not be collected. Social media platforms can reduce the frequency of mentions by stakeholders, which would consequently reduce the level of public attention, so they still represent an effective reflection of the actual situation regarding public attention. Although social media data are a great and reliable resource to reflect public opinions on SDGs, along with the previous social media data-based studies, it is essential to integrate the advantages of traditional surveys and social media data to comprehensively reflect the perspectives of the public regarding the SDGs. In specific, future studies could collect individual-level data on the attention each SDG receives from members within the stakeholder groups through surveys. This would provide a more detailed and statistically robust understanding of their perspectives. Secondly, Weibo, China News, and Toutiao are not the only three representatives of Chinese social media platforms. Recently, two short video platforms, TikTok and Kuaishou, have attracted many users. Whether our findings can be extended to short video platforms will hopefully be addressed in further research. Finally, our study only focused on the two main stakeholders, the public and the government. Future research studies could be further expanded to other stakeholders.

Author Contributions: Conceptualization: H.F., L.F., K.Z., Q.Z. and L.D.D.; methodology: H.F. and L.F.; writing—original draft preparation: H.F., L.F., K.Z. and Q.Z.; writing—review and editing: H.F., L.F., K.Z., Q.Z. and L.D.D.; visualization: K.Z., Q.Z. and L.D.D.; supervision: K.Z., Q.Z. and L.D.D. All authors have read and agreed to the published version of the manuscript.

Funding: This research received funding from the Fundamental Research Funds for the Central Universities of Central South University (2024ZZTS340).

Data Availability Statement: Data developed in this study will be made available on request to the corresponding authors.

Conflicts of Interest: The authors declare no conflicts of interest.

References

1. Sachs, J.D. From Millennium Development Goals to Sustainable Development Goals. *Lancet* **2012**, *379*, 2206–2211. [CrossRef] [PubMed]
2. Huck, W. Transforming Our World: The 2030 Agenda for Sustainable Development. In *Sustainable Development Goals*; Springer: New York, NY, USA, 2023.
3. Bain, P.G.; Kroonenberg, P.M.; Johansson, L.O.; Milfont, T.L.; Crimston, C.R.; Kurz, T.; Bushina, E.; Calligaro, C.; Demarque, C.; Guan, Y.; et al. Public Views of the Sustainable Development Goals across Countries. *Nat. Sustain.* **2019**, *2*, 819–825. [CrossRef]
4. Sachs, J.; Schmidt-Traub, G.; Kroll, C.; Durand-Delacre, D.; Teksoz, K. *SDG Index and Dashboards Report 2017*; SDSN: New York, NY, USA, 2017.
5. Hák, T.; Janoušková, S.; Moldan, B. Sustainable Development Goals: A Need for Relevant Indicators. *Ecol. Indic.* **2016**, *60*, 565–573. [CrossRef]
6. Li, Z.; Zheng, K.; Zhong, Q. Comprehensive Evaluation and Spatial-Temporal Pattern of Green Development in Hunan Province, China. *Sustainability* **2022**, *14*, 6819. [CrossRef]
7. Zhong, Q.; Li, Z. Long-Term Trends of Vegetation Greenness under Different Urban Development Intensities in 889 Global Cities. *Sustain. Cities Soc.* **2024**, *106*, 105406. [CrossRef]
8. Fu, H.; Liu, J.; Dong, X.; Chen, Z.; He, M. Evaluating the Sustainable Development Goals within Spatial Planning for Decision-Making: A Major Function-Oriented Zone Planning Strategy in China. *Land* **2024**, *13*, 390. [CrossRef]

9. Ormanovic, S.; Ciric, A.; Talovic, M.; Alic, H.; Jeleskovic, E.; Causevic, D. Importance-Performance Analysis: Different Approaches. *Acta Kinesiol.* **2017**, *11*, 58–66.
10. George, T.E.; Karatu, K.; Edward, A. An Evaluation of the Environmental Impact Assessment Practice in Uganda: Challenges and Opportunities for Achieving Sustainable Development. *Heliyon* **2020**, *6*, e04758. [CrossRef] [PubMed]
11. Meuleman, L. Public Administration and Governance for the Sdgs: Navigating between Change and Stability. *Sustainability* **2021**, *13*, 5914. [CrossRef]
12. Zhao, Z.; Pan, Y.; Zhu, J.; Wu, J.; Zhu, R. The Impact of Urbanization on the Delivery of Public Service–Related SDGs in China. *Sustain. Cities Soc.* **2022**, *80*, 103776. [CrossRef]
13. Wang, B.; Xu, S.; Sun, K.; Chang, X.; Wang, Z.; Zhao, W. Government Responsive Selectivity and Public Limited Mediation Role in Air Pollution Governance: Evidence from Large Scale Text Data Content Mining. *Resour. Conserv. Recycl.* **2022**, *187*, 106553. [CrossRef]
14. Van Tulder, R.; Rodrigues, S.B.; Mirza, H.; Sexsmith, K. The UN's Sustainable Development Goals: Can Multinational Enterprises Lead the Decade of Action? *J. Int. Bus. Policy* **2021**, *4*, 1–21. [CrossRef]
15. De Guimarães, J.C.F.; Severo, E.A.; Júnior, L.A.F.; Da Costa, W.P.L.B.; Salmoria, F.T. Governance and Quality of Life in Smart Cities: Towards Sustainable Development Goals. *J. Clean. Prod.* **2020**, *253*, 119926. [CrossRef]
16. Kanie, N.; Biermann, F. *Governing through Goals: Sustainable Development Goals as Governance Innovation*; MIT Press: Cambridge, MA, USA, 2017.
17. Shulla, K.; Leal Filho, W.; Sommer, J.H.; Salvia, A.L.; Borgemeister, C. Channels of Collaboration for Citizen Science and the Sustainable Development Goals. *J. Clean. Prod.* **2020**, *264*, 121735. [CrossRef]
18. Dalampira, E.; Nastis, S.A. Mapping Sustainable Development Goals: A Network Analysis Framework. *Sustain. Dev.* **2020**, *28*, 46–55. [CrossRef]
19. Tiefenbeck, V.; Staake, T.; Roth, K.; Sachs, O. For Better or for Worse? Empirical Evidence of Moral Licensing in a Behavioral Energy Conservation Campaign. *Energy Policy* **2013**, *57*, 160–171. [CrossRef]
20. Ilyas, S.; Hu, Z.; Wiwattanakornwong, K. Unleashing the Role of Top Management and Government Support in Green Supply Chain Management and Sustainable Development Goals. *Environ. Sci. Pollut. Res.* **2020**, *27*, 8210–8223. [CrossRef]
21. Forestier, O.; Kim, R.E. Cherry-picking the Sustainable Development Goals: Goal Prioritization by National Governments and Implications for Global Governance. *Sustain. Dev.* **2020**, *28*, 1269–1278. [CrossRef]
22. Le Blanc, D. Towards Integration at Last? The Sustainable Development Goals as a Network of Targets. *Sustain. Dev.* **2015**, *23*, 176–187. [CrossRef]
23. Bruns, A.; Burgess, J. Crisis Communication in Natural Disasters: The Queensland Floods and Christchurch Earthquakes. *Twitter Soc.* **2014**, *89*, 373–384.
24. Tufekci, Z. Engineering the Public: Big Data, Surveillance and Computational Politics. *First Monday* **2014**, *19*, 7. [CrossRef]
25. Stieglitz, S.; Mirbabaie, M.; Ross, B.; Neuberger, C. Social Media Analytics—Challenges in Topic Discovery, Data Collection, and Data Preparation. *Int. J. Inf. Manag.* **2018**, *39*, 156–168. [CrossRef]
26. Saura, J.R.; Ribeiro-Soriano, D.; Palacios-Marqués, D. From User-Generated Data to Data-Driven Innovation: A Research Agenda to Understand User Privacy in Digital Markets. *Int. J. Inf. Manag.* **2021**, *60*, 102331. [CrossRef]
27. Ostherr, K.; Borodina, S.; Bracken, R.C.; Lotterman, C.; Storer, E.; Williams, B. Trust and Privacy in the Context of User-Generated Health Data. *Big Data Soc.* **2017**, *4*, 1–11. [CrossRef]
28. Zhang, B.; Yin, J.; Jiang, H.; Chen, S.; Ding, Y.; Xia, R.; Wei, D.; Luo, X. Multi-Source Data Assessment and Multi-Factor Analysis of Urban Carbon Emissions: A Case Study of the Pearl River Basin, China. *Urban Clim.* **2023**, *51*, 101653. [CrossRef]
29. Zheng, Y.; Du, S.; Zhang, X.; Bai, L.; Wang, H. Estimating Carbon Emissions in Urban Functional Zones Using Multi-Source Data: A Case Study in Beijing. *Build. Environ.* **2022**, *212*, 108804. [CrossRef]
30. Zhou, Y.; Yi, P.; Li, W.; Gong, C. Assessment of City Sustainability from the Perspective of Multi-Source Data-Driven. *Sustain. Cities Soc.* **2021**, *70*, 102918. [CrossRef]
31. Esmailpour, J.; Aghabayk, K.; Vajari, M.A.; De Gruyter, C. Importance–Performance Analysis (IPA) of Bus Service Attributes: A Case Study in a Developing Country. *Transp. Res. Part A Policy Pract.* **2020**, *142*, 129–150. [CrossRef]
32. Yang, T.; Dang, Y.; Wu, J. Dynamic Perceived Quality Analysis Using Social Media Data at Macro-and Micro-Levels. *Ind. Manag. Data Syst.* **2023**, *123*, 1465–1495. [CrossRef]
33. Watanabe, K.; Zhou, Y. Theory-Driven Analysis of Large Corpora: Semisupervised Topic Classification of the UN Speeches. *Soc. Sci. Comput. Rev.* **2022**, *40*, 346–366. [CrossRef]
34. Karan, M.; Vulić, I.; Korhonen, A.; Glavaš, G. *Classification-Based Self-Learning for Weakly Supervised Bilingual Lexicon Induction*; Association for Computational Linguistics: Stroudsburg, PA, USA, 2020.
35. Husain, F. OSACT4 Shared Task on Offensive Language Detection: Intensive Preprocessing-Based Approach. *arXiv* **2020**, arXiv:2005.07297.
36. Viswanathan, R. Word Relationships Are Not Created Equally. In Proceedings of the 2021 IEEE Pune Section International Conference (PuneCon), Pune, India, 16–19 December 2021; IEEE: Piscataway, NJ, USA, 2021; pp. 1–6.
37. Fouad, M.M.; Mahany, A.; Aljohani, N.; Abbasi, R.A.; Hassan, S.-U. ArWordVec: Efficient Word Embedding Models for Arabic Tweets. *Soft. Comput.* **2020**, *24*, 8061–8068. [CrossRef]

38. Chen, M.; Mussalli, G.; Amel-Zadeh, A.; Weinberg, M.O. NLP for SDGs: Measuring Corporate Alignment with the Sustainable Development Goals. *J. Impact ESG Invest.* **2022**, *2*, 61–81. [CrossRef]
39. Ling, M.; Chen, Q.; Sun, Q.; Jia, Y. Hybrid Neural Network for Sina Weibo Sentiment Analysis. *IEEE Trans. Comput. Soc. Syst.* **2020**, *7*, 983–990. [CrossRef]
40. Chen, Q.; Min, C.; Zhang, W.; Wang, G.; Ma, X.; Evans, R. Unpacking the Black Box: How to Promote Citizen Engagement through Government Social Media during the COVID-19 Crisis. *Comput. Human. Behav.* **2020**, *110*, 106380. [CrossRef] [PubMed]
41. Chen, H.; Huang, X.; Li, Z. A Content Analysis of Chinese News Coverage on COVID-19 and Tourism. *Curr. Issues Tour.* **2022**, *25*, 198–205. [CrossRef]
42. Kuai, J.; Lin, B.; Karlsson, M.; Lewis, S.C. From Wild East to Forbidden City: Mapping Algorithmic News Distribution in China through a Case Study of Jinri Toutiao. *Digit. J.* **2023**, *11*, 1521–1541. [CrossRef]
43. Chang, L.-C.; Wang, X.; Zhang, S.-Q.; Chen, C.-B.; Yuan, Z.-H.; Zeng, X.-J.; Chu, J.-Q.; Tsai, S.-B. User-Driven Business Model Innovation: An Ethnographic Inquiry into Toutiao in the Chinese Context. *Asia Pac. Bus. Rev.* **2021**, *27*, 359–377. [CrossRef]
44. Wagner, T.; Guhl, D.; Langhals, B. The Impact of Data Preparation and Model Complexity on the Natural Language Classification of Chinese News Headlines. *Algorithms* **2024**, *17*, 132. [CrossRef]
45. United Nations. The Sustainable Development Goals Report 2023. 2023. Available online: https://sdgs.un.org/documents/sustainable-development-goals-report-2023-53220 (accessed on 10 April 2024).
46. Osman, A.I.A.; Ahmed, A.N.; Chow, M.F.; Huang, Y.F.; El-Shafie, A. Extreme Gradient Boosting (Xgboost) Model to Predict the Groundwater Levels in Selangor Malaysia. *Ain. Shams Eng. J.* **2021**, *12*, 1545–1556. [CrossRef]
47. Fang, Z.; Yang, S.; Lv, C.; An, S.; Wu, W. Application of a Data-Driven XGBoost Model for the Prediction of COVID-19 in the USA: A Time-Series Study. *BMJ Open* **2022**, *12*, e056685. [CrossRef] [PubMed]
48. Amjad, M.; Ahmad, I.; Ahmad, M.; Wróblewski, P.; Kamiński, P.; Amjad, U. Prediction of Pile Bearing Capacity Using XGBoost Algorithm: Modeling and Performance Evaluation. *Appl. Sci.* **2022**, *12*, 2126. [CrossRef]
49. Zamzuri, M.H.A.; Sofian, N.; Hassan, R. Forecasting of Poverty Using the Ensemble Learning Classification Methods. *Int. J. Perceptive Cogn. Comput.* **2023**, *9*, 1. [CrossRef]
50. Martilla, J.A.; James, J.C. Importance-Performance Analysis. *J. Mark.* **1977**, *41*, 77–79. [CrossRef]
51. Jou, R.-C.; Day, Y.-J. Application of Revised Importance–Performance Analysis to Investigate Critical Service Quality of Hotel Online Booking. *Sustainability* **2021**, *13*, 2043. [CrossRef]
52. Jin, C.; Zhu, D.; Chen, X.; He, J. Evaluation of the Tourists' Satisfaction of Smart Tourist Attractions Using Importance-Performance Analysis: —Taking Jiuzhaigou as an Example. In Proceedings of the Distributed, Ambient and Pervasive Interactions: 8th International Conference, DAPI 2020, Held as Part of the 22nd HCI International Conference, HCII 2020, Copenhagen, Denmark, 19–24 July 2020; Proceedings 22. Springer: Berlin/Heidelberg, Germany, 2020; pp. 391–399.
53. Saransi, F.; Erlinda, N. Assessing Service Quality and Passenger Satisfaction: A Quantitative Study of Sea Transportation on the Padang Mentawai. *Asian J. Econ. Bus. Account* **2024**, *24*, 347–355. [CrossRef]
54. Ulkhaq, M.M.; Wijayanti, W.R.; Dilaga, D.A.; Qadarullah, A.A.; Arfi, F. Combining the Analytic Hierarchy Process and Importance-Performance Analysis to Assess Service Quality of m-Commerce: A Case of Indonesian m-Commerce. In Proceedings of the 2021 IEEE 8th International Conference on Industrial Engineering and Applications (ICIEA), Chengdu, China, 23–26 April 2021; IEEE: Piscataway, NJ, USA, 2021; pp. 6–10.
55. The SDG Transformation Center for the United Nations. Sustainable Development Report 2023. 2023. Available online: https://sdgtransformationcenter.org/reports/sustainable-development-report-2023 (accessed on 10 April 2024).
56. Shuai, C.; Yu, L.; Chen, X.; Zhao, B.; Qu, S.; Zhu, J.; Liu, J.; Miller, S.A.; Xu, M. Principal Indicators to Monitor Sustainable Development Goals. *Environ. Res. Lett.* **2021**, *16*, 124015. [CrossRef]
57. Giles-Corti, B.; Lowe, M.; Arundel, J. Achieving the SDGs: Evaluating Indicators to Be Used to Benchmark and Monitor Progress towards Creating Healthy and Sustainable Cities. *Health Policy* **2020**, *124*, 581–590. [CrossRef] [PubMed]
58. Sachs, J.D.; Kroll, C.; Lafortune, G.; Fuller, G.; Woelm, F. *Sustainable Development Report 2022*; Cambridge University Press: Cambridge, UK, 2022; ISBN 1009210084.
59. Sebestyén, V.; Abonyi, J. Data-Driven Comparative Analysis of National Adaptation Pathways for Sustainable Development Goals. *J. Clean. Prod.* **2021**, *319*, 128657. [CrossRef]
60. Fekih Zguir, M.; Dubis, S.; Koç, M. Embedding Education for Sustainable Development (ESD) and SDGs Values in Curriculum: A Comparative Review on Qatar, Singapore and New Zealand. *J. Clean. Prod.* **2021**, *319*, 128534. [CrossRef]
61. Fowler, A.; Biekart, K. Multi-Stakeholder Initiatives for Sustainable Development Goals: The Importance of Interlocutors. *Public Adm. Dev.* **2017**, *37*, 81–93. [CrossRef]
62. Yamane, T.; Kaneko, S. Impact of Raising Awareness of Sustainable Development Goals: A Survey Experiment Eliciting Stakeholder Preferences for Corporate Behavior. *J. Clean. Prod.* **2021**, *285*, 125291. [CrossRef]
63. Maslow, A.; Lewis, K.J. Maslow's Hierarchy of Needs. *Salenger Inc.* **1987**, *14*, 987–990.
64. Frey, D.F. Economic Growth, Full Employment and Decent Work: The Means and Ends in SDG 8. *Int. J. Hum. Rights* **2017**, *21*, 1164–1184. [CrossRef]
65. Leal Filho, W.; Raath, S.; Lazzarini, B.; Vargas, V.R.; de Souza, L.; Anholon, R.; Quelhas, O.L.G.; Haddad, R.; Klavins, M.; Orlovic, V.L. The Role of Transformation in Learning and Education for Sustainability. *J. Clean. Prod.* **2018**, *199*, 286–295. [CrossRef]
66. Somin, I. Rational Ignorance. In *Routledge International Handbook of Ignorance Studies*; Routledge: London, UK, 2022; pp. 313–322.

67. Bromberger, S. Rational Ignorance. *Synthese* **1988**, 47–64. [CrossRef]
68. Krek, A. Rational Ignorance of the Citizens in Public Participatory Planning. In Proceedings of the 10th symposium on Information- and Communication Technologies (ICT) in Urban Planning and Spatial Development and Impacts of ICT on Physical Space, CORP, Vienna, Austria, 22–25 February 2005; Volume 5, p. 420.
69. DeMarrais, E.; Earle, T. Collective Action Theory and the Dynamics of Complex Societies. *Annu. Rev. Anthropol.* **2017**, *46*, 183–201. [CrossRef]
70. Ostrom, E. Collective Action Theory. In *The Oxford Handbook of Comparative Politics*; Oxford Academic: Oxford, UK, 2009.
71. Holcombe, R.G. A Theory of the Theory of Public Goods. *Rev. Austrian Econ.* **1997**, *10*, 1–22. [CrossRef]
72. Oakland, W.H. Theory of Public Goods. In *Handbook of Public Economics*; Elsevier: Amsterdam, The Netherlands, 1987; Volume 2, pp. 485–535, ISBN 978-0-444-87908-0.
73. Howlett, M. Governance Modes, Policy Regimes and Operational Plans: A Multi-Level Nested Model of Policy Instrument Choice and Policy Design. *Policy Sci.* **2009**, *42*, 73–89. [CrossRef]
74. Rabie, A. Government Policy Reviews and Reforms Relating to the Environment. *S. Afr. J. Environ. Law Policy* **1999**, *6*, 121–148.
75. Kejžar, K.Z.; Ponikvar, N. Global Public Goods in Light of the New Paradigm of Sustainable Development under the Challenges Facing Today's World. In *Global Public Goods and Sustainable Development in the Practice of International Organizations*; Brill: Leiden, The Netherlands, 2023; pp. 21–41, ISBN 9004687262.
76. Olusa, A. Public Goods in Attainment of Sustainable City Growth in Ibadan, Nigeria. Available online: https://www.researchgate.net/profile/Adekemi-Olusa/publication/353606541_Public_Goods_in_Attainment_of_Sustainable_City_Growth_in_Ibadan_Nigeria/links/61057793169a1a0103cc7bfd/Public-Goods-in-Attainment-of-Sustainable-City-Growth-in-Ibadan-Nigeria.pdf (accessed on 6 June 2024).
77. Sæbø, J.I.; Nicholson, B.; Nielsen, P.; Sahay, S. Digital Global Public Goods. *arXiv* **2021**, arXiv:2108.09718.
78. Anwar, M.S.; Ruhtiani, M.; Hendriana, R. Blue Carbon: Integrative Management of Coastal Ecosystems Based on Regional Autonomy. *Jambe Law J.* **2023**, *6*, 185–204. [CrossRef]
79. Zhong, Q.; Fu, H.; Yan, J.; Li, Z. How Does Energy Utilization Affect Rural Sustainability Development in Traditional Villages? Re-Examination from the Coupling Coordination Degree of Atmosphere-Ecology-Socioeconomics System. *Build. Environ.* **2024**, 111541. [CrossRef]
80. Beck, D.; Ferasso, M. How Can Stakeholder Capitalism Contribute to Achieving the Sustainable Development Goals? A Cross-Network Literature Analysis. *Ecol. Econ.* **2023**, *204*, 107673. [CrossRef]
81. Glass, L.M.; Newig, J.; Ruf, S. MSPs for the SDGs—Assessing the Collaborative Governance Architecture of Multi-Stakeholder Partnerships for Implementing the Sustainable Development Goals. *Earth Syst. Gov.* **2023**, *17*, 100182. [CrossRef]
82. Guan, T.; Meng, K.; Liu, W.; Xue, L. Public Attitudes toward Sustainable Development Goals: Evidence from Five Chinese Cities. *Sustainability* **2019**, *11*, 5793. [CrossRef]
83. Trust Barometer. 2019 Edelman Trust Barometer-Global Report. 2019. Available online: https://www.edelman.com/sites/g/files/aatuss191/files/2019-03/2019_Edelman_Trust_Barometer_Global_Report.pdf?utm_source=website&utm_medium=global_report&utm_campaign=downloads/ (accessed on 10 April 2024).

land

Article

Impact of Free Trade (Pilot) Zone Establishment on Urban Land Use Efficiency—Empirical Evidence from Cities in China

Siwen Xia [1], Jiaying Han [2], Anglu Li [2], Penghao Ye [2,*] and Huarong Zhang [1]

[1] School of Business Administration, Zhongnan University of Economics and Law, Wuhan 430073, China; xiasiwen@stu.zuel.edu.cn (S.X.); zhrzuel@126.com (H.Z.)
[2] International Business School, Hainan Open Economy Research Institute, Hainan University, Haikou 570228, China; 20213003006@hainanu.edu.cn (J.H.); 20213003072@hainanu.edu.cn (A.L.)
* Correspondence: paulyph@hainanu.edu.cn

Abstract: The establishment of the Free Trade (Pilot) Zone (FTZ) is a new attempt by China to embark on modern sustainable urban planning and governmental territorial management at the development stage of a high-quality economy. As urban lands serve as the limited resource foundation for civic production and livelihoods, enhancing land use efficiency becomes a key measure to facilitate metropolitan economies of high quality. Utilizing panel data from prefecture-level cities spanning from 2005 to 2021, this study constructs a multi-period difference-in-differences model to explore the impact of FTZs on urban land use efficiency (ULUE). The results indicate that FTZs can significantly enhance urban land use efficiency, with a more pronounced policy effect observed in central cities, inland cities, and cities with higher urbanization rates. Furthermore, the FTZs also demonstrate spatial spillover effects on urban land use efficiency. This study holds great significance for relevant government departments in formulating policies to optimize land resource allocation, promote FTZ development, and foster high-quality urban planning and territorial management.

Keywords: free trade (pilot) zone; urban land use efficiency; multi-period difference-in-differences; heterogeneity analysis; spatial spillover effect

1. Introduction

Since the reform and opening up, particularly after joining the World Trade Organization (WTO) in 2001, China's economy has experienced sustained rapid growth for many years. However, starting in 2008, China's economic growth rate has gradually slowed down, partly due to external shocks from the economic crisis and challenging external demand conditions; on the other hand, it is also closely related to internal factors such as unbalanced economic development structures and an over-reliance on external demand. Profound changes in both domestic and international environments have compelled China to proactively move from a focus on quantity to quality in its transformation. High-quality urban economic development requires the transformation and upgrading of traditional industries, optimizing the layout of emerging industries, and, simultaneously, leveraging institutional reforms to advance comprehensive reforms in urban planning, construction, and management [1,2]. The establishment of Free Trade (Pilot) Zones (FTZs) is an important measure for China to deepen reforms comprehensively and seek high-quality economic development under new circumstances.

Since the reform and opening up in 1978, China has established various special zones in response to domestic and international situations, including Special Economic Zones (SEZs), High-Tech Industrial Development Zones, National New Areas, and Free Trade (Pilot) Zones. High-Tech Industrial Development Zones and National New Areas focus on promoting high-tech products or regional development through policy incentives provided by the government. In contrast, SEZs and FTZs drive comprehensive development across China by actively opening up and integrating with international standards. However, SEZs were initially founded under a planned economy framework, aiming at facilitating the flow of goods, capital, talent, and technology through preferential policies, essentially constituting a form of "border opening". Free Trade (Pilot) Zones, on the other hand, are set up under a market economy framework, aiming to actively benchmark against international rules and create a "soft environment" that attracts, accumulates, and cultivates high-end production factors, fundamentally constituting a form of "domestic opening". Establishing FTZs is widely recognized as one of the key strategies for becoming a global hub port [3]. However, the size, geographic location, and other characteristics of FTZs vary by country or region, leading to differences in the form, positioning, and level of development of FTZs. For example, an FTZ could be as small as a duty-free shop at an airport or as large as the entire city of Hong Kong [4].

Most studies on Free Trade (Pilot) Zones focus on foreign investment management systems [5], trade supervision systems [6], financial service systems [7], etc. Within these zones, however, during the research process in the Free Trade Pilot Zones, we discovered that in addition to defining the scope and subdividing the areas of the Free Trade Pilot Zones, there are also regulations concerning the development and utilization of land and space in these zones. For instance, the "China (Shanghai) Pilot Free Trade Zone Lingang New Area Territorial Space Master Plan (2021–2035)" issued in 2021, and the "Notice on Printing and Distributing the Overall Plan of Beijing, Hunan, and Anhui (Pilot) Free Trade Zone and the Regional Expansion Plan of Zhejiang Pilot Free Trade Zone" issued by The State Council of China in 2020, all emphasize the need for the development and utilization of the free trade zone to comply with land use, ecological environmental protection, and planning regulations, aligning with territorial space planning and the requirements of economizing and intensive land use. Some free trade zones have specific strategies for intensive and sustainable urban land use. For example, the Shanghai FTZ has pioneered mixed-use land utilization, the Qingdao FTZ has implemented a flexible supply of industrial land and the 1.5-level land development model, and the Yingkou Free Trade Zone has piloted the "standard land +" system, among others.

The regulations on land development and the pioneering land use models in Free Trade Pilot Zones naturally raise questions: Can the establishment of these zones affect the land use efficiency of their host cities? If so, are there regional differences in the effects? Is there a spillover or suction effect on the land use efficiency of neighboring cities? Answering these questions not only helps in a more comprehensive evaluation and understanding of the construction effectiveness of the Free Trade Pilot Zones but also holds significant practical importance for China in further improving the development of these zones.

Based on this, the paper first uses the Super-Efficiency Slacks-Based Measure (Super-SBM) method to calculate the urban land use efficiency of 297 Chinese cities from 2005 to 2021. Then, using a multi-period Difference-in-Differences (DID) approach, it focuses on examining the impact of China's Free Trade Pilot Zones on urban land use efficiency. Furthermore, the paper discusses the heterogeneous effects and spatial effects of the Free Trade Pilot Zones on urban land use efficiency (ULUE). The marginal contribution of this paper is primarily twofold. Firstly, while most studies on pilot free trade zones focus on trade and economic development, few scholars delve into the efficiency of land resource utilization. This paper bridges this gap by examining the construction of FTZs in conjunction with urban land use efficiency, thus enriching the research perspective on FTZs and factors influencing ULUE. Secondly, the research significance is underscored by China's economic transition from high-speed growth to high-quality growth. Rational allocation of land resources becomes imperative for urban economic transformation. Therefore, exploring the impact of pilot free trade zone policies on urban land use efficiency holds great importance in strengthening policy evaluation and design by relevant government departments, thereby enhancing the government's role as the "visible hand" in the market economy.

The structure of the remaining sections of the paper is organized as follows: Section 2 is the literature review and research hypothesis; Section 3 describes the policy background, research design, and data description; Section 4 presents the analysis of empirical results and robustness tests; Section 5 conducts further analysis; Section 6 draws conclusions and discusses policy implications.

2. Literature Review and Research Hypothesis

2.1. Literature Review

2.1.1. Research on FTZ Policy

Free Trade Zones (FTZs) serve as significant mechanisms for countries to adapt to globalization and cultivate comparative advantages in global trade [8]. Ghosh et al. [9] conducted an empirical test on the geographical and economic spillover effects of free trade zones in the United States using the propensity score matching method. They found that FTZs contribute to the growth of non-manufacturing industries, with the positive impact exhibiting spillover effects, particularly within a 5-mile radius of the FTZs. Similarly, some scholars conducted a comparative study of the Masan Port Free Trade Zone and the Customs Free Trade Zone (CFZ) in South Korea, revealing that FTZs outperform CFZs in terms of economic development strategy, industrial structure, and investment promotion [10]. Since the establishment of the Shanghai FTZ in 2013, China's pilot FTZ policy has not only expanded trade openness and attracted foreign investment but has also undertaken key tasks such as fostering further reforms and institutional innovation [11]. Several studies have indicated that this policy can enhance trade efficiency by reducing trade costs [12] and, to some extent, improve trade structure by increasing the share of general trade while reducing the share of processing trade [13].

In addition, pilot free trade zones can effectively facilitate in the attraction of international capital and the expansion of domestic capital abroad [14]. Further analysis of the heterogeneous effects of FTZs reveals that cities that implemented special economic zone policies early tend to experience greater success in attracting foreign capital. This phenomenon may be attributed to the lagged effects of FTZ implementation and supportive policies [15]. It is noteworthy that studies have demonstrated that FTZs can generate external agglomeration effects, leading to knowledge spillovers and technological innovation through collaboration and exchange with neighboring cities [16,17].

An important advantage of FTZs in fostering knowledge spillovers lies in their conducive institutional environment [18], which provides a solid foundation for enhancing innovation capacity [19]. While the environment is a crucial aspect of FTZs, research on the impact of FTZ policies on the environment has not yielded a unified conclusion. Some scholars argue that trade liberalization can enhance environmental quality through technological progress and effectively address the tension between demand and resources [20]. Wang et al. [21] found that the pilot free trade zone policy effectively improves urban green total factor productivity, with this impact gradually increasing over time. Hu et al. [22] also support this view. However, some scholars argue that the establishment of free trade zones can lead to an environmental "policy trap", exacerbating environmental constraints [23]. Grossman and Krueger [24] identified that due to the interaction of scale, structural, and technological effects, environmental quality can significantly decrease with the promotion of free trade zones. Similarly, Liu et al. [25] confirmed that trade liberalization could harm the environment in developing countries.

Apart from the macro perspective mentioned above, a small number of scholars have examined the policy effects of free trade zones from a micro perspective. Li et al. [26] explored the influence of policies in pilot free trade zones on the development performance of port-listed companies within these zones. They concluded that port-listed companies effectively leveraged the agglomeration effect of free trade zones to access high-quality resources such as capital, management, and talent, thereby realizing long-term and rapid development. Jiang et al. [27] found that free trade zones have significantly enhanced the green total factor productivity of enterprises, with this improvement strengthening year by year.

2.1.2. The Measurement of Urban Land-Use Efficiency

Urban land serves as the spatial carrier for a city's economic, social, and environmental aspects, and its utilization efficiency is directly linked to urban socio-economic development and living environments [28]. Consequently, countries and regions around the world place great emphasis on the efficient use of land. For example, the United States has initiated smart growth and urban growth boundaries; the United Kingdom and Germany focus on compact city development [29]; the Netherlands and the Hong Kong region of China implement diversified intensive use [30,31]; and Singapore imposes land use structural constraints [32]. The measurement of land use efficiency has been continuously optimized, undergoing a transformation from single-index measurement to multi-index comprehensive measurement and then to non-parametric measurement [33–35]. Most traditional methods of measuring land use efficiency only consider a single index, such as economic aspects like land use density and unit land yield [36] or ecological [37] and social aspects [38]. Obviously, a single index cannot fully capture the relationship between various inputs and outputs in urban land use processes. Therefore, scholars have proposed multi-index comprehensive evaluation methods [39–41].

However, the multi-index comprehensive measurement method still faces challenges such as subjective index selection and weight determination [42]. As a result, non-parametric methods based on comprehensive evaluation of multiple inputs and outputs have become widely adopted. Gao et al. [43] constructed a two-stage Data Envelopment Analysis (DEA) method to measure urban construction land use efficiency, finding that efficiency during the construction phase was higher than during the production phase. Other scholars have also used DEA methods to measure urban land use efficiency from different perspectives, such as adjusting agriculture [44] and input-output structures [45], without considering non-expected outputs.

In addition to economic output, it has become consensus that non-expected outputs such as waste gas and wastewater discharge should be included in land use efficiency calculations. The Slack-Based Measure (SBM) model for land use efficiency measurement is an improvement over the traditional DEA model, as it accounts for non-expected outputs of land use, making it the mainstream method. Zhou et al. [46] and Xiao et al. [47] considered environmental pollution and carbon emissions as non-expected outputs when measuring land use efficiency. The Super-SBM model, based on non-expected output, addresses the issue of decomposing the efficiency value of effective decision-making units, preventing the loss of decision-making information in practical applications [48,49]. Furthermore, Huang et al. [50] proposed an SBM model considering global reference to address the challenge of comparing efficiency across time.

2.1.3. Factors Influencing Urban Land-Use Efficiency

Scholars have also delved into the factors influencing urban land use efficiency [51–54]. Existing research primarily examines these factors from the perspectives of economic development [55,56], market openness [57], industrial structure [58], technological innovation ability [59], city size [60] and characteristics [61], public infrastructure level [62], government behavior and policy [63,64], land system [65], and land finance [66], among others. For instance, Fan et al. [67] explored the main driving factors of land use efficiency using the grey correlation model, revealing that neglecting negative ecological impacts would lead to an overestimation of land use efficiency. Zhou et al. [68] conducted an empirical assessment of the level of land-intensive use in the Shanghai Pilot Free Trade Zone using the Delphi method and entropy value method, identifying development time, land use intensity, differences in leading functions, industrial land use structure, and diversification of employment structure as significant influencing factors.

It is evident from the existing literature that most research on Free Trade Pilot Zones (FTZs) focuses on their economic impacts, such as trade, investment, total factor productivity, and business environment. Studies on the factors affecting urban land use efficiency are also abundant. However, research specifically examining the relationship between FTZs and urban land use efficiency is very scarce, with few in-depth studies on the heterogeneity of FTZ impacts across different regions and city sizes. This paper considers the literature on measures of land use efficiency and uses the Super-SBM method to initially measure urban land use efficiency. It then treats the establishment of FTZs as a quasi-natural experiment to explore their impact on urban land use efficiency (ULUE) through a multi-period DID analysis, thus further extending the existing research findings.

2.2. Research Hypothesis

Enhancing the efficiency of land resource allocation is a crucial element in the development strategies of each Free Trade Pilot Zone. For instance, "In the Free Trade Pilot Zones, land is supplied based on the industrial chain, with integrated supply for multiple plots of land involved in the key links and core projects of the industrial chain. There is also support for localities in prioritizing the reasonable land use needs of the Free Trade Pilot Zones when arranging land use plans". Specifically, the Free Trade Pilot Zones have achieved success in models of efficient land use. For example, the Shanghai Free Trade Pilot Zone pioneered a pilot for mixed-use development, innovatively planning the unified construction of research and development facilities and production plants after land acquisition by businesses within the zone, thus achieving efficient land use [69]. Additionally, the Free Trade Pilot Zones explore the implementation of a "negative list", improve legal regulations such as intellectual property protection and arbitration judicial review, continually expand the areas of openness, and enhance the level of trade and investment facilitation, which benefits the improvement of the business environment and leads to an upgrade in the industrial structure within the zone [70], thereby enhancing urban land use efficiency. Accompanying the institutional innovations of the Free Trade Pilot Zones are reductions in institutional transaction costs and the inflow of high-quality investments and high-level talents, which favor technological innovation and, in turn, boost urban land use efficiency [71].

Based on the aforementioned analysis, this paper proposes Hypothesis 1:

H1: *Pilot free trade zone policies can enhance urban land-use efficiency within the zone.*

The policy of establishing pilot free trade zones does not solely aim at the development of the city where they are located; it also considers whether the free trade zone impacts surrounding cities. If so, further investigation is needed to determine whether it is a spillover effect or a siphon effect. Upon reviewing the overall plans for the construction of pilot trade zones, many emphasize the goal of forming high-standard and high-quality pilot free trade zones with outstanding radiation-driving effects within three to five years of reform and exploration. Existing research also indicates that Free Trade Pilot Zones indeed can produce significant positive spillover effects [72–74].

Based on this, it is inferred that the improvement effect of pilot free trade zones on urban land use efficiency can extend to surrounding cities, exerting a positive spillover effect on their land use efficiency. Therefore, Hypothesis 2 is proposed:

H2: *The construction of pilot free trade zones can enhance the land-use efficiency of surrounding cities through spillover effects.*

Due to differences in resource endowment, geographical location, and policy environment, cities will inevitably experience varying levels of development. Since China's reform and opening up, cities in coastal areas and the eastern regions have vigorously developed export-oriented economies, capitalizing on their geographical advantages and policy incentives. They have consequently accumulated significant capital, technology, and high-quality labor. With the initiatives of "Developing the West" and the "Rise of the Central Region", the central and western regions of China have seized the opportunity to achieve relatively rapid development. The differences in industrial structures and economic development gaps between regions have gradually narrowed.

However, there are still significant disparities in development between regions. Cities in the central and western regions of China have relatively weaker geographical conditions, resource endowments, and policy environments, which make them less attractive for factor agglomeration. However, the construction of Free Trade Pilot Zones (FTZs) facilitates the revitalization of production factors and optimizes resource allocation [75]. Consequently, this paper posits that the impact of FTZ policies varies among cities in the eastern, central, and western regions, as well as between coastal and inland areas, with potentially greater policy marginal utility for cities in the central and western regions, as well as for inland cities. Furthermore, considering that land is a core element of urbanization, the policy effects of FTZs on Urban Land Use Efficiency (ULUE) may differ depending on the level of urbanization. During urbanization, land use methods are optimized, and the extensive use of land shifts toward conservation and intensification, thereby enhancing land use efficiency [76,77]. Therefore, this paper argues that the establishment of FTZs has a more pronounced policy effect on land use efficiency in cities with higher rates of urbanization.

Building on this, Hypothesis 3 is proposed:

H3: *The Free Trade Pilot Zone policy has a more significant impact on cities in the central and western regions, inland cities, and cities with high urbanization rates.*

Figure 1 describes the theoretical hypotheses framework.

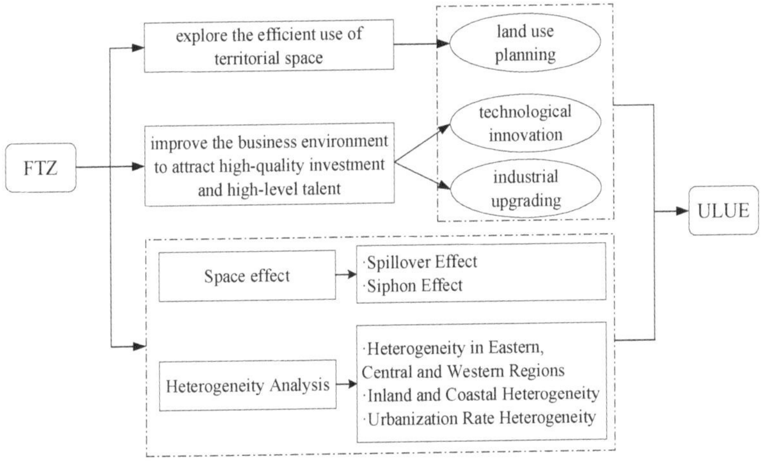

Figure 1. The theoretical hypotheses.

3. Policy Background and Research Design

3.1. Policy Background

Since the official launch of the China (Shanghai) Pilot Free Trade Zone in September 2013, by the first half of 2024, China has essentially established a new pattern of regional opening up. This pattern is comprehensive and high-level, coordinated across eastern, central, and western regions, and integrates both land and maritime strategies, with 22 Free Trade Zones forming the main framework. Specifically, in April 2015, three Chinese provinces—Guangdong, Tianjin, and Fujian—were approved as pilot free trade zones. In March 2017, seven provinces—Liaoning, Zhejiang, Henan, Hubei, Chongqing, Sichuan, and Shaanxi—were officially established as the second batch of pilot free trade zones, marking a deeper expansion of FTZs from coastal areas to inland regions. In October 2018, the China (Hainan) Pilot Free Trade Zone was established, designating the entire island of Hainan as a pilot zone. Subsequently, in August 2019, six provinces—Shandong, Jiangsu, Guangxi, Hebei, Yunnan, and Heilongjiang—were approved as pilot free trade zones. In September 2020, Beijing, Hunan, and Anhui were newly established as pilot free trade zones.

Finally, in October 2023, the first pilot free trade zone in China's northwest border region, the Xinjiang Pilot Free Trade Zone, was established. At this juncture, the network of pilot free trade zones covering coastal, border, and inland areas was formally established, becoming an essential platform for the development of China's open economy and global free trade.

The geographical distribution of China's Free Trade Zones and their founding years is shown in Figure 2.

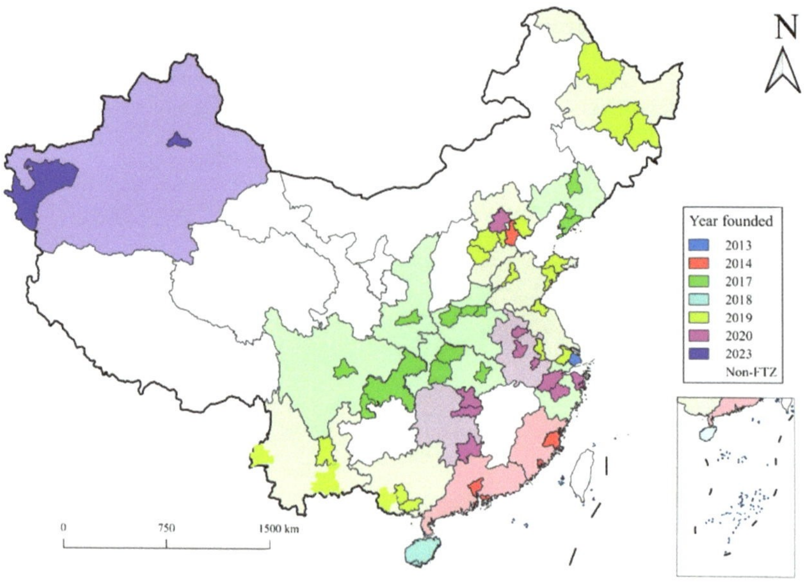

Figure 2. The Map of the Geographical Distribution of China's Free Trade Zones from 2013 to 2023. (Source: Plotted by authors based on official authoritative documents).

3.2. Model Setting

After the official establishment of the Shanghai Free Trade Zone in 2013, China established a total of 22 free trade zones in batches in 2015, 2017, 2018, 2019, 2020, and 2023, providing a favorable "quasi-natural experiment" scenario for this study. Due to data availability issues, the Xinjiang Free Trade Zone, officially inaugurated in November 2023, is not included in the analysis. Therefore, drawing from existing literature, this paper constructs a multi-period difference-in-differences model based on the implementation timing of the six batches of pilot free trade zone policies in 2013, 2015, 2017, 2018, 2019, and 2020. It establishes virtual variables for pilot free trade zone policies and analyzes their impact on urban land use efficiency. Given that urban land use efficiency in different regions is influenced by factors such as economic development level and industrial structure, and to address the problem of endogeneity arising from missing variables, we construct the following bidirectional fixed multi-period Difference-in-Differences (DID) model:

$$ULUE_{it} = \alpha + \beta \cdot DID_{it} + \sum_{j}^{n} (\gamma_j \cdot Control_{jt}) + \mu_i + \lambda_t + \eta_{pt} + \varepsilon_{it} \quad (1)$$

where $ULUE_{it}$ represents the urban land use efficiency of the city i in year t, and α is the constant term; DID_{it} is the policy dummy variable. $Control_{it}$ represents a series of control variables that may affect urban land use efficiency independently of FTZ policy, all of which are control variables of baseline regression. μ_i is the individual (city) fixed effect, indicating the characteristics of the city level that do not change with time, such as the city's terrain, climate, and other natural conditions. λ_t is the time fixed effect. In addition, after considering that the influence of province characteristics on land use efficiency has time-varying characteristics, the interactive fixed effect of province and year is added to the equation, i.e., η_{pt}. ε_{it} is the classic random error term. This paper is mainly concerned with the coefficient β in the equation, which is the impact of the implementation of the FTZ policy on urban land use efficiency. The following analysis defaults to clustering standard errors at the city level.

3.3. Variable Selection

3.3.1. Explained Variables

Urban land use efficiency (*ULUE*): Based on the literature review of urban land use efficiency measurements previously discussed, and considering the impacts of urban land use on the economy, society, and the environment—especially the high environmental costs incurred—this paper posits that urban land use efficiency should not only consider the inputs and expected outputs from production and living activities on the land but also the unintended outputs resulting from these activities. Therefore, this paper, referencing Tone (2001) [78], employs the Super-SBM model that accounts for undesirable outputs to measure urban land use efficiency. The advantage of this model lies in its ability to address the oversight of relaxation variables in the efficiency evaluation process within the radial model.

Let us consider a set of cities (n = 1, 2, ..., N), where each city serves as a production decision-making unit (DMU). Each DMU comprises an input, expected output, and unexpected output, denoted as m, l, and h, respectively. The calculation formula is as follows:

$$min\theta^* = \frac{1 + \frac{1}{m}\sum_{m=1}^{M}\left(s_m^x / s_{jm}^t\right)}{1 - \frac{1}{l+h}\left[\sum_{l=1}^{L}\left(s_l^y / s_{jl}^t\right) + \sum_{h=1}^{H}\left(s_h^b / b_{jh}^t\right)\right]} \tag{2}$$

$$s.t.\begin{cases} x_{jm}^t \geq \sum\limits_{j=1,\, j\neq 0}^{n} \lambda_j^t x_{jm}^t + s_m^x \\ y_{jl}^t \geq \sum\limits_{j=1,\, j\neq k}^{n} \lambda_j^t y_{jl}^t - s_l^y \\ b_{jh}^t \geq \sum\limits_{j=1,\, j\neq k}^{n} \lambda_j^t b_{jh}^t + s_h^b \\ \lambda_j^t \geq 0,\, s_m^x \geq 0,\, s_l^y \geq 0,\, j = 1, 2, \ldots, n \end{cases} \tag{3}$$

where θ^* represents the urban land use efficiency value; x_j^t, y_j^t and b_j^t respectively denote the input, expected output, and non-expected output values of the DMU_j at time t; s_m^x, s_l^y, and s_h^b represent the relaxation vectors for input, expected output, and non-expected output; and λ is the weight vector.

Following classical economic thought, this paper selects the most fundamental production inputs—land, capital, and labor—as factors influencing land use efficiency. The corresponding indicators are represented respectively by the built-up area, fixed capital stock, and the number of employed individuals in the secondary and tertiary industries of each city. The fixed capital stock is calculated using the perpetual inventory method, with a depreciation rate of 9.6% based on the year 2000, as per the practices of scholars such as Hall and Jones (1999) [79] and Young (2003) [80].

Economic benefit output is measured by the added value of the secondary and tertiary industries, while the general public budget revenue of local governments represents the social benefit output. The ecological environment benefit output is indicated by the green coverage rate of built-up areas. To comprehensively assess the non-expected output outputs stemming from various types of urban land use, three commonly used indicators—industrial sulfur dioxide emissions, industrial wastewater emissions, and industrial smoke and dust emissions—are selected. Additionally, carbon dioxide emissions, introduced by

Glaeser and Kahn (2008) [81], are used to characterize environmental pollution arising from commercial service land and residential land.

The evaluation indicators for *ULUE* are presented in Table 1.

Table 1. *ULUE* Evaluation Indicators.

Input and Output	Index	Variables	Unit
Inputs	Land	Built-up area	km^2
	Capital	Stock of fixed capital	Ten thousand yuan
	Labor force	Population employed in secondary and tertiary industries	Ten thousand people
Expected output	Economic benefits	Added value of secondary and tertiary industries	One hundred million yuan
	Social benefits	Local general public budget revenue	Ten thousand yuan
	Ecological and environmental benefits	Green coverage rate of built-up area	%
Non-expected output Inputs	Negative environmental benefits	Industrial wastewater	Ten thousand tons
		Industrial smoke dust	Tons
		Industrial sulfur dioxide	Tons
		CO_2 emissions	One million tons

3.3.2. Core Explanatory Variables

The dummy variable (*DID*) for the Free Trade Zone policy equals the interaction term between the treatment group dummy variable (*FTZ*) and the post-period dummy variable (*Post*).

$$DID_{it} = FTZ_i \times Post_{it} \quad (4)$$

where *FTZ* represents whether a city i is designated as a free trade zone, with a value of 1 if the city is approved as a free trade zone and 0 otherwise. *Post* represents the dummy variable for the policy implementation period, with a value of 1 for the year t when the free trade zone in city i is approved and subsequent years, and 0 otherwise. Since the establishment of the Shanghai FTZ in 2013, China has established 22 provincial FTZs in seven batches, covering four municipalities directly under the central government (Shanghai, Tianjin, Chongqing, and Beijing) and eighteen provinces.

Due to data availability constraints, we will exclude discussion on the Xinjiang FTZ, which was officially inaugurated on 1 November 2023, and focus solely on the policy effects of the first six batches of FTZs. Furthermore, it is important to note that FTZs in the four municipalities directly under the Central Government (Beijing, Shanghai, Tianjin, Chongqing) actually consist of multiple sub-zones. For ease of analysis, the baseline model assumes the existence of only one free trade zone per municipality. Hainan Province, on the other hand, is designated as a free trade port for the entire province. Given administrative consistency, the study focuses only on three prefecture-level cities in Hainan Province: Haikou, Sanya, and Danzhou. Additionally, due to data limitations and statistical considerations at the county level, if the free trade pilot area is a district or county unit, the corresponding city is taken as the study sample. Please refer to Table A1 in Appendix A for details on the establishment of the respective pilot trade zones.

3.3.3. Control Variables

In addition to being influenced by policies of the pilot free trade zone, urban land use efficiency in the FTZ is also closely associated with factors such as the level of social and economic development, regional industrial structure, population size, and local financial support. To minimize estimation bias resulting from missing variables, this paper incorporates a set of city-level control variables into the empirical model, drawing from existing literature on urban land use efficiency and pilot free trade zone policies. Specifically: (1) Level of economic development ($lnAGDP_{pc}$): The natural logarithm of per capita GDP is used to represent the level of economic development. Generally, cities with stronger economic strength tend to have higher scientific and technological investment, greater human resource capacity to utilize resources, and, thus, higher urban land use efficiency. (2) Level of industrial structure (*ind*): The level of industrial structure is measured by the ratio of the added value of the tertiary industry to the added value of the secondary industry. Changes in industrial structure directly or indirectly affect land use structure and patterns, thereby influencing land use efficiency. (3) Population size (*lnPop*): Population size is represented by the logarithm of the total population of the city. Generally, the crowding effect resulting from population agglomeration reduces the land's carrying capacity, leading to a decline in urban land use efficiency. (4) Government intervention (*gov*): The degree of government intervention is reflected through the proportion of fiscal expenditure to GDP. Generally, government intervention, to a certain extent, improves urban land use efficiency because the government considers both efficiency and equity to ensure the region's sustainable development. (5) Level of human capital (*hum*): The level of urban human capital is represented by the number of college students per capita. Human capital is also a crucial factor influencing land use efficiency. (6) Level of openness to the outside world (*open*): The proportion of foreign direct investment in GDP is used to reflect the city's level of openness to the outside world. Generally, a higher level of openness allows the region to understand and adopt foreign advanced experience and technology earlier, resulting in relatively higher initial urban land use efficiency.

Table 2 presents the definitions of all variables.

Table 2. Definition and description of variables.

Type	Symbol	Definition
Explained variable	*ULUE*	Urban land-use efficiency measured using the Super-SBM model considering undesired outputs
Explanatory variable	*DID*	Interaction term for area and time of implementation of free trade pilot zones
Control variable	$lnAGDP_{pc}$	Logarithmic value of GDP per capita
	ind	Ratio of value added of tertiary sector to value added of secondary sector
	lnpop	Logarithmic value of total urban population
	gov	Ratio of fiscal expenditure to GDP
	human	Value of university students per capita in cities
	open	Ratio of foreign direct investment to GDP

3.4. Sample and Data Description

This paper selects data from prefecture-level cities spanning from 2005 to 2021 as samples to investigate the impact of pilot free trade zone policies on urban land use efficiency. The relevant data primarily originate from the "China City Statistical Yearbook", "China Urban Construction Statistical Yearbook", statistical yearbooks and bulletins of individual cities, "China Statistical Yearbook on Environment", and the China Carbon Emission Accounts and Datasets (CEADs). To ensure the accuracy of the research conclusions, interpolation methods were utilized to address missing data, and samples with significant missing data were excluded during the study period. The descriptive statistics of the main variables are presented in Table 3.

Table 3. Descriptive statistics.

Variables	Observed Quantity	Mean Value	Standard Deviation	Min	Max
ULUE	4932	0.40	0.25	0.000	1.000
DID	4932	0.039	0.194	0.000	1.000
$\ln AGDP_{pc}$	4932	10.43	0.76	4.595	13.056
ind	4932	0.98	0.56	0.094	5.929
ln*pop*	4932	5.83	0.79	−1.609	8.141
gov	4932	0.19	0.14	0.043	2.349
hum	4932	4.69	1.01	0.000	7.165
open	4932	0.14	0.93	0.000	57.694

4. Result Analysis and Robustness Test

4.1. Baseline Regression Results

To test Hypothesis 1, regression was performed on Model (1), and the results are shown in Table 4. Without including control variables, column (1) shows that the regression coefficient for the construction of Free Trade Pilot Zones on urban land use efficiency is 0.1735, significant at the 1% level as positive, suggesting that the establishment of Free Trade Pilot Zones improves urban land use efficiency, preliminarily validating Hypothesis 1. After adding control variables and fixing individual effects (City FE) and time effects (Year FE), the coefficient of the core explanatory variable in column (2) is 0.0375, significant at the 5% level as positive. This implies that, with all other conditions being constant, cities with Free Trade Pilot Zones have increased their land use efficiency by 0.0375 compared to cities without such zones, which is equivalent to 9.375% of the average land use efficiency among the sample cities, with both statistical and economic significance. Furthermore, to control for the annual changes in the impact of the Free Trade Pilot Zone policy on ULUE, column (3) further controls for province-year interaction fixed effects (Province*Year FE), and column (4) performs a province-clustered analysis based on column (3), finding that the coefficient of the core explanatory variable is significantly positive at the 1% level. These results indicate that the Free Trade Pilot Zone policy significantly enhances urban land use efficiency, thus validating Hypothesis 1.

Table 4. Baseline result.

	(1)	(2)	(3)	(4)
	ULUE	*ULUE*	*ULUE*	*ULUE*
DID	0.1735 *** (9.4665)	0.0375 ** (2.5073)	0.0592 *** (4.0201)	0.0592 ** (2.4514)
ln$AGDP_{pc}$		0.1759 *** (5.8811)	0.1726 *** (3.4519)	0.1726 *** (3.2838)
ind		−0.0637 *** (−4.5244)	−0.0124 (−0.8299)	−0.0124 (−0.5840)
ln*pop*		0.1422 *** (4.6358)	0.1579 *** (4.2481)	0.1579 *** (4.2787)
gov		−0.0779 ** (−2.0508)	−0.0709 (−1.0789)	−0.0709 (−0.7052)
hum		−0.0127 ** (−2.2572)	−0.0027 (−0.4286)	−0.0027 (−0.4975)
open		−0.0025 (−0.8424)	−0.0015 (−0.4982)	−0.0015 * (−2.0551)
City FE	/	√	√	√
Year FE	/	√	√	√
Province*Year FE	/	/	√	√
Constant	0.3947 *** (111.7285)	−2.1265 *** (-5.3851)	−2.2817 *** (-3.3854)	−2.2817 *** (−4.5713)
N	4932	4932	4845	4845
R^2	0.0188	0.7089	0.7581	0.7581

Note: Robust standard errors in parentheses. * $p < 0.1$, ** $p < 0.05$, *** $p < 0.01$.

4.2. Robustness Test

4.2.1. Parallel Trend Test

The prerequisite for using the difference-in-differences model is that the treatment group and the control group should adhere to the parallel trend assumption before the policy shock. That is, before the implementation of the free trade zone policy, the urban land use efficiency of the treatment group and the control group should ideally exhibit similar trends. Only under these circumstances can the policy effect coefficient more accurately reflect the impact of the policy.

This study employs an event analysis approach, where the year a city is established as a free trade zone is assigned a value of 0. The year following establishment (Year 1 or Year 2) is assigned a value of +1 (+2), while the year preceding establishment (Year −1 or Year −2) is assigned a value of −1 (−2), and so forth. The sample range in this study is [−15, 8]. Following the approach of Fajgelbaum et al. (2020) [82] and considering the scarcity of observations on both sides, this study sets relative years that are less than or equal to −8 uniformly to −8 and relative years that are greater than or equal to 6 all to 6. Additionally, the year preceding the city's designation as a free trade zone (−1 period) is considered the base period, from which corresponding regression coefficients are obtained.

As illustrated in Figure 3, prior to the implementation of the pilot free trade zone policy, the regression coefficients of both the treatment and control groups were not significant and fluctuated around 0. This suggests that there was no significant difference between the regression coefficients of the two groups before the policy implementation, thus satisfying the parallel trend hypothesis. Furthermore, examining the dynamic effect after the implementation of the pilot free trade zone policy, in the first year after policy implementation, the estimated coefficient $\hat{\beta}$ may not be significant due to lag effects, while in the second year after policy implementation, $\hat{\beta}$ is positive and significant. But in the third year, it becomes insignificant suddenly, which may be affected by the unexpected epidemic outbreak. Subsequently, the coefficient $\hat{\beta}$ is gradually and significantly positive, showing an upward trend, indicating that the pilot free trade zone policy can improve urban land use efficiency, consistent with the baseline regression results.

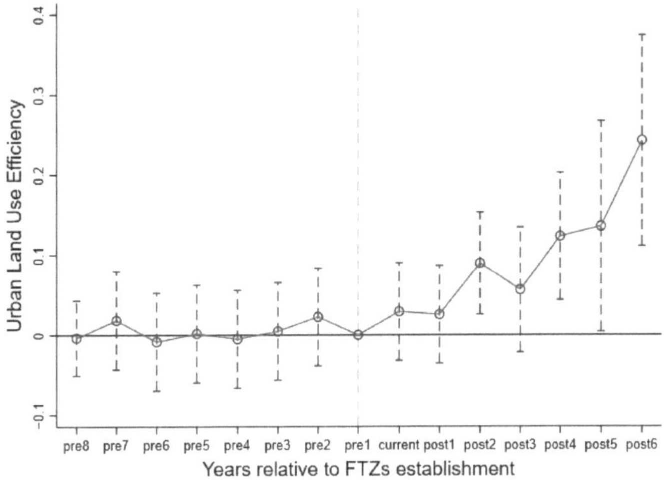

Figure 3. Results of Parallel Trend Test.

4.2.2. Placebo Test

The baseline regression results confirm that FTZ policy can improve urban land use efficiency. However, this result may be influenced by other random factors, and the regression results above could be biased due to other confounding policies or unobserved factors. Therefore, a random policy treatment group was obtained by randomly sampling samples and randomly generating the policy implementation year. Meanwhile, interaction terms of pseudo-virtual policies were generated for placebo tests. In order to enhance the effectiveness of the placebo test, this paper randomly selected 1000 samples and conducted 1000 regression iterations according to model (1). Figure 4 illustrates the coefficient kernel density obtained from regression and the corresponding p-value distribution. The centers of distribution of the coefficients from the random sampling test are all near zero, significantly different from the estimated coefficient of the baseline regression (dashed line value = 0.0592). Most of the p-values are greater than 0.1, indicating that the coefficients are not statistically significant, at least at the 10% level. This suggests that the baseline regression results are robust and less affected by other policies or unobservable factors.

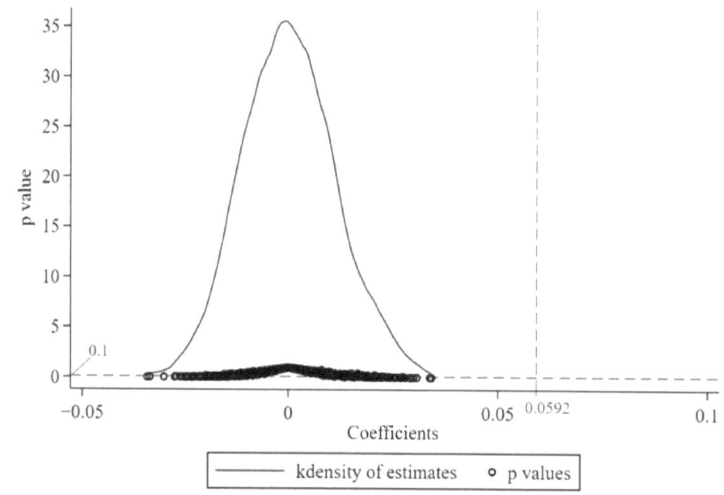

Figure 4. Results of the Placebo Test.

4.2.3. PSM-DID

Since cities with FTZs may have different initial conditions compared to those without, there could be endogeneity issues. Additionally, cities establishing FTZs may exhibit better economic development levels, industrial structures, and policy environments, leading to a potential "self-selection effect". Consequently, direct regression on all samples of cities with and without FTZs may introduce "sample selection bias", resulting in biased research conclusions. To address this potential "self-selection effect", Propensity Score Matching with Difference-in-Differences (PSM-DID) was employed to identify comparable control group cities for those with FTZs. Subsequently, matched samples were utilized for regression analysis.

In this paper, the core explanatory variable (DID) is treated as the treatment variable, while six control variables in the baseline regression are selected as first-order covariates. Additionally, 17 interaction terms, such as gov#gov, pop#ind, and ind#ind, are chosen as second-order covariates. The logit model is employed, and the default kernel function and bandwidth are utilized for kernel matching. The matching results are presented in Table 5. The average treatment effect on the treated (ATT) is calculated as 0.028447, with an absolute T-value of 3.29. This T-value exceeds the absolute value of 2.56 at the 5% significance level, indicating that the average treatment effect is statistically significant at the 1% level. Therefore, the treatment variable significantly impacts the outcome variable.

Table 5. Average Treatment Effect.

Variables	Samples	Treatment Group	Control Group	Differences	Residuals	T-Value
ULUE	Unmatched	0.568218452	0.394671119	0.173547334	0.017876457	9.71
	ATT	0.56962411	0.663307424	−0.093683314	0.028447391	−3.29
	ATU	0.541227468	0.439847344	−0.101380124		
	ATE			−0.100082037		

Before sample matching, the *p*-values of control variables in both the treatment group and the control group were significant at the 1% level, indicating a significant difference between the two groups. After propensity score matching, the absolute values of standard deviations of each covariate were all less than 20%, and the *p*-values after matching were all greater than 1%, indicating no significant difference between the treatment group and the control group post-matching. Propensity score matching effectively corrects the distribution bias of propensity score values, rendering the matched data in this paper valid and credible. The balance test of propensity score matching is presented in Table A2 in Appendix A.

In order to clearly explain the impact of establishing the pilot free trade zone on urban land use efficiency, this paper employs data after kernel matching for regression analysis to underscore the robustness of the regression results. As depicted in Table 6, in the presence of city fixed effects (City FE), time fixed effects (Year FE), and province-year fixed effects (Province*Year FE), columns (1) and (2) show the regression results without and with control variables, respectively, where the estimated coefficients of the core explanatory variables are significant at the 5% and 10% levels, respectively. Columns (3) and (4) show the regression results without and with control variables, respectively, under further province-clustered analysis, where the estimated coefficients of the core explanatory variables are 0.0687 and 0.0520 and are significant at the 5% and 10% levels, respectively.

Table 6. PSM-DID Estimation.

	(1)	(2)	(3)	(4)
	ULUE	*ULUE*	*ULUE*	*ULUE*
DID	0.0687 **	0.0520 *	0.0687 **	0.0520 *
	(2.2282)	(1.7916)	(2.3960)	(1.7597)
Constant	0.5344 ***	−0.7317	0.5344 ***	−0.7317
	(79.7026)	(−0.6451)	(107.8556)	(−0.5687)
Controls	/	√	/	√
City FE	√	√	√	√
Year FE	√	√	√	√
Province*Year FE	√	√	√	√
N	949	949	949	949
R^2	0.8432	0.8466	0.8432	0.8466

Note: Robust standard errors in parentheses. * $p < 0.1$, ** $p < 0.05$, *** $p < 0.01$.

Observing the regression results post data matching, it becomes evident that the coefficient of the pilot free trade zone variable is significantly positive, signifying a notable positive impact of establishing the pilot free trade zone on urban land use efficiency. The significance of each variable remains largely consistent before and after matching, in comparison with the baseline regression results, thereby further affirming the robustness and reliability of the baseline regression model.

4.2.4. Entropy Balance

Compared to PSM matching, entropy balance matching can further ensure a multidimensional balance of covariates between the treatment group and the control group. This involves narrowing the sample moment difference of covariates between the two groups, thereby minimizing the influence of selection bias without losing samples. In this paper, six control variables from the baseline regression were selected as covariates. Entropy balance matching was applied, and the method of moment balance was adopted to align the distribution of first-order, second-order, and third-order moments of covariates between the control group and the treatment group, achieving inter-group equilibrium.

As depicted in Tables 7 and 8, the mean, variance, and kurtosis of the six covariates in both the treatment and control groups are nearly identical. Subsequently, regression analysis was conducted on the samples following entropy balance matching. The outcomes are illustrated in Table 9. Controlling for city fixed effects (City FE), time fixed effects (Year FE), and province-year fixed effects (Province*Year FE), columns (1) and (2) respectively show the regression results without and with control variables, where the estimated coefficients of the core explanatory variables are 0.1012 and 0.0946, both significant at the 1% level. Columns (3) and (4) involve further province-clustered analysis, displaying the regression results without and with control variables, where the estimated coefficients of the core explanatory variables are 0.1012 and 0.0946, respectively, with corresponding standard errors of 2.9898 and 3.2605, both significant at the 1% level. Overall, the estimated coefficients of the core explanatory variables after entropy balancing are consistently significant at the 1% level and positive, confirming the reliability of the baseline regression results.

Table 7. Characteristics of Covariates of Treatment Group and Control Group Before Entropy Matching.

	Treatment			Control		
	Mean	Variance	Skewness	Mean	Variance	Skewness
$\ln AGDP_{pc}$	11.4	0.1905	−0.4032	10.39	0.5532	−0.3684
ind	1.798	0.9312	2.047	0.9492	0.2627	2.474
ln*pop*	6.295	0.6997	−0.6473	5.814	0.6196	−2.242
gov	0.1723	0.003718	1.548	0.1909	0.02048	5.695
hum	5.78	0.635	−0.4511	4.641	0.9813	−0.01645
open	0.2062	0.03595	1.033	0.1367	0.8894	52.43

Table 8. Characteristics of Covariables in the Treatment Group and Control Group after Entropy Matching.

	Treat			Control		
	Mean	Variance	Skewness	Mean	Variance	Skewness
$\ln AGDP_{pc}$	11.4	0.1905	−0.4032	11.4	0.1905	−0.4031
ind	1.798	0.9312	2.047	1.797	0.9311	2.047
ln*POP*	6.295	0.6997	−0.6473	6.295	0.6997	−0.6473
gov	0.1723	0.003718	1.548	0.1723	0.003718	1.548
hum	5.78	0.635	−0.4511	5.78	0.635	−0.4512
open	0.2062	0.03595	1.033	0.2063	0.03597	1.034

Table 9. Entropy-balancing Estimation.

	(1)	(2)	(3)	(4)
	ULUE	*ULUE*	*ULUE*	*ULUE*
DID	0.1012 ***	0.0946 ***	0.1012 ***	0.0946 ***
	(3.6955)	(3.3388)	(2.9898)	(3.2605)
Constant	0.5364 ***	−0.1496	0.5364 ***	−0.1496
	(36.3990)	(−0.1383)	(30.3554)	(−0.0970)
Controls	/	√	/	√
City FE	√	√	√	√
Year FE	√	√	√	√
Province*Year FE	√	√	√	√
N	4841	4841	4841	4841
R^2	0.9197	0.9220	0.9197	0.9220

Note: Robust standard errors in parentheses. *** $p < 0.01$.

4.2.5. Expected Effect Test

The premise of considering the policy shock of joining the pilot free trade zone as a quasi-natural experiment is that this policy shock exists in randomness. Drawing from Lu and Yu's method [83], this paper tests the expected effect of the policy shock from the pilot free trade zone. By adding dummy variables for the pre-implementation period of the pilot cities (represented respectively as one year before, two years before, and three years before) and their interactions with "treatment" in the baseline regression model, we examine whether cities approved as pilot zones have already formed adjustment expectations in the one, two, or three years prior to the policy shock. If the coefficients of these interaction terms are significantly non-zero, it indicates that cities approved as pilot zones had anticipatory effects on the policy before the implementation of the pilot free trade zone. In this case, the outcome variables of the treatment and control groups in the difference-in-differences (DID) estimation in this paper are not comparable before the policy shock of the pilot free trade zone, resulting in bias in the estimation results of DID.

The test results, as shown in Table 10, indicate that the estimated coefficients for the three interaction terms are 0.0000, 0.0371, and 0.0116, none of which passed the significance test. After including the aforementioned interaction terms in the regression, the estimated coefficients of the core explanatory variables are significantly positive and not much different from the baseline regression result of 0.1735. This indicates that cities approved as FTZs did not anticipate adjustments before the impact of FTZ policies, suggesting a strong policy externality. Therefore, the DID estimation results described above are deemed more credible.

Table 10. Expected Effect Test.

	(1)	(2)	(3)
	ULUE	*ULUE*	*ULUE*
DID	0.0573 ***	0.0604 ***	0.0582 ***
	(3.7987)	(3.9821)	(3.8403)
treat × one year before	0.0000		
	(0.0018)		
treat × two year before		0.0371	
		(1.5576)	
treat × three year before			0.0116
			(0.4786)
_cons	0.3964 ***	0.3959 ***	0.3962 ***
	(202.3755)	(201.8074)	(201.7808)
N	4845	4845	4845
R^2	0.7490	0.7491	0.7490

Note: Robust standard errors in parentheses. *** $p < 0.01$.

4.2.6. Excluding Interference from Other Relevant Policies

As national or regional central cities, the cities approved as pilot free trade zones are often affected by various location-oriented policies at the national level. In order to exclude the influence of other location-oriented policies, this paper focuses on the following two types of major location-oriented policies at the national level: ① The Industrial Structure Transformation (Pilot) Zone (ISTZ). The main goal of the ISTZ is to accelerate the transformation and upgrading of old industrial cities and resource-based cities by promoting urban renewal and industrial structure adjustment, which undoubtedly affects the ULUE of relevant cities. China has supported the construction of two batches of the ISTZ in 2017 and 2019, which intersected with the FTZ policy. In view of the above reasons, it is necessary to exclude the policy interference of the ISTZ. As shown in column (1) of Table 11, after controlling for the ISTZ policy, the estimated coefficient of the core explanatory variable, DID, is 0.0597, significant at the 5% level. ② The Ecological Conservation (Pilot) Zone (ECZ). According to a document released by the Ministry of Ecology and Environment of the People's Republic of China, a total of seven ECZ zones were established between 2017 and 2023 at a rate of one batch per year. The ECZ encompasses administrative units at various levels, including provinces, cities, districts, and counties. However, this paper focuses on the ecological civilization demonstration zones at the prefecture-level city level. It is important to note that due to the intersecting nature of the list of ecological civilization demonstration zones, districts or counties within the same city are considered as research samples if they are designated as demonstration zones. In cases where different districts and counties within the same city are designated as demonstration zones at different times, they are categorized into batches based on their implementation dates. In addition, this paper's study period spans from 2005 to 2021. Given that the fifth batch of the ECZ was established in October 2021, its policy effects are challenging to investigate within this timeframe. Consequently, this paper only considers the policy effects of the first four batches of the ECZ. The results are presented in column (2) of Table 11. In column (2), the estimated coefficient for the ECZ is 0.0346, and the estimated coefficient for the core explanatory variable DID is 0.0715, significant at the 5% level. Column (3) of Table 11 shows the regression results when controlling for both the ISTZ and ECZ policies, where the estimated coefficient for the core explanatory variable DID is 0.0720, also significant at the 5% level. These results indicate that whether controlling separately or jointly for the ISTZ and ECZ, the estimated coefficients of the core explanatory variable are significantly positive at the 5% level. This demonstrates that, even after excluding interference from related policies, the positive impact of the Free Trade Zone policy on ULUE remains evident.

Table 11. Excluding the Interference of Other Relevant Policies.

	(1)	(2)	(3)
	ULUE	*ULUE*	*ULUE*
DID	0.0597 **	0.0715 **	0.0720 **
	(2.4579)	(2.4881)	(2.5151)
ISTZ	−0.0224		0.0345 **
	(−0.9962)		(2.7893)
ECZ		0.0346 **	−0.0176
		(2.7872)	(−0.4269)
_cons	−2.2543 ***	−3.0777 ***	−3.0469 ***
	(−4.4662)	(−3.6379)	(−3.5924)
Controls	√	√	√
City FE	√	√	√
Year FE	√	√	√
Province*Year FE	√	√	√
N	4846	2156	2156
R^2	0.7582	0.8198	0.8198

Note: Robust standard errors in parentheses. ** $p < 0.05$, *** $p < 0.01$.

5. Further Analysis

5.1. Spillover Effect or Siphon Effect

Based on Model (1) discussed earlier, it has been found that the establishment of Free Trade Pilot Zones can significantly enhance urban land use efficiency. Since the Free Trade Zones carry the significant mission of exploring new paths and accumulating new experiences for comprehensively deepening reforms and expanding openness, this paper intends to explore, based on the fundamental conclusions, whether the Free Trade Pilot Zones have produced spatial effects on neighboring cities. If so, the question remains whether these are spillover effects or siphon effects. Based on this, and referring to Model (1), the following model is constructed:

$$ULUE_{it} = \alpha + \beta_1 \cdot FTZ_{it} + \sum_{j}^{n} (\gamma_j \cdot Control_{jt}) + \mu_i + \lambda_t + \eta_{pt} + \varepsilon_{it} \quad (5)$$

In Equation (5), FTZ_{it} represents whether city i has a neighboring city with a pilot free trade zone in year t; if it does, it is assigned a value of 1; otherwise, 0. Take Xiamen, Fujian Province, as an example. Xiamen is adjacent to Quanzhou and Zhangzhou, and Xiamen was approved as an FTZ in 2015. Therefore, before 2015, there were no neighboring cities with FTZs for Xiamen, so the value was 0; from 2015 onwards, the value is assigned as 1. If a city is adjacent to multiple pilot free trade zone cities, it is considered to be adjacent to the city that established the FTZ earlier. If a city's neighboring cities have never established an FTZ during the sample period, the value of FTZ_{it} is consistently assigned as 0. Other variables and parameters in the model are the same as in Model (1), with β_1 serving as the test basis for the spillover effect of FTZs. If β_1 is significantly positive, it indicates that the establishment of FTZs can drive the improvement of urban land use efficiency in surrounding neighboring cities, indicating a radiation effect. If β_1 is significantly negative, it indicates that the establishment of FTZs will reduce the urban land use efficiency of surrounding neighboring cities, indicating a siphoning effect. It should be noted that this model primarily examines the impact of pilot free trade zones on surrounding cities, so the model estimation does not include samples from the cities within the pilot free trade zone itself. Table 12 presents the estimation results of Model (5). Specifically, column (1) controls for city fixed effects (City FE), time fixed effects (Year FE), and province-year fixed effects (Province*Year FE) but does not include control variables. The estimated coefficient of the core explanatory variable is 0.0351, which is significant at the 1% level. Column (2) builds on column (1) by introducing control variables, with the estimated coefficient of the core explanatory variable being 0.0365, which is also significant at the 1% level. The regression results indicate that the establishment of pilot free trade zones has a positive and significant effect on the urban land use efficiency of surrounding neighboring cities. This suggests a clear radiating effect of pilot free trade zone construction on surrounding cities. Hypothesis 2 is supported.

Table 12. Analysis of Spatial Effects of Free Trade Pilot Zones.

	(1)	(2)
	ULUE	*ULUE*
Space Effect	0.0351 ***	0.0365 ***
	(3.9439)	(3.5396)
_cons	0.3826 ***	−1.9652 ***
	(160.5378)	(−6.3669)
Controls	/	√
City FE	√	√
Year FE	√	√
Province*Year FE	/	√
N	4144	4115
R^2	0.6915	0.7669

Note: Robust standard errors in parentheses. *** $p < 0.01$.

5.2. Heterogeneity Analysis

5.2.1. Heterogeneity in Eastern, Central, and Western Regions

As mentioned earlier, pilot free trade zones are dispersed across various regions of China, each with significant differences in development level, stage of development, and resource endowment. Therefore, it is necessary to assess the heterogeneity of pilot free trade zone policies on urban land use efficiency using traditional methods for Eastern, Central, and Western China. Utilizing data from the National Bureau of Statistics of China and standard classification criteria, this study categorizes sample cities into eastern, central, and western regions and conducts empirical regression analysis. Each control variable is included, and the regression results, including fixed effects and province cluster analysis, are presented in Table 13.

Table 13. Heterogeneity in Eastern, Central, and Western China.

	(1)	(2)	(3)
	East	Central	West
DID	0.0385	0.0326	0.1732 *
	(0.8212)	(1.3883)	(1.8310)
Constant	−2.0308 *	−3.3349 ***	−1.7914 ***
	(−2.3877)	(−8.5998)	(−4.8042)
Controls	√	√	√
City FE	√	√	√
Year FE	√	√	√
Province*Year FE	√	√	√
N	1394	1512	1481
R^2	0.7779	0.6986	0.6981

Note: Robust standard errors in parentheses. * $p < 0.1$, *** $p < 0.01$.

In the eastern and central regions, the estimated coefficients for the core explanatory variables are 0.0385 and 0.0326, but they are not significant, indicating that the pilot free trade zone policy has no significant impact on the land use efficiency of cities in these regions. This may be attributed to the fact that the eastern and central regions already possess a solid development foundation and well-established development plans, resulting in a limited spatial effect of the policy and making it challenging to demonstrate its true impact. In the western region, the estimated coefficient for the core explanatory variable is 0.1732, which is significant at the 10% level. A possible explanation is that the economic resources and development systems in the western region are relatively weak. The implementation of the Free Trade Pilot Zone policy in the western region can initially release greater innovation and opening benefits, fully leveraging its "latecomer advantage" to enhance urban land use efficiency. Thus, the pilot free trade zone policy not only contributes to China's sustained economic growth but also facilitates the balanced development of China's eastern, central, and western regions.

5.2.2. Inland and Coastal Heterogeneity

Considering the differences in the functional orientation of each Free Trade Zone and the resource endowments of their respective regions, this paper categorizes the sample cities into coastal and inland cities and investigates their respective differences in urban land use efficiency, as presented in Table 14. The impact coefficient of the Free Trade Zone policy on the land use efficiency of coastal cities is 0.0354, but it is not significant; the impact coefficient of the Free Trade Zone policy on the land use efficiency of inland cities is 0.0854, which is significant at the 10% level, indicating that, compared to coastal cities, inland cities lack advantages in economic and trade development and factor flow. Consequently, pilot free trade zone policies are likely to have a more significant impact on improving land use efficiency in inland cities.

Table 14. Coastal and Inland Heterogeneity.

	(1)	(2)
	Coastal	**Inland**
DID	0.0354	0.0854 *
	(0.9902)	(2.0103)
Constant	−2.2535 **	−2.1847 ***
	(−3.3513)	(−3.0703)
Controls	√	√
City FE	√	√
Year FE	√	√
Province*Year FE	√	√
N	1903	2942
R^2	0.7967	0.6967

Note: Robust standard errors in parentheses. * $p < 0.1$, ** $p < 0.05$, *** $p < 0.01$.

5.2.3. Urbanization Rate Heterogeneity

Under the context of China vigorously coordinating the advancement of new urbanization, the urbanization rate has a direct or indirect impact on urban land use efficiency. To investigate this, the study divides the sample into two categories—high urbanization rate and low urbanization rate—using the median urbanization rate of the overall sample. Population regression analysis is then conducted according to Equation (1). The regression results are presented in Table 15. The estimated coefficient of the Free Trade Pilot Zone policy on the land use efficiency of cities with high urbanization rates is 0.0564, which is significant at the 5% level. For cities with low urbanization rates, the estimated coefficient of the core explanatory variable is 0.0037, which is not significant. This indicates that the Free Trade Pilot Zone policy has a more significant impact on the land use efficiency of cities with high urbanization rates. This discrepancy may be attributed to the fact that, compared to areas with low urbanization rates characterized by extensive land use, regions with higher urbanization rates have more optimized urban spatial layouts and intensive land use. As a result, the free trade zone policy has a more substantial impact on urban land use efficiency in these areas.

Table 15. Heterogeneity of Urbanization Rate.

	(1)	(2)
	High Urbanization Rate	**Low Urbanization Rate**
DID	0.0564 **	0.0037
	(2.1871)	(0.1017)
Constant	−3.1166 ***	−1.0748 *
	(−5.1512)	(−1.7768)
Controls	√	√
City FE	√	√
Year FE	√	√
Province*Year FE	√	√
N	2390	2344
R^2	0.8224	0.7161

Note: Robust standard errors in parentheses. * $p < 0.1$, ** $p < 0.05$, *** $p < 0.01$.

6. Conclusions and Policy Implications

Pilot Free Trade Zones (FTZs) represent special economic zones established independently by China. With institutional innovation as their core principle and replicability as a fundamental requirement, FTZs conduct pilot trials to accelerate the transformation of government functions, explore institutional and mechanism innovations, and facilitate investment and trade, thereby paving the way for comprehensive reform and expanded opening up. Given that land serves as a crucial asset for the establishment and development of FTZs, it is imperative to examine the relationship between FTZ construction and urban land use efficiency. However, as mentioned in the literature review section, existing studies have paid little attention to the impact of Free Trade Pilot Zones on urban land use efficiency. In addressing this, the present paper treats the FTZ policy as a quasi-natural experiment. Leveraging panel data from 297 cities in China spanning from 2005 to 2021, we employ a multi-period differential method to empirically investigate the relationship between FTZ construction and urban land use efficiency. The research findings are as follows: Firstly, pilot free trade zone policies can significantly enhance urban land use efficiency, offering practical insights and a solid foundation for both established and prospective pilot free trade zones. The expansion of pilot free trade zones should be further encouraged to fully leverage and amplify the positive impact of this policy. Secondly, this study conducted parallel trend tests and placebo tests, employing methods such as PSM-DID, entropy balance, and control for other relevant policies of the same period. The results consistently support the beneficial effect of FTZs on urban land use efficiency. Thirdly, the impact of FTZ policies on urban land use efficiency varies across regions. Through heterogeneity analysis, we found that compared to central and eastern cities, as well as coastal cities and those with a high urbanization rate, the effect of this policy is more pronounced in western regions, inland cities, and cities with lower urbanization rates. It is noteworthy that pilot free trade zones also exhibit spatial spillover effects on the land use efficiency of neighboring cities.

Based on the aforementioned research findings, this paper proposes the following policy implications: Firstly, it is crucial to further promote the establishment of pilot free trade zones. The fundamental conclusion that pilot free trade zone policies can significantly enhance urban land use efficiency provides a solid rationale for advancing the implementation of such policies and refining urban land planning strategies. Secondly, in the endeavor to promote pilot free trade zones, due consideration should be given to the heterogeneity of cities in terms of their geographical location and urbanization rate. Tailored policy combinations should be devised for pilot free trade zones situated in eastern and central regions, as well as in inland and less urbanized areas, to maximize the effectiveness of policy implementation. Thirdly, there is a need to bolster inter-regional cooperation and development efforts, fostering synergies between local and neighboring regions and fully harnessing the positive spatial spillover effects of pilot free trade zone policies.

During the implementation of the free trade zone policy, government departments should also be aware of the potential negative impacts associated with free trade zones. For instance, in countries or regions with relatively insufficient technological advantages, more liberalized trade may reduce economic welfare, gradually exacerbating regional development disparities [84]. In industries with more severe pollution, stricter environmental regulations could reduce the likelihood and volume of exports [85]. Furthermore, heightened levels of trade liberalization might increase pollution [86], intensifying the globalization of pollution [87]. In response, regions should develop preemptive strategies to maximize the effectiveness of policy implementation.

However, this paper still has several shortcomings including the following: First, due to the lack of relevant data, the time span of the study sample covers 2005–2021, which makes the empirical part of this paper relatively weak. Second, since Free Trade Pilot Zones are actually divided into districts, this paper uses prefecture-level city data for analysis, which may bias the empirical results. Further research on this aspect requires additional verification and discussion in the future.

Author Contributions: Conceptualization, S.X., P.Y. and H.Z.; methodology, S.X.; software, S.X. and P.Y.; validation, J.H. and A.L.; formal analysis, S.X.; investigation, S.X. and P.Y.; resources, S.X.; data curation, S.X. and P.Y.; writing—original draft preparation, S.X.; writing—review and editing, S.X., P.Y., J.H., A.L. and H.Z.; visualization, S.X. and P.Y.; supervision, H.Z.; project administration, S.X.; funding acquisition, P.Y. All authors have read and agreed to the published version of the manuscript.

Funding: This research was funded by The Education Department of Hainan Province (Project No. Hnky2024-8), 2024 Hainan Provincial Philosophy and Social Sciences Planning Project (Project No. HNSK(JD)24-09), 2022 Hainan Provincial Philosophy and Social Sciences Planning Project (Project No. HNSK(JD)22-04), and Hainan Provincial Natural Science Foundation of China (Grant No. 723QN224, 722MS045, 722QN293, 721QN0873 and 721MS026).

Data Availability Statement: The original contributions presented in the study are included in the article. Further inquiries can be directed to the authors.

Conflicts of Interest: The authors declare no conflicts of interest.

Appendix A

Table A1. The Establishment of China's Pilot Free Trade Zones.

Year	Province/Municipality	Prefecture-Level Cities
2013	Shanghai	
2015	Canton	Guangzhou, Shenzhen, Zhuhai
	Tianjin	
	Fujian	Fuzhou, Xiamen
2017	Liaoning	Yingkou, Shenyang, Dalian
	Zhejiang	Zhoushan (In 2020, Hangzhou, Ningbo, and Jinhua were added.)
	Henan	Luoyang, Kaifeng, Zhengzhou
	Hubei	Yichang, Xiangyang, Wuhan
	Chongqing	
	Sichuan	Chengdu
	Shaanxi	Xi'an
2018	Hainan	Sanya, Haikou, Danzhou
2019	Shandong	Jinan, Qingdao, Yantai
	Jiangsu	Nanjing, Suzhou, Lianyungang
	Guangxi	Chongzuo, Qinzhou, Nanning
	Hebei	Baoding, Shijiazhuang, Langfang, Tangshan
	Yunnan	Kunming
	Heilongjiang	Harbin, Heihe, Mudan River
2020	Beijing	
	Hunan	Yueyang, Chenzhou, Changsha
	Anhui	Bengbu, Wuhu, Hefei
2023	Xinjiang	Khorgos, Kashgar, Urumqi

Source: Ministry of Commerce, People's Republic of China.

Table A2. Propensity Score Matching Balance Test.

Variables	Samples	Mean: Treatment Group	Mean: Control Group	Deviation %	Deviation Reduction Rate %	T-Value	P > |t|
ln $AGDP_{pc}$	Before matching	11.404	10.387	166.8	98.7	18.87	0
	After matching	11.398	11.385	2.1		0.28	0.779
ind	Before matching	1.7975	0.94918	109.8	90.1	21.5	0
	After matching	1.7502	1.8342	−10.9		−0.86	0.393
ln *pop*	Before matching	6.2954	5.8139	59.3	90.2	8.31	0
	After matching	6.3164	6.3636	−5.8		−0.58	0.563
gov	Before matching	0.17226	0.19094	−17	88.9	−1.81	0.071
	After matching	0.16923	0.1713	−1.9		−0.32	0.746
open	Before matching	0.20623	0.13668	10.2	75.4	1.02	0.306
	After matching	0.20608	0.18899	2.5		0.87	0.383
hum	Before matching	5.7799	4.6409	126.7	92.2	15.77	0
	After matching	5.7751	5.6859	9.9		1.08	0.282
gov#gov	Before matching	0.03337	0.05693	−16.5	96	−1.63	0.103
	After matching	0.03226	0.0332	−0.7		−0.32	0.748
ln *pop#ind*	Before matching	11.13	5.5208	128.5	87.2	24.48	0
	After matching	10.966	11.684	−16.4		−1.15	0.252
ind#ind	Before matching	4.1574	1.1635	72.9	83.4	20.1	0
	After matching	3.847	4.3453	−12.1		−0.98	0.33
ln $AGDP_{pc}$#ln $AGDP_{pc}$	Before matching	130.24	108.44	169.5	98.7	19.66	0
	After matching	130.11	129.82	2.3		0.29	0.775
hum#ind	Before matching	10.51	4.4516	129.8	94.6	27.41	0
	After matching	10.24	10.565	−7		−0.54	0.592
gov#ind	Before matching	0.33229	0.20707	44.4	76.8	6.05	0
	After matching	0.31503	0.34409	−10.3		−1.01	0.315
hum#gov	Before matching	0.98036	0.85595	23.5	97.1	2.52	0.012
	After matching	0.96076	0.9571	0.7		0.11	0.91
*hum#*ln $AGDP_{pc}$	Before matching	66.106	48.517	153.9	93.8	19.72	0
	After matching	66.025	64.942	9.5		0.98	0.326
hum#hum	Before matching	34.039	22.519	126.2	90.8	16.89	0
	After matching	34.001	32.936	11.7		1.16	0.247
*ind#*ln $AGDP_{pc}$	Before matching	20.423	9.866	124.1	91.1	25.08	0
	After matching	19.892	20.829	−11		−0.85	0.395
*open#*ln *pop*	Before matching	1.3335	0.74707	25.5	82.3	2.71	0.007
	After matching	1.3297	1.226	4.5		0.83	0.41
open#open	Before matching	0.07829	0.90788	−2.4	98.9	−0.23	0.815
	After matching	0.07873	0.06941	0		0.64	0.521
*hum#*ln *pop*	Before matching	36.481	27.019	131.1	97.5	18.19	0
	After matching	36.576	36.337	3.3		0.32	0.753
hum#open	Before matching	1.2371	0.69449	15.7	77.1	1.58	0.114
	After matching	1.2366	1.1124	3.6		1.06	0.289
ln *pop#*ln *pop*	Before matching	40.328	34.421	64.5	91.4	9.84	0
	After matching	40.549	41.057	−5.5		−0.51	0.609
ln *pop#*ln $AGDP_{pc}$	Before matching	71.844	60.328	119.2	95.9	17.26	0
	After matching	72.042	72.514	−4.9		−0.47	0.639
*gov#*ln *pop*	Before matching	1.0657	1.0833	−3.2	−40.9	−0.35	0.724
	After matching	1.0519	1.0767	−4.6		−0.68	0.497

References

1. Yigitcanlar, T. Smart Cities: An Effective Urban Development and Management Model? *Aust. Plan.* **2015**, *52*, 27–34. [CrossRef]
2. Chakrabarty, B. Urban Management: Concepts, Principles, Techniques and Education. *Cities* **2001**, *18*, 331–345. [CrossRef]
3. Lee, S.; Lee, C. The Perception of Port Logistics Related Firms According to the Adoption of Busan Free Trade Zone. *J. Korean Navig. Port Res.* **2003**, *27*, 563–568.
4. Grubel, H.G. Towards a Theory of Free Economic Zones. *Weltwirtschaftliches Arch.* **1982**, *118*, 39–61. [CrossRef]
5. Zhao, A.; Wang, J.; Guan, H. Has the Free Trade Zone Construction Promoted the Upgrading of the City's Industrial Structure? *Sustainability* **2022**, *14*, 5482. [CrossRef]
6. On the Issues Concerning Intellectual Property in Transit Goods and Parallel Imports in China (Shanghai) Pilot Free Trade Zone-All Databases. Available online: https://webofsciencehtbprolclarivatehtbprolcn-s.libvpn.zuel.edu.cn/wos/alldb/full-record/KJD:ART001918390 (accessed on 23 December 2023).
7. Wan, Z.; Zhang, Y.; Wang, X.; Chen, J. Policy and Politics behind Shanghai's Free Trade Zone Program. *J. Transp. Geogr.* **2014**, *34*, 1–6. [CrossRef]
8. Akbari, M.; Azbari, M.E.; Chaijani, M.H. Performance of the Firms in a Free-Trade Zone: The Role of Institutional Factors and Resources. *Eur. Manag. Rev.* **2019**, *16*, 363–378. [CrossRef]
9. Ghosh, S.; Reynolds, C.L.; Rohlin, S.M. The Spillover Effects of United States Foreign Trade Zones. *Appl. Econ.* **2016**, *48*, 4112–4130. [CrossRef]
10. Su, K.Y.; Dae-cheol, J. The development of Masan Port through comparison of Free Trade Zone. *J. Korea Port Econ. Assoc.* **2003**, *19*, 161–188.
11. Yao, D.; Whalley, J. The China (Shanghai) Pilot Free Trade Zone: Background, Developments and Preliminary Assessment of Initial Impacts. *World Econ.* **2016**, *39*, 2–15. [CrossRef]
12. Wang, L.; Shao, J. Can China's Pilot Free Trade Zone Improve Trade Efficiency? *J. Int. Commer. Econ. Policy* **2022**, *13*, 2250006. [CrossRef]
13. Chen, W.; Hu, Y.; Liu, B.; Wang, H.; Zheng, M. Does the Establishment of Pilot Free Trade Test Zones Promote the Transformation and Upgradation of Trade Patterns? *Econ. Anal. Policy* **2022**, *76*, 114–128. [CrossRef]
14. Zhang, X.; Lu, F.; Xue, D. Does China's Carbon Emission Trading Policy Improve Regional Energy Efficiency? —An Analysis Based on Quasi-Experimental and Policy Spillover Effects. *Environ. Sci. Pollut. Res.* **2022**, *29*, 21166–21183. [CrossRef] [PubMed]
15. Chen, H.; Yuan, B.; Cui, Q. Does the Pilot Free Trade Zone Policy Attract the Entering of Foreign-Invested Enterprises? The Evidence from China. *Appl. Econ. Lett.* **2021**, *28*, 1162–1168. [CrossRef]
16. Hausmann, R.; Obach, J.; Santos, M.A. Special Economic Zones in Panama: Technology Spillovers from a Labor Market Perspective. Center for International Development at Harvard University. 2016. Available online: https://growthlab.hks.harvard.edu/publications/special-economic-zones-panama (accessed on 28 June 2024).
17. Li, H.; Chen, J.; Wan, Z.; Zhang, H.; Wang, M.; Bai, Y. Spatial Evaluation of Knowledge Spillover Benefits in China's Free Trade Zone Provinces and Cities. *Growth Chang.* **2020**, *51*, 1158–1181. [CrossRef]
18. Seyoum, B.; Ramirez, J. Foreign Trade Zones in the United States: A Study with Special Emphasis on the Proposal for Trade Agreement Parity. *J. Econ. Stud.* **2012**, *39*, 13–30. [CrossRef]
19. Diez-Vial, I.; Fernandez-Olmos, M. Knowledge Spillovers in Science and Technology Parks: How Can Firms Benefit Most? *J. Technol. Transf.* **2015**, *40*, 70–84. [CrossRef]
20. Erdogan, A.M. Bilateral Trade and the Environment: A General Equilibrium Model Based on New Trade Theory. *Int. Rev. Econ. Financ.* **2014**, *34*, 52–71. [CrossRef]
21. Wang, A.; Hu, Y.; Li, Y.; Rao, S.; Lin, W. Do Pilot Free Trade Zones Improve the Green Total Factor Productivity? Evidence from a Quasi-Natural Experiment in China. *Environ. Sci. Pollut. Res.* **2022**, *29*, 75307–75321. [CrossRef]
22. Hu, Z.; Lan, F.; Xu, H. Green TFP Heterogeneity in the Ports of China's Pilot Free Trade Zone under Environmental Constraints. *Int. J. Environ. Res. Public Health* **2021**, *18*, 12910. [CrossRef]
23. Zhuo, C.; Mao, Y.; Rong, J. Policy Dividend or "Policy Trap"? Environmental Welfare of Establishing Free Trade Zone in China. *Sci. Total Environ.* **2021**, *756*, 143856. [CrossRef] [PubMed]
24. Grossman, G.M.; Krueger, A.B. Environmental Impacts of a North American Free Trade Agreement. National Bureau of Economic Research Working Paper Series. 1991. Available online: https://www.nber.org/papers/w3914 (accessed on 28 June 2024). [CrossRef]
25. Liu, M.; Zhang, B.; Liao, X. Can Trade Liberalization Promote Green Production? Evidence from China's Manufacturing Enterprises. *J. Asian Econ.* **2022**, *79*, 101369. [CrossRef]
26. Li, S.; Liu, J.; Kong, Y. Pilot Free Trade Zones and Chinese Port-Listed Companies Performance: An Empirical Research Based on Quasi-Natural Experiment. *Transp. Policy* **2021**, *111*, 125–137. [CrossRef]
27. Jiang, Y.; Wang, H.; Liu, Z. The Impact of the Free Trade Zone on Green Total Factor Productivity—Evidence from the Shanghai Pilot Free Trade Zone. *Energy Policy* **2021**, *148*, 112000. [CrossRef]
28. Makowski, D.; Hendrix, E.M.T.; van Ittersum, M.K.; Rossing, W.A.H. A Framework to Study Nearly Optimal Solutions of Linear Programming Models Developed for Agricultural Land Use Exploration. *Ecol. Model.* **2000**, *131*, 65–77. [CrossRef]
29. Howley, P. Attitudes towards Compact City Living: Towards a Greater Understanding of Residential Behaviour. *Land Use Policy* **2009**, *26*, 792–798. [CrossRef]

30. Geurs, K.T. Job Accessibility Impacts of Intensive and Multipleland-Use Scenarios for the Netherlands' Randstad Area. *J. Hous. Built Environ.* **2006**, *21*, 51–67. [CrossRef]
31. Lau, S.S.Y.; Giridharan, R.; Ganesan, S. Policies for Implementing Multiple Intensive Land Use in Hong Kong. *J. Hous. Built Environ.* **2003**, *18*, 365–378. [CrossRef]
32. Ming, Y.S.; Hin, H.K. Planned Urban Industrialization and Its Effect on Urban Industrial Real Estate Valuation: The Singapore Experience. *Habitat Int.* **2006**, *30*, 509–539. [CrossRef]
33. Chen, Y.; Chen, Z.; Xu, G.; Tian, Z. Built-up Land Efficiency in Urban China: Insights from the General Land Use Plan (2006–2020). *Habitat Int.* **2016**, *51*, 31–38. [CrossRef]
34. Martinho, V.J.P.D. Efficiency, Total Factor Productivity and Returns to Scale in a Sustainable Perspective: An Analysis in the European Union at Farm and Regional Level. *Land Use Policy* **2017**, *68*, 232–245. [CrossRef]
35. Kaur, H.; Garg, P. Urban Sustainability Assessment Tools: Review. *J. Clean. Prod.* **2019**, *210*, 146–158. [CrossRef]
36. Hui, E.C.M.; Wu, Y.; Deng, L.; Zheng, B. Analysis on Coupling Relationship of Urban Scale and Intensive Use of Land in China. *Cities* **2015**, *42*, 63–69. [CrossRef]
37. Wang, Z.; Chen, J.; Zheng, W.; Deng, X. Dynamics of Land Use Efficiency with Ecological Intercorrelation in Regional Development. *Landsc. Urban Plan.* **2018**, *177*, 303–316. [CrossRef]
38. Desiere, S.; Jolliffe, D. Land Productivity and Plot Size: Is Measurement Error Driving the Inverse Relationship? *J. Dev. Econ.* **2018**, *130*, 84–98. [CrossRef]
39. Guastella, G.; Pareglio, S.; Sckokai, P. A Spatial Econometric Analysis of Land Use Efficiency in Large and Small Municipalities. *Land Use Policy* **2017**, *63*, 288–297. [CrossRef]
40. Ferreira, M.D.P.; Feres, J.G. Farm Size and Land Use Efficiency in the Brazilian Amazon. *Land Use Policy* **2020**, *99*, 104901. [CrossRef]
41. Jin, G.; Deng, X.; Zhao, X.D.; Guo, B.; Yang, J. Spatio-Temporal Patterns of Urban Land Use Efficiency in the Yangtze River Economic Zone during 2005–2014. *Acta Geogr. Sin* **2018**, *73*, 1242–1252.
42. Cheng, H.; Zhang, Y.; Li, H. Effects of Free Trade Zone Policy on China's Income Disparity. *Singap. Econ. Rev.* **2022**, *67*, 709–732. [CrossRef]
43. Jingxin, G.; Jinbo, S.; Lufang, W. A New Methodology to Measure the Urban Construction Land-Use Efficiency Based on the Two-Stage DEA Model. *Land Use Policy* **2022**, *112*, 105799. [CrossRef]
44. Yang, C.-H.; Wu, L.; Lin, H.-L. Analysis of Total-Factor Cultivated Land Efficiency in China's Agriculture. *Agric. Econ.—Czech* **2010**, *56*, 231–242. [CrossRef]
45. Gao, X.; Zhang, A.; Sun, Z. How Regional Economic Integration Influence on Urban Land Use Efficiency? A Case Study of Wuhan Metropolitan Area, China. *Land Use Policy* **2020**, *90*, 104329. [CrossRef]
46. Zhou, X.; Wu, D.; Li, J.; Liang, J.; Zhang, D.; Chen, W. Cultivated Land Use Efficiency and Its Driving Factors in the Yellow River Basin, China. *Ecol. Indic.* **2022**, *144*, 109411. [CrossRef]
47. Xiao, Y.; Zhong, J.-L.; Zhang, Q.-F.; Xiang, X.; Huang, H. Exploring the Coupling Coordination and Key Factors between Urbanization and Land Use Efficiency in Ecologically Sensitive Areas: A Case Study of the Loess Plateau, China. *Sustain. Cities Soc.* **2022**, *86*, 104148. [CrossRef]
48. Lu, X.; Chen, D.; Kuang, B.; Zhang, C.; Cheng, C. Is High-Tech Zone a Policy Trap or a Growth Drive? Insights from the Perspective of Urban Land Use Efficiency. *Land Use Policy* **2020**, *95*, 104583. [CrossRef]
49. Zhu, X.; Li, Y.; Zhang, P.; Wei, Y.; Zheng, X.; Xie, L. Temporal–Spatial Characteristics of Urban Land Use Efficiency of China's 35mega Cities Based on DEA: Decomposing Technology and Scale Efficiency. *Land Use Policy* **2019**, *88*, 104083. [CrossRef]
50. Wang, P.; Shao, Z.; Wang, J.; Wu, Q. The Impact of Land Finance on Urban Land Use Efficiency: A Panel Threshold Model for Chinese Provinces. *Growth Chang.* **2021**, *52*, 310–331. [CrossRef]
51. Paulsen, K. Geography, Policy or Market? New Evidence on the Measurement and Causes of Sprawl (and Infill) in US Metropolitan Regions. *Urban Stud.* **2014**, *51*, 2629–2645. [CrossRef]
52. Barbosa, J.A.; Braganca, L.; Mateus, R. Assessment of Land Use Efficiency Using BSA Tools: Development of a New Index. *J. Urban Plan. Dev.* **2015**, *141*, 04014020. [CrossRef]
53. Osman, T.; Divigalpitiya, P.; Arima, T. Driving Factors of Urban Sprawl in Giza Governorate of Greater Cairo Metropolitan Region Using AHP Method. *Land Use Policy* **2016**, *58*, 21–31. [CrossRef]
54. Caputo, P.; Pasetti, G.; Ferrari, S. Implementation of an Urban Efficiency Index to Comprehend Post-Metropolitan Territories—The Case of Greater Milan in In Italy. *Sustain. Cities Soc.* **2019**, *48*, 101565. [CrossRef]
55. Yu, J.; Zhou, K.; Yang, S. Land Use Efficiency and Influencing Factors of Urban Agglomerations in China. *Land Use Policy* **2019**, *88*, 104143. [CrossRef]
56. Chen, W.; Chen, W.; Ning, S.; Liu, E.; Zhou, X.; Wang, Y.; Zhao, M. Exploring the Industrial Land Use Efficiency of China's Resource-Based Cities. *Cities* **2019**, *93*, 215–223. [CrossRef]
57. Huang, Z.; He, C.; Zhu, S. Do China's Economic Development Zones Improve Land Use Efficiency? The Effects of Selection, Factor Accumulation and Agglomeration. *Landsc. Urban Plan.* **2017**, *162*, 145–156. [CrossRef]
58. Li, P.; Lu, Y.; Wang, J. Does Flattening Government Improve Economic Performance? Evidence from China. *J. Dev. Econ.* **2016**, *123*, 18–37. [CrossRef]

59. Xie, H.; Chen, Q.; Lu, F.; Wang, W.; Yao, G.; Yu, J. Spatial-Temporal Disparities and Influencing Factors of Total-Factor Green Use Efficiency of Industrial Land in China. *J. Clean. Prod.* **2019**, *207*, 1047–1058. [CrossRef]
60. Yan, S.; Peng, J.; Wu, Q. Exploring the Non-Linear Effects of City Size on Urban Industrial Land Use Efficiency: A Spatial Econometric Analysis of Cities in Eastern China. *Land Use Policy* **2020**, *99*, 104944. [CrossRef]
61. He, S.; Yu, S.; Li, G.; Zhang, J. Exploring the Influence of Urban Form on Land-Use Efficiency from a Spatiotemporal Heterogeneity Perspective: Evidence from 336 Chinese Cities. *Land Use Policy* **2020**, *95*, 104576. [CrossRef]
62. Sun, Y.; Ma, A.; Su, H.; Su, S.; Chen, F.; Wang, W.; Weng, M. Does the Establishment of Development Zones Really Improve Industrial Land Use Efficiency? Implications for China's High-Quality Development Policy. *Land Use Policy* **2020**, *90*, 104265. [CrossRef]
63. Wang, A.; Lin, W.; Liu, B.; Wang, H.; Xu, H. Does Smart City Construction Improve the Green Utilization Efficiency of Urban Land? *Land* **2021**, *10*, 657. [CrossRef]
64. Liu, Z.; Lai, B.; Wu, S.; Liu, X.; Liu, Q.; Ge, K. Growth Targets Management, Regional Competition and Urban Land Green Use Efficiency According to Evidence from China. *Int. J. Environ. Res. Public Health* **2022**, *19*, 6250. [CrossRef] [PubMed]
65. Ye, L.; Huang, X.; Yang, H.; Chen, Z.; Zhong, T.; Xie, Z. Effects of Dual Land Ownerships and Different Land Lease Terms on Industrial Land Use Efficiency in Wuxi City, East China. *Habitat Int.* **2018**, *78*, 21–28. [CrossRef]
66. Zhong, S.; Li, X.; Ma, J. Impacts of Land Finance on Green Land Use Efficiency in the Yangtze River Economic Belt: A Spatial Econometrics Analysis. *Environ. Sci. Pollut. Res.* **2022**, *29*, 56004–56022. [CrossRef] [PubMed]
67. Differential Characteristics and Driving Factors of Land Use Efficiency in Different Functional Cities Based on Undesirable Outputs. Available online: http://www.resci.cn/EN/10.18402/resci.2018.05.08 (accessed on 14 November 2023).
68. Zhou, L.; Shi, Y.; Cao, X. Evaluation of Land Intensive Use in Shanghai Pilot Free Trade Zone. *Land* **2019**, *8*, 87. [CrossRef]
69. Feng, Y.; Li, Y.; Nie, C. The Effect of Place-Based Policy on Urban Land Green Use Efficiency: Evidence from the Pilot Free-Trade Zone Establishment in China. *Land* **2023**, *12*, 701. [CrossRef]
70. Hu, H.; Wang, S.; He, J. Comparative Advantages of Free Trade Port Construction in Shanghai under the Belt and Road Initiative. *Int. J. Financ. Stud.* **2020**, *8*, 6. [CrossRef]
71. Jiang, H.; Liang, Y.; Pan, S. Foreign Direct Investment and Regional Innovation: Evidence from China. *World Econ.* **2022**, *45*, 1876–1909. [CrossRef]
72. Guan, H.; Wang, J.; Zhao, A. Free Trade Zone Policies and Green Development: An Empirical Examination Based on China's Free Trade Zone Cities. *Environ. Dev. Sustain.* **2024**. [CrossRef]
73. Fan, D.; He, M.; Yu, Y.; Liu, Y. Has the Establishment of a FTZ Promoted the High-Quality Development of the Pharmaceutical Manufacturing Industry?—Quasi Natural Experiments Based on the Establishment of the First Three Batches of Pilot FTZ. *J. Pharm. Innov.* **2024**, *19*, 30. [CrossRef]
74. Wang, H.; Zhang, Y.; Liu, Z.; Liu, R.; Li, K. The Impact and Mechanisms of the Shanghai Pilot Free-Trade Zone on the Green Total Factor Productivity of the Yangtze River Delta Urban Agglomeration. *Environ. Sci. Pollut. Res.* **2022**, *29*, 40997–41011. [CrossRef]
75. Venables, A.J. Winners and Losers from Regional Integration Agreements. *Econ. J.* **2003**, *113*, 747–761. [CrossRef]
76. Lu, X.; Ke, S. Evaluating the Effectiveness of Sustainable Urban Land Use in China from the Perspective of Sustainable Urbanization. *Habitat Int.* **2018**, *77*, 90–98. [CrossRef]
77. Tang, M.; Li, Z.; Hu, F.; Wu, B. How Does Land Urbanization Promote Urban Eco-Efficiency? The Mediating Effect of Industrial Structure Advancement. *J. Clean. Prod.* **2020**, *272*, 122798. [CrossRef]
78. Tone, K. A Slacks-Based Measure of Efficiency in Data Envelopment Analysis. *Eur. J. Oper. Res.* **2001**, *130*, 498–509. [CrossRef]
79. Hall, R.E.; Jones, C.I. Why Do Some Countries Produce So Much More Output Per Worker than Others?*. *Q. J. Econ.* **1999**, *114*, 83–116. [CrossRef]
80. Young, A. Gold into Base Metals: Productivity Growth in the People's Republic of China during the Reform Period. *J. Political Econ.* **2003**, *111*, 1220–1261. [CrossRef]
81. Glaeser, E.L.; Kahn, M.E. The greenness of cities: Carbon dioxide emissions and urban development. *J. Urban Econ.* **2010**, *67*, 404–418. [CrossRef]
82. Fajgelbaum, P.D.; Goldberg, P.K.; Kennedy, P.J.; Khandelwal, A.K. The Return to Protectionism*. *Q. J. Econ.* **2020**, *135*, 1–55. [CrossRef]
83. Lu, Y.; Yu, L. Trade Liberalization and Markup Dispersion: Evidence from China's WTO Accession. *Am. Econ. J. Appl. Econ.* **2015**, *7*, 221–253. [CrossRef]
84. Gilbert, J.; Beladi, H.; Oladi, R. North-South Trade Liberalization and Economic Welfare. *Rev. Dev. Econ.* **2015**, *19*, 1006–1017. [CrossRef]
85. Shi, X.; Xu, Z. Environmental Regulation and Firm Exports: Evidence from the Eleventh Five-Year Plan in China. *J. Environ. Econ. Manag.* **2018**, *89*, 187–200. [CrossRef]
86. Copeland, B.R.; Scott, T.M. North-South Trade and the Environment. *Q. J. Econ.* **1994**, *109*, 755–787. [CrossRef]
87. Zhang, Q.; Jiang, X.; Tong, D.; Davis, S.J.; Zhao, H.; Geng, G.; Feng, T.; Zheng, B.; Lu, Z.; Streets, D.G.; et al. Transboundary Health Impacts of Transported Global Air Pollution and International Trade. *Nature* **2017**, *543*, 705. [CrossRef]

land

MDPI

Article

Regional Coexistence in the Digital Era: Spatial–Temporal Evolution and Sustainable Strategies of the Coupled System in the Yangtze River Basin, China

Tianxiang Long [1,2,3], Yuxin Liu [4] and Qikang Zhong [5,*]

1 College of Architecture and Urban Planning, Hunan City University, Yiyang 413000, China
2 Key Laboratory of Urban Planning Information Technology of Hunan Provincial Universities, Yiyang 413000, China
3 Key Laboratory of Digital Urban and Rural Spatial Planning of Hunan Province, Yiyang 413000, China
4 School of Economics and Management, Tongji University, Shanghai 200092, China
5 School of Architecture and Art, Central South University, Changsha 410083, China
* Correspondence: 201311063@csu.edu.cn

Abstract: Against the backdrop of globalization and ecological civilization, this study aims to analyze the patterns of system coupling coordination development in the Yangtze River Basin under the interacting influences of population growth, ecological conservation, energy utilization, and digital economic development. Using a multisource model, this paper explores the state of coordinated development, spatial–temporal evolution characteristics, and influencing factors in the Yangtze River Basin from 2011 to 2020. The results indicate the following: (1) The overall degree of coupling coordination in the Yangtze River Basin shows better performances in the eastern coastal areas compared to the central and western regions. Over time, the spatial autocorrelation of coupling and coordination increases, exhibiting a significant spatial clustering trend. (2) The Moran's I index increased from 0.327 to 0.370, with high–high clusters primarily distributed in economically developed coastal provinces, while low–low clusters were observed in remote provinces in the central and western regions, revealing regional development imbalance issues. (3) The driving force analysis shows that green coverage and GDP are the core factors influencing the spatial differentiation of coupling coordinated development. Factors such as the urbanization rate, nighttime light index, and energy consumption had significant impacts in certain years but are generally considered minor factors. The results of this study not only contribute to understanding the dynamic mechanisms of regional coupling and development but also provide a scientific basis for formulating regional coordinated development policies, promoting the achievement of win–win goals of economic growth and ecological civilization in the Yangtze River Basin and similar regions.

Keywords: population-ecology-energy-digital economy; coupling coordination development; spatial–temporal evolution; spatial autocorrelation; driving factors; Yangtze River Basin

1. Introduction

Population, ecology, energy, and digital economy (PEED) are core drivers of economic growth and social stability. According to the World Population Prospects 2024 report, the global population is expected to increase by 2 billion over the next 60 years, peaking at approximately 10.3 billion by the mid-2080s. This rapid population growth will lead to a surge in resource and energy demands, intensifying dependence on various energy sources and exacerbating environmental pollution and ecological degradation [1–3]. Data from the International Energy Agency indicate that oil accounts for 33% of global energy consumption, while renewable energy, despite significant growth over the past decade, constitutes only 10% of total energy consumption [4]. The United Nations Convention to Combat Desertification (UNCCD) reports that 12 million hectares of land are degraded annually, affecting the lives of over 1.5 billion people. However, population structure and distribution are also heavily influenced by ecological quality and energy supply, as regions with a healthy environment and abundant energy resources tend to attract population inflows.

In recent years, the development of information and communication technologies has given rise to the digital economy, which mitigates ecological pressures by enhancing resource utilization efficiency and optimizing energy management [5]. Additionally, the digital economy has created numerous employment opportunities, facilitating population mobility. As of 2020, approximately 53.6% of global households have internet access, a significant increase from about 20% in 2005. The proliferation of digital technologies has improved the convenience of household life, enabling better access to efficient urban management and high-quality public services. As the demand for high-quality growth intensifies, competition for resources among PEED has become increasingly fierce. Achieving relatively balanced and coordinated coupling among these four elements is essential to meet the needs of sustainable development and realize the harmonious coexistence of the economy, society, and environment.

As the world's second-largest economy and largest primary energy consumer, China has increasingly emphasized the environmental pollution and sustainable development challenges arising from energy production and consumption in recent years. Particularly in the Yangtze River Basin, which contributes over 40% of China's GDP and plays a critical role in the Belt and Road Initiative as a key corridor for regional economic integration, balancing the interplay among PEED has become an ongoing challenge. The basin is a significant center for energy production and consumption in China, housing world-class hydropower facilities such as the Three Gorges Dam, which is the largest hydropower station globally. In 2020, the dam generated over 110 billion kWh of electricity, providing substantial clean energy to downstream regions [6]. Concurrently, the Yangtze River Basin boasts advanced digital infrastructure, positioning it at the forefront of China's digital economy development [7]. However, the basin is home to a large population, and the high population density has led to urban expansion and land resource constraints, causing ecological degradation, such as wetland loss and soil erosion, which pose severe threats to the environment. On the one hand, the challenges highlight the tight supply–demand tensions of regional resources, necessitating urgent resource management and optimization. On the other hand, the high-quality development of the basin relies on the harmonious advancement of population management, ecological protection, energy transition, and digital innovation. Successfully coordinating these factors will enhance the overall competitiveness and sustainability of the entire basin. Therefore, studying the coupling coordination development of PEED in the region is crucial not only for understanding and addressing the complex interactions and challenges in these areas but also for promoting balanced economic, social, and environmental development. This research will provide policy recommendations for the sustainable development of the region and offer insights that can be applied to other regions.

In recent years, significant progress has been made in research on the coupled coordination development of systems such as population, ecology, and energy. Most studies have focused on exploring the interactions between environmental sustainability and economic development. For example, Wang et al. (2019) examined the coupling relationship between energy efficiency and population in Guangdong Province [8], while Fu et al. (2022) analyzed the coordinated development of natural resources and financial systems [9]. However, existing studies often concentrate on simple coupling relationships between two or three systems, overlooking the important influence of the digital economy as an emerging economic system. Moreover, there is a lack of research on the spatiotemporal evolution characteristics of specific regions, as most studies focus on the national or provincial level, neglecting an in-depth exploration of spatiotemporal differences within regions.

Given this context, this paper first analyzes the mechanisms of coupling coordination relationships among PEED in the Yangtze River Basin and constructs a corresponding composite evaluation index system. Second, the coupling coordination degree model is employed to measure the development status of PEED coupling coordination in the basin from 2011 to 2020, with a detailed analysis conducted from both temporal and spatial perspectives. Third, the spatial autocorrelation model is used to analyze the spatial correlations of this coupling coordination. Finally, geographic detectors and correlation analysis are utilized to explore external influencing factors. This study not only reveals the dynamic changes in the coupling development within regions but also provides a quantitative assessment of regional development imbalances. It offers theoretical support for achieving the dual goals of economic growth and ecological protection. It provides reliable research results for a comprehensive understanding of the current status and potential of coupled and coordinated development in the Yangtze River Basin.

2. Literature Review

2.1. State of the Research on Coupled Coordination Systems

The coupling relationship among ecology, energy, population, and the digital economy is a complex scientific issue involving long-term time series. Research on the relationship among these four subsystems, both domestically and internationally, has mainly focused on three aspects. First, from the perspective of content, most studies concentrate on constructing and calculating coupling coordination development systems. Existing research typically evaluates the coordination degree of complex systems by establishing indicator systems. For the study of the coordination degree of the two subsystems, Wang et al. (2019) developed a coupling coordination model to study the relationship between energy efficiency and population in Guangdong Province [8]. Ariken et al. (2020) used multi-source remote sensing data to construct a remote sensing ecological index and coupling coordination model to comprehensively evaluate the coupling coordination relationship between urbanization and ecological environment in Yanqi Basin from 2000 to 2018 [10]. Subsequently, Ariken, Zhang, and weng Chan (2021) conducted a coupling assessment of urbanization and ecological environment development along the Silk Road Economic Belt from 2004 to 2018, integrating the evaluation model with the coupling coordination model and the GTWR model [11]. Zheng, Khan, and Abbas (2022) used the coupling coordination model to examine the relationship between urbanization and ecological environment in China's central region, from comprehensive evaluation analysis, coupling coordination, and spatiotemporal evolution characteristics to decoupling path exploration and influencing factors analysis [12]. Fu et al. (2022) employed the entropy method and coupling coordination development model to measure the coupling development level of the digital economy and ecological environment [9]. Other scholars have applied the coupling coordination model to quantitatively analyze the urbanization and land-use efficiency in China's Loess Plateau, aiming to evaluate the comprehensive development level and coordination of this unique region [13]. Regarding three subsystems, Chen et al. (2019) used data from 36 major Chinese cities from 2003 to 2016 to establish an entropy-weighted TOPSIS model to evaluate urban environment–economic–population systems [14]. Zameer et al. (2020) evaluated the coupling coordination degree of natural resources, financial development, and ecological efficiency in China's eastern, central, and western provinces from a regional perspective, finding that financial development and natural resource levels in the central and western regions are higher than in the east, while utilization efficiency is lower [15]. Dong et al. (2021) used data from the Yangtze River Delta urban agglomeration from 2010 to 2018 to construct an economic–energy–environment system evaluation index and apply the coupling coordination model for a comprehensive assessment [16]. Liu et al. (2022) developed an energy–economic–environment indicator system from the perspective of green development and ecological integrity, using a distance-based coupling coordination model and dynamic comprehensive coordination model to evaluate 11 provinces along the Yangtze River Economic Belt from 2008 to 2017 [17]. These studies mainly focus on calculating the coordination degree of various combinations of social, economic, environmental, and energy systems, indicating that different complex systems have certain impacts on coupled development. However, research on the coordination development level of PEED remains limited, and the interactions and influencing factors among these four elements have yet to be fully explored. Therefore, exploring the coupling relationship among PEED can provide a more comprehensive understanding of its potential impact on overall societal development. In-depth analysis of the interactions among these complex systems can not only optimize resource allocation and utilization, enhancing system resilience, but also provide crucial insights for developing more effective and sustainable development policies, achieving multi-objective coordinated development.

2.2. State of the Research Method

Regarding research methods, data envelopment analysis (DEA), structural equation modeling (SEM), multi-indicator comprehensive evaluation methods, system dynamics models, and coupling coordination models are widely used [13,18–21]. Scholars have utilized the super-efficiency DEA from a coupling perspective to quantify urban land-use efficiency and ecological carrying capacity in Jilin Province, proposing a multistage coupling mechanism corresponding to different stages of urban development [22]. Xing, Xue, and Hu (2019) employed a system dynamics model to establish a system model that integrates the interactions among economic, resource, and environmental subsystems, clearly revealing the coupling relationships among the elements [20]. Wang et al. (2019) based on sustainable development data and the coupling coordination model from 2010 to 2016 in Shandong Province, quantitatively analyzed regional sustainable development trends and the coupling coordination state of various subsystems [23]. Bahtebay et al. (2021) constructed a coordination development index of urbanization and resource environment in Xinjiang from 2004 to 2018, using the Kaya identity, logarithmic mean Divisia index model, and regression analysis to measure the turning point of urbanization rate on resource and environmental pressure [24]. Yang et al. (2023) employed the coupling coordination model and spatiotemporal evolution laws to study the coupling coordination development status and spatiotemporal characteristics of China's digital economy and carbon environment governance. This provides some support for applying models in studying the coupled and coordinated development of PEED in the Yangtze River Basin [25].

2.3. State of the Research Scale

In terms of research scale, the literature mainly focuses on national, provincial, or urban levels, with relatively few studies directly targeting the Yangtze River Basin. For example, Tomal (2021) took Poland as the research area, identifying barriers to development at the local coupling level for each municipality [26]. Li et al. (2021) analyzed the coupling coordination development status of social security and economic development systems nationwide from 2002 to 2018 and highlighted regional development imbalances [27]. Sun and Cui (2018) used four autonomous regions in China as case studies to analyze the coupling coordination of the economic, social, and environmental benefits of urban infrastructure [21]. Abulimiti et al. (2022) explored the coordination development status of population, economy, society, and human habitat factors in Xinjiang's county-level urbanization process from 1996 to 2018, across time and space scales [28]. Feng et al. (2022) studied the Yellow River Basin, discussing how to balance water resource utilization, industrial development, and ecological protection [29]. Zhong, Li, and He (2023) analyzed how urban policies affect the coordination development of coupled systems on an urban scale [30]. Cui et al. (2019) used Kunming as a case study to explore the coupling coordination relationship between social economy and water environment on an urban scale [31]. Although there is extensive research on national provinces, cities, and other river basins, studies that comprehensively reveal the coordination development status of the Yangtze River Basin from a more holistic perspective are lacking. The Yangtze River Basin holds significant importance in China's economic and ecological systems, and in-depth exploration of the interactions among PEED and their impact on overall development from a more comprehensive and systematic perspective is of great significance.

3. Data and Methods

3.1. Study Area

The Yangtze River Basin spans 19 provinces across China (as is shown in Figure 1), extending from the western highlands to the eastern coastal regions. The selection of the regional scale for this analysis, specifically focusing on the Yangtze River Basin, is based on both the ecological and economic significance of the region and the research objectives. The Yangtze River Basin plays a pivotal role in China's economy, accounting for over 40% of the national GDP, and is a key area for both economic development and ecological conservation under national policies such as the Yangtze River Economic Belt Development Plan. This makes it an ideal case study for examining the coupled dynamics of PEED.

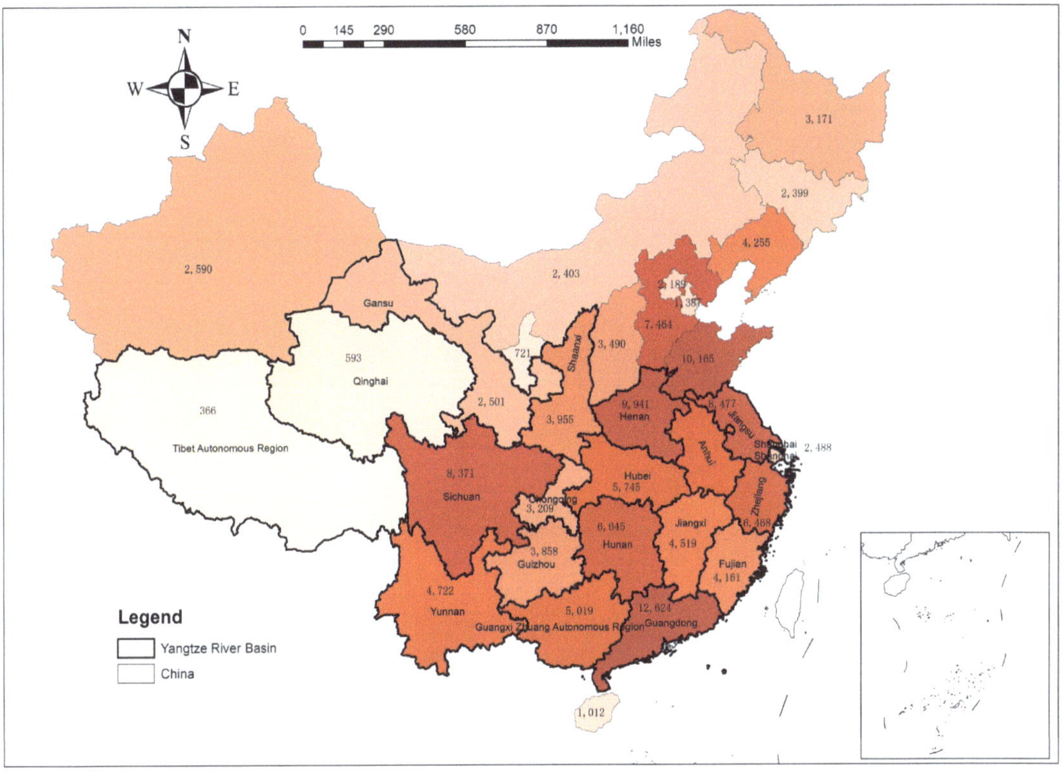

Figure 1. Geographical Distribution of the Yangtze River Basin Across 19 Provinces in China. Note: number means the population of 31 provinces in 2020. Note: The unit of population is in tens of thousands.

Additionally, the regional scale is chosen because it allows for a more detailed examination of spatial heterogeneity across provinces within the basin. Unlike studies at the national level that may overlook local variations or city-level studies that may be too localized to capture broader trends, the regional approach enables us to identify distinct spatial clustering patterns and development imbalances between the eastern, central, and western parts of the basin. This is particularly important in understanding how different regions contribute to the overall coordination of PEED systems, which are influenced by both regional policies and local socio-economic conditions.

Moreover, this scale allows for the application of spatial autocorrelation methods, such as Moran's I, which are crucial for detecting spatial dependencies and clustering in the interaction of these systems. This spatial analysis further reinforces the necessity of studying regional dynamics on this scale, as it provides insights into the geographical patterns of development that are not apparent in more aggregate analyses.

Lastly, the regional scale has direct policy relevance. The Chinese government's development strategies for the Yangtze River Basin emphasize regional coordination and balancing economic growth with ecological sustainability. Therefore, this scale of analysis is particularly useful for producing insights that can inform regional planning and policy making aimed at promoting sustainable development across diverse geographic areas.

3.2. Remote Sensing Data

3.2.1. Entropy-TOPSIS Method

The Entropy-TOPSIS method was selected for this study as it provides a robust approach for evaluating the multidimensional aspects of the coupled system across the PEED subsystems. The entropy weight method is particularly effective in assigning objective weights to indicators based on their variability, reducing the subjectivity that often arises from arbitrary weighting. This ensures a more reliable and data-driven evaluation of system performance, which is crucial given the diverse nature of the subsystems under consideration.

While the TOPSIS method has been widely applied in previous studies for multicriteria decision analysis, the novelty of this study lies in its application within the context of four interacting subsystems in the Yangtze River Basin, incorporating the digital economy as a key system component. Existing studies that employ the Entropy-TOPSIS method typically focus on simpler interactions, such as economic-environmental or population-ecological relationships [17]. However, few studies have applied this method to explore the dynamic coupling of four complex systems simultaneously, especially in a regional context where spatial heterogeneity and temporal evolution play significant roles.

(1) The standardization is publicized as follows:

$$Z_{ij} = \frac{x_{ij} - \min x_{ij}}{\max x_{ij} - \min x_{ij}} \quad (i = 1, 2, \cdots, m;\ j = 1, 2, \cdots, n) \tag{1}$$

and the negative indicators as follows:

$$Z_{ij} = \frac{\max x_{ij} - x_{ij}}{\max x_{ij} - \min x_{ij}} \quad (i = 1, 2, \cdots, m;\ j = 1, 2, \cdots, n) \tag{2}$$

where the j index value of the i evaluation sample is recorded as x_{ij}.

(2) Calculate the weight of the i evaluated object on the j indicator as follows:

$$Y_{ij} = Z_{ij} / \sum Z_{ij} \tag{3}$$

Calculate the entropy of the j indicator as follows:

$$e_j = -k \sum_{i=1}^{m} Y_{ij} \ln Y_{ij} \tag{4}$$

Determine the j indicator's variability coefficient as follows:

$$g_i = 1 - e_j \tag{5}$$

Determine the weighting coefficients as follows:

$$w_j = g_j / \sum_{j=1}^{n} g_j \tag{6}$$

where Y_{ij} represents the weight value of the j indicator of the i evaluation sample; e_j represents the information entropy value of the j indicator; g_j represents the information entropy redundancy of the j indicator; and w_j represents the weight coefficient value of the j indicator.

(3) The following is the formula for the comprehensive evaluation index:

$$S_{IJ} = w_j Z_{ij} \tag{7}$$

where S_{IJ} is the weighted comprehensive value, w_j is the weight of the th indicator, and Z_{ij} is the standardized data.

(4) The calculation formulas for the positive ideal solution (Z_j^+) and negative ideal solution (Z_j^+) of the TOPSIS model are as follows:

$$Z_j^+ = \max(S_{1j}, S_{2j}, \cdots S_{ij}) \tag{8}$$

$$Z_j^- = \max(S_{1j}, S_{2j}, \cdots S_{ij}) \tag{9}$$

(5) Calculate the distance between each decision unit and the optimal solution (D_j^+) and the distance between it and the worst solution (D_j^-) as follows:

$$D_j^+ = \sqrt{\sum_{j=1}^{n}\left(Z_j^+ - r_{ij}\right)^2} \tag{10}$$

$$D_j^- = \sqrt{\sum_{j=1}^{n}\left(Z_j^- - r_{ij}\right)^2} \tag{11}$$

(6) Calculate the relative progress (C_j) of each decision-making unit and obtain the comprehensive evaluation results of each decision-making unit as follows:

$$C_j = \frac{D_j^-}{D_j^- + D_j^+} \tag{12}$$

where $C_j \in [0,1]$ The closer C_j is to 1, the better the comprehensive evaluation of each decision-making unit.

3.2.2. Coupling Coordination Degree Model

This study uses the coupling coordination model to calculate the coupling coordination degree, which is convenient for measuring the strength of the interaction between systems and provides a scientific basis for sustainable development. The model calculation formula is as follows:

$$D = \sqrt{C * T} \tag{13}$$

$$C = \left\{\frac{E_1 E_2 E_3 E_4}{[(E_1 + E_2 + E_3 + E_4)/4]^4}\right\}^{1/4} \tag{14}$$

$$T = aE_1 + bE_2 + cE_3 + dE_4 \tag{15}$$

where D is the CCD, $D \in [0,1]$; C represents the coupling degree between the PEED systems $C \in [0,1]$. The larger C indicates a higher level of coupling coordination development and better development of the four subsystems. Referring to previous studies, the coupling degree is divided into five stages according to the numerical range (as is shown in Supplementary Table S1). E_1 represents the population system evaluation value; E_2 represents the ecosystem evaluation value; E_3 represents the energy system evaluation value; E_4 represents the digital economy system evaluation value. T represents the coordination value of the four systems; a, b, c, d are all coefficients to be determined. Here, we consider the four subsystems—PEED—to be of equal importance, and therefore assign them equal weights in the coupling coordination degree model. This assumption is based on the need to provide a balanced evaluation of the interactions among these subsystems, each of which plays a critical and interdependent role in regional sustainable development. The equal weighting approach is widely used in similar studies of coupled systems [32–34], where the goal is to avoid subjective bias in assigning differential importance to each subsystem. By treating the subsystems as equally important, we aim to maintain an objective framework that reflects the integrated nature of the development challenges in the Yangtze River Basin, i.e., $a = b = c = d = 1/4$. Classification of coupling coordination degree is presented in Supplementary Table S2.

3.2.3. Exploratory Spatial Data Analysis

In the test stage, this study used the Global Moran's I index and the Local Moran's I, to analyze the overall and local spatial autocorrelation characteristics of the resilience of the human settlement environment in the study area.

Calculation of the Global Moran's I index: $G_0 = \sum_{i=1}^{n}\sum_{j=1}^{n} w_{ij}$, where n is the total number of spatial units; y_i and y_j represent the attribute values of the i spatial unit and the j spatial unit, respectively; $\overline{y}$ is the mean of the attribute values of all spatial units; and w_{ij} is the spatial weight value. The calculation formula is as follows:

$$I = \frac{n}{G_0} \times \frac{\sum_{i=1}^{n}\sum_{j=1}^{n} w_{ij}(y_i - \overline{y})\left(y_j - \overline{y}\right)}{\sum_{i=1}^{n}(y_i - \overline{y})^2} \tag{16}$$

Calculation of the Local Moran's I index: $F_i = y_i - \overline{y}$, $F_j = y_j - \overline{y}$, where w_{ij} is the spatial weight value; n is the total number of all regions in the study area; and I_i represents the Local Moran's I of the i region. The calculation formula is as follows:

$$I_i = \frac{F_i}{G_2}\sum_{j \neq 1}^{n} w_{ij}F_j \tag{17}$$

3.2.4. Geographic Detector Model

This study mainly used the geographic detector model to reveal the impact of dynamic factors on different scales for the coupling coordination of the human settlement environment system in each region. The significance of the main effects can help reveal causal relationships. The model is as follows:

$$q = 1 - \frac{1}{N\sigma^2}\sum_{h-1}^{L} N_h\sigma_h^2 \tag{18}$$

where q is the influence of the dynamic factor on the coupling coordination system, $q \in [0, 1]$, with the larger q represents the stronger explanatory power of the dynamic factor on the coupling coordination degree; h represents the stratification of the dynamic factor, with N_h and N represent the number of units in layer h and the whole area, respectively; and σ_h^2 and σ^2 represent the variance of y in layer h and the whole area, respectively.

3.3. *Indicators and Data Sources*

3.3.1. Coupling Coordination System Index Selection

Population dynamics have a direct impact on a country's socioeconomic structure. The ecological environment is closely tied to human well-being and national sustainable development. Energy serves as the cornerstone of modern economies, so energy policy and sustainability are critical for long-term development. The digital economy growth is a hallmark of modern socioeconomic progress, playing a crucial role in enhancing societal and economic vitality. Building upon existing research [8,35–37], this study develops an evaluation system for the coupled coordination development of PEED in the Yangtze River Basin. Table 1 below shows the evaluation index system.

Table 1. Coupling coordination classification.

Subsystem	Evaluation Index	Unit	Attribute	Weight (%)	References
Population	Permanent population at year-end	10,000 people	-	21.038	[35,38,39]
	Birth rate	%	+	25.941	
	Natural population growth rate	‰	+	13.827	
	Urban employed population	10,000 people	+	18.845	
	Urban registered unemployment rate	%	-	20.349	
Ecology	Sulfur dioxide emissions	10,000 tons	-	15.720	[1,2,40]
	Particulate matter, dust emissions	10,000 tons	-	10.084	
	Sewage treatment rate	%	+	15.037	
	Comprehensive utilization rate of solid waste	%	+	19.741	
	Forest coverage rate	%	+	20.727	
	Per capita park green space area	Square meters	+	18.691	
Energy	Electricity generation	100 million kWh	+	22.668	[2,41,42]
	Per capita LPG supply	m^3/person	+	26.007	
	Per capita natural gas supply	m^3/person	+	21.274	
	Electricity consumption	100 million kWh	−	15.354	
	Gas penetration rate	Tons standard coal/10,000 CNY	+	14.697	
Digital economy	Mobile phone penetration rate	%	+	14.973	[25,32,43,44]
	Number of internet access ports	10,000 units	+	18.144	
	Level of mobile online payment	/	+	23.154	
	Breadth of digital financial coverage	%	+	28.805	
	Degree of digitalization in finance	%	+	14.924	

3.3.2. Selection of Driving Factors

To comprehensively assess the external influencing factors on the level of coupled coordination development, the following six key variables are selected (as presented in Table 2): GDP, nighttime light index, urbanization rate, greening coverage rate, energy consumption intensity, and CO_2 emission intensity. GDP, as a macroeconomic indicator, reflects the economic scale and growth rate of the Yangtze River Basin, making it a crucial factor in measuring economic development levels [45]. Nighttime light index which is a novel tool for measuring economic activity, provides more intuitive and real-time information on economic activities [46]. Urbanization rate not only reflects changes in population structure and distribution but also relates closely to regional economic development, ecological environment, and resource consumption [13]. The combined analysis of these variables helps reveal the interaction between economic growth and population dynamics. Greening coverage rate, an indicator of ecological environment quality, reflects the health of ecosystems and the expansion of green ecological spaces within the basin, It is crucial for evaluating ecosystem services and contributions to human well-being [47]. Additionally, energy consumption intensity and CO_2 emission intensity are directly related to energy efficiency and environmental pressure. They are important indicators for evaluating the effectiveness of energy policies and environmental protection, providing a critical perspective for understanding the region's sustainable development capacity.

Moreover, the relationship between Tables 1 and 2 lies in the fact that while the subsystems (Table 1) are the core components of the coupling coordination system, the external factors (Table 2) moderate or influence how these subsystems interact over time. In other words, Table 1 focuses on the internal structure of the system, while Table 2 considers how external forces shape or influence this structure. Together, these two sets of factors provide a comprehensive view of both the internal and external dynamics affecting regional development. By analyzing both sets of factors, we can gain deeper insight into how the internal performance of the subsystems is influenced by external conditions, allowing for a more holistic understanding of the driving forces behind the regional coupling coordination process.

Table 2. Driving factors.

Variable Type	Variable Name	Unit	Code
Explained Variable	Coupling Coordination Degree	/	D
Explanatory Variables	GDP	100 million CNY	X1
	Nighttime Light Index	/	X2
	Urbanization Rate	%	X3
	Greening Coverage Rate	%	X4
	Energy Consumption Intensity	Tons standard coal/10,000 CNY	X5
	CO_2 Emission Intensity	t/10,000 CNY	X6

3.3.3. Data Resources

The study examined panel data from the Yangtze River Basin from 2011 to 2020. Administrative boundary data were obtained from the Resource and Environmental Science Data Platform, Chinese Academy of Sciences (https://www.resdc.cn/, accessed on 1 May 2024). The indicators and driving factors are based on data from the "China Urban Statistical Yearbook", "China Energy Statistical Yearbook", "Statistical Bulletin of National Economic and Social Development", and National Bureau of Statistics. Missing data were filled using the interpolation method [48].

4. Results

4.1. Spatiotemporal Evolution of Regional Coupling Coordination Systems

4.1.1. Temporal Evolution Characteristics of Coupling Coordination Degree

This study first calculated the comprehensive evaluation indices and coupling degree of the PEED subsystems in the Yangtze River Basin from 2011 to 2020. To better reflect the temporal variations in the coupling coordination development across the region, the 19 provinces in the Yangtze River Basin were treated as a whole, and the annual results for each province are visualized (as presented in Figures 2–4).

The empirical results indicate that from 2011 to 2020, the overall coupling coordination degree in the basin showed an upward trend. This indicates that the interactions among population, ecological protection, energy consumption, and digital economy growth became increasingly harmonious.

However, from 2016 to 2017, the overall growth in the coupling coordination slowed down. This may be related to increased economic pressures, adjustments in traditional industrial structures, and the intensification of environmental protection policies during that period. Although economic growth slowed, the implementation of environmental policies promoted long-term sustainable development.

Moreover, provinces such as Guangdong, Fujian, and Zhejiang exhibited strong coupling coordination degrees, likely due to more active economic activities, particularly the rapid development of digital economy, which has better balanced the dual goals of ecological protection and economic development. In contrast, provinces like Sichuan and Yunnan showed relatively low coupling coordination degrees, which may be attributed to their regional industrial structures and geographical conditions.

Finally, provinces like Tibet, Yunnan, Guizhou, and Gansu experienced significant fluctuations in their coupling coordination degrees. This suggests that these regions might be significantly influenced by policy adjustments, industrial transitions, or changes in the external economic environment when facing the challenges of balancing economic development and ecological protection.

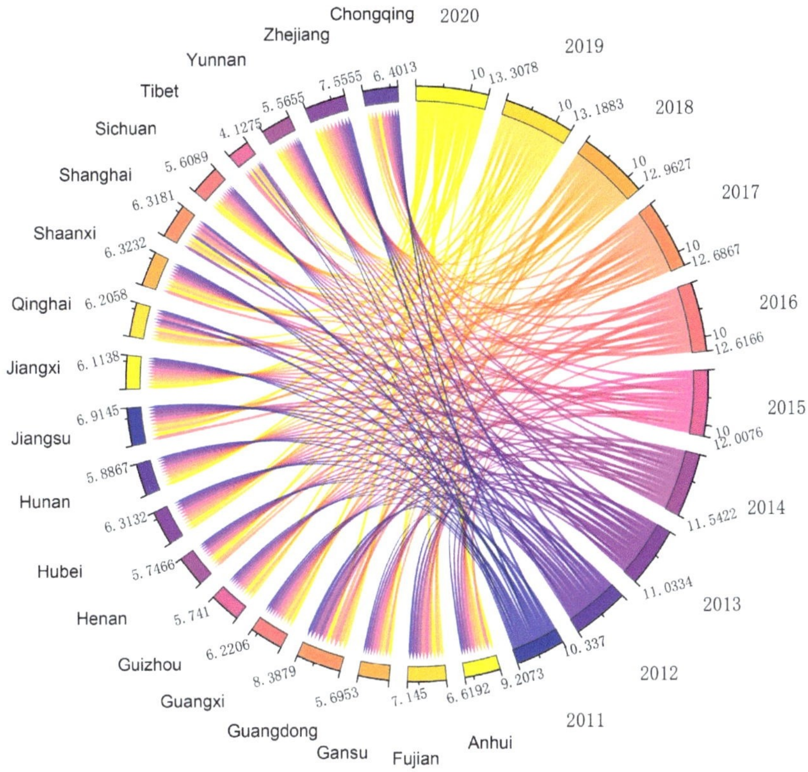

Figure 2. Time series changes in the coupling coordination average index for the Yangtze River Basin from 2011 to 2020.

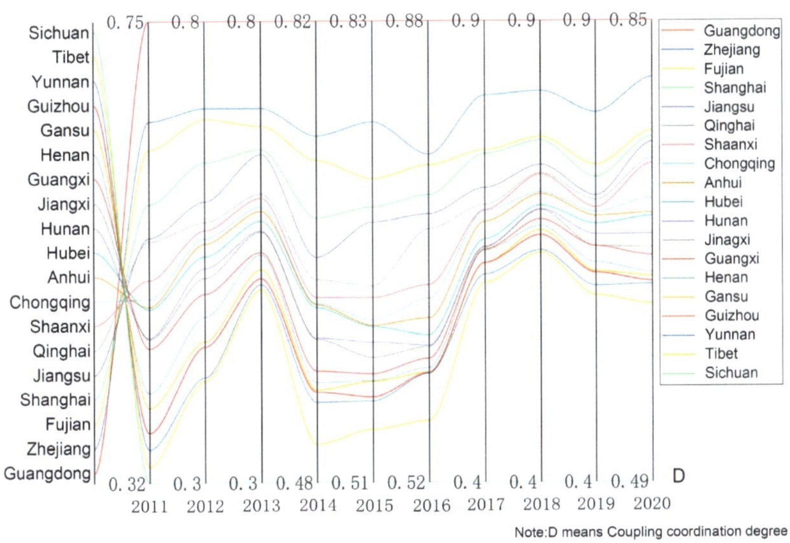

Figure 3. Time series changes in the coupling coordination index for the Yangtze River Basin from 2011 to 2020.

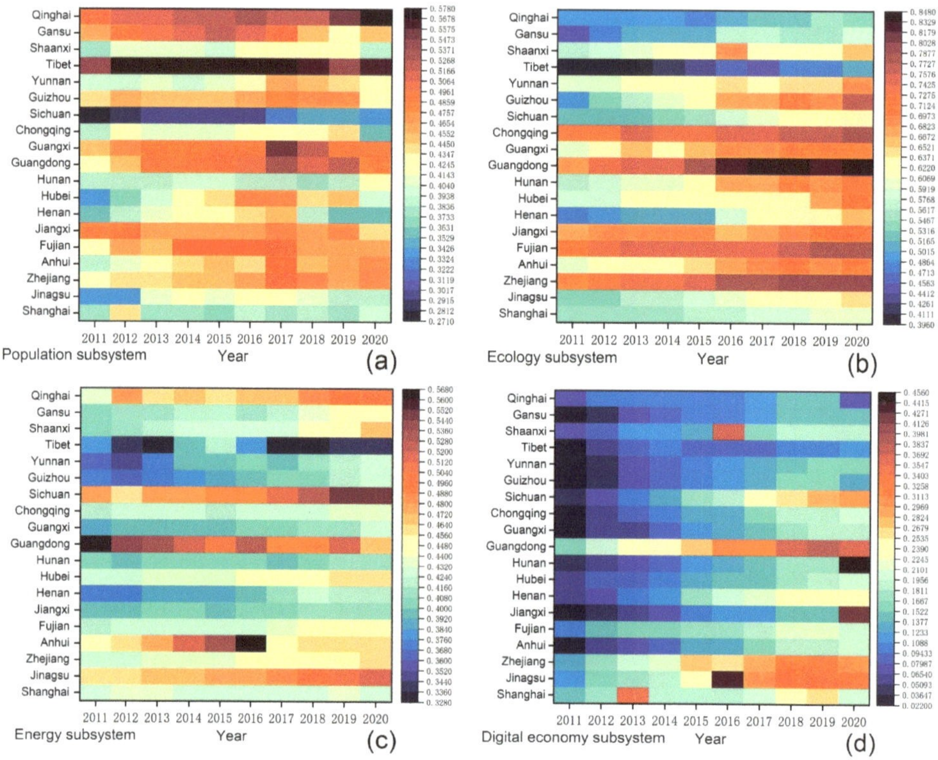

Figure 4. Time series changes in subsystem indices for the Yangtze River Basin from 2011 to 2020. Note: (**a**–**d**) means the order.

The changes in the population subsystem typically reflect shifts in population growth, urbanization levels, population policies, and labor market (as shown in Figure 4a). Overall, the scores for most provinces showed an upward trend over the decade, despite some fluctuations. However, some provinces experienced a decline or fluctuation. Additionally, lower scores in eastern cities could be associated with high population density, elevated living costs, and aging.

The ecological subsystem scores reflect the levels of environmental quality, pollution control, and ecological protection (as shown in Figure 4b). Provinces like Fujian, Zhejiang, Chongqing, and Guangdong achieved high scores, possibly due to the implementation of stricter environmental protection measures and a balanced approach between environmental conservation and economic development. In contrast, the ecological challenges faced by Gansu, Tibet, and Qinghai might be tied to their fragile ecosystems and the pressures of resource exploitation. Provinces such as Guangxi and Jiangxi scored moderately, indicating stable development within their ecological subsystems, though there remains room for improvement.

The strong performances of Sichuan and Guangdong in the energy subsystem may be related to their optimized energy structures and improved energy efficiencies (as shown in Figure 4c). The lower scores in Tibet, Yunnan, and Guizhou might have resulted from insufficient energy infrastructure and low utilization rates for renewable energy. The fluctuations observed in Jiangsu and Qinghai could reflect the impact of industrial restructuring and changes in energy policies. Moreover, provinces like Hubei, Chongqing, and Shanghai displayed relatively stable but lower energy scores during this period.

The overall score for the digital economy subsystem exhibited a continuous upward trend (as shown in Figure 4d), reflecting the rapid development of China's digital economy and the national emphasis on this sector. Guangdong, Jiangsu, and Zhejiang led in the digital economy. Notably, the significant score increases in Shaanxi, Zhejiang, and Jiangsu between 2015 and 2017 suggest effective policy adjustments or the initiation of key projects during this period. Meanwhile, although Yunnan, Gansu, and Guizhou started with lower scores, they have recently shown remarkable growth.

4.1.2. Spatial Evolution Characteristics of Coupling Coordination Degree

To capture the spatial evolution of coupling coordination development in the Yangtze River Basin, this study visualizes the coupling coordination degree and the evaluation indices of the four subsystems from 2011 to 2020 (as presented in Supplementary Figure S1a–e).

Overall, the coupling coordination degree in the Yangtze River Basin exhibits a spatial pattern of higher values in the eastern regions and lower values in the western regions (Supplementary Figure S1a). The eastern provinces demonstrated a higher coupling coordination degree, while the western provinces showed lower levels. The southern provinces such as Fujian and Guangdong scored higher than the northern provinces like Qinghai and Tibet.

The eastern region (Supplementary Figure S1b–e), which includes provinces such as Zhejiang, Jiangsu, Shanghai, Guangdong, and Fujian, consistently exhibited higher coupling coordination degrees. These provinces benefit from strong economic bases, advanced digital infrastructures, and significant policy support that prioritize regional development and sustainability. The southern provinces, particularly Fujian and Guangdong, scored higher compared to the northern provinces because of their openness to trade and robust economic growth, driven by coastal advantages and industrial diversity. In the eastern region, provinces such as Jiangsu and Zhejiang maintained high population subsystem scores, reflecting their dense populations, high rates of urbanization, and population mobility. These provinces benefit from abundant educational and healthcare resources, which support their large populations and create a dynamic labor market. The ecological subsystem scores are generally higher in eastern provinces, particularly in Zhejiang and Fujian. The success of these policies in the coastal areas is evident in their higher ecological subsystem scores, as these regions prioritize environmental governance and sustainable development. The energy subsystem in the eastern region showed a tri-core pattern connecting Sichuan, Anhui, and Guangdong, with Jiangsu, Zhejiang, and Fujian playing key roles in energy utilization and improvement over time. These provinces form a network that fosters energy cooperation and technological innovation, contributing to a spillover effect across neighboring regions. The digital economy is a significant driver of the eastern region's higher coupling coordination degree. Provinces such as Zhejiang, Shanghai, Jiangsu, and Guangdong have emerged as core hubs of the digital economy. These provinces lead in e-commerce, artificial intelligence, and digital financial services, making them central to China's digital transformation.

The western region (Supplementary Figure S1b–e), which includes provinces such as Qinghai, Tibet, Sichuan, and Yunnan, generally exhibited lower coupling coordination degrees compared to the east. These provinces face challenges related to economic underdevelopment, infrastructure gaps, and limited policy intervention. However, there have been improvements in some areas due to national initiatives, such as the Western Development Strategy, which aims to enhance economic growth and infrastructure in these provinces. The population subsystem scores in the western region show an interesting pattern, with upstream provinces like Sichuan and Yunnan demonstrating notable increases in their scores. These increases were largely driven by new urbanization initiatives and the government's focus on population redistribution to less developed areas. Nevertheless, population growth remained limited compared to the eastern provinces, reflecting the region's slower pace of urban development. The ecological subsystem scores in the western region were lower overall, particularly in provinces such as Tibet and Qinghai. These regions face significant ecological challenges due to their harsh environmental conditions and lack of infrastructure for comprehensive ecological protection. However, policies like the "Overall Plan for the Reform of the Ecological Civilization System" have started to improve the situation, as seen in provinces like Sichuan and Yunnan, where scores have gradually improved because of ecological restoration projects. Energy development in the western region is centered around provinces like Sichuan, which acts as a key hub for energy production. However, provinces such as Qinghai and Tibet still lag behind in energy infrastructure development, leading to lower energy subsystem scores. Despite this, there is potential for improvement through renewable energy initiatives, particularly in the solar and hydropower sectors, which have been prioritized in recent government strategies. The western region lags behind the east in terms of digital economy development. However, certain provinces, such as Chongqing and Sichuan, are showing momentum in digital transformation, supported by government initiatives aimed at improving digital infrastructure. While there are still significant disparities between the east and west, the digital economy in the western region is gradually improving, with a focus on bridging the digital divide through investments in broadband and e-commerce platforms.

4.2. Spatial Autocorrelation Analysis

The spatial autocorrelation test results, analyzed by Geoda software 1.14, are shown in Figure 5, indicating the spatial autocorrelation characteristics of the coupling coordination degree of PEED in the Yangtze River Basin. The results reveal statistically significant values for Moran's I at the 5% significance level across ten periods. From 2010 to 2019, the Global Moran's I increased from 0.327 to 0.370; the Global Moran's I was the same as the Local Moran's I. Moreover, the confidence percentage of over 80% of Local Moran is over 95%, suggesting that the spatial autocorrelation of the coupling coordination degree has strengthened, with clustering patterns emerging among similar provinces.

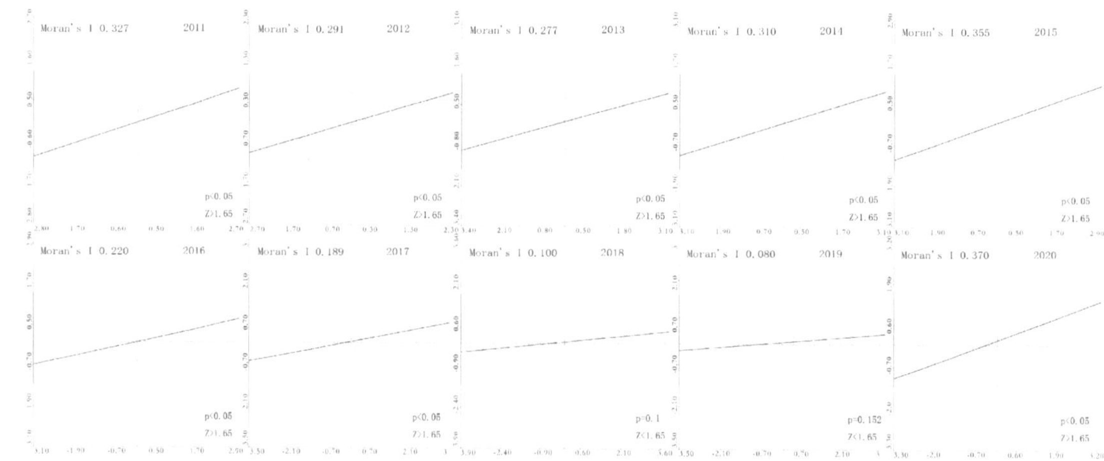

Figure 5. Moran's I calculation results for the coupling coordination degree in Yangtze River Basin from 2011 to 2020.

Additionally, this study applied LISA cluster mapping (as presented in Figure 6) to categorize the local spatial characteristics of the four subsystems' coupling coordination into four types of spatial associations, thereby clarifying the coordination development patterns and evolutionary trends across provinces. High–high clusters (areas where both the region and its neighbors show high coupling coordination degrees) were primarily concentrated in the eastern coastal provinces, including Zhejiang, Fujian, and Guangdong. Over time, these policies created an environment conducive to greater coordination among the four subsystems (PEED), resulting in stable high–high clusters in these areas. In contrast, the central and western regions, including Sichuan, Yunnan, and Tibet, consistently displayed low–low clusters, reflecting their slower economic development, weaker infrastructure, and limited policy support. These areas, particularly Tibet, are geographically remote and face challenges in attracting investment and fostering technological innovation, leading to persistent gaps in development compared to the eastern provinces. The presence of high–low clusters (regions with high coupling coordination degrees but surrounded by areas with lower degrees) often reflects localized development successes, where specific provinces have implemented successful regional policies that have not yet spread to neighboring regions. Overall, there is a clear distinction in the coupling coordination degree of the PEED systems between the eastern coastal regions and the central–western regions.

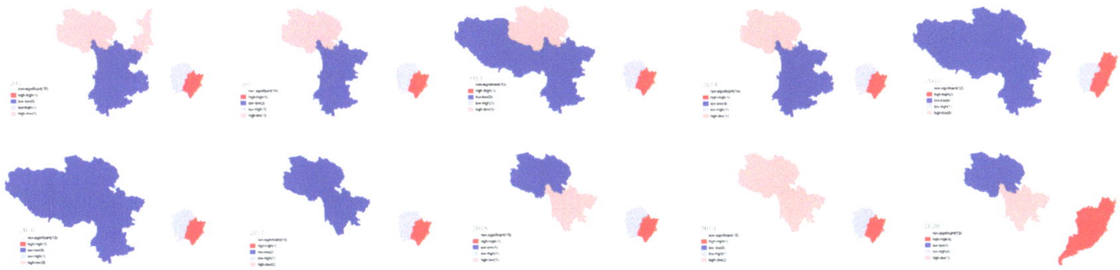

Figure 6. LISA clustering results of the coupling coordination degrees in the Yangtze River Basin from 2011 to 2020.

4.3. Analysis of Influencing Factors

4.3.1. Detection Result Analysis

To further analyze the external factors influencing the spatiotemporal evolution of coupling coordination development in the Yangtze River Basin.

As shown in Table 3, the dominant factors contributing to spatial differentiation in coupling coordination development vary across different periods. The study conducted a frequency analysis of factors with q-values greater than 0.5 and *p*-values less than 0.05 for each period, identifying the most frequently occurring factors as core influences. Between 2011 and 2016, greening coverage rate emerged as a core factor significantly impacting the coupling coordination development. Additionally, GDP played a significant role from 2014 to 2017, which may have contributed positively to the region's coupling coordination development. The influence of the urbanization rate highlights the economic agglomeration effects and the impact of population dynamics on resource allocation and environmental quality. Meanwhile, changes in the night light index indicate fluctuations in regional economic activity levels. Notably, energy consumption and CO_2 emission intensity did not show significant effects on coupling coordination development, possibly due to effective carbon reduction strategies or shifts in energy policies.

As illustrated in Supplementary Figure S2, this study also reveals the interaction relationships among driving factors. Overall, the interactions among these factors had significantly intensified. Strong interactions were observed between urbanization rate and CO_2 emission intensity, greening coverage rate∩energy consumption, energy consumption∩nighttime light index, and urbanization rate∩nighttime light index. This indicates that the interplay among these factors can have a more pronounced impact on the spatial differentiation of coupling coordination development in the Yangtze River Basin.

Table 3. Factor detection results. Note: * means $p < 0.05$, ** means $p < 0.01$, *** means $p < 0.001$.

	GDP	Urbanization Rate	Greening Coverage Rate	Energy Consumption	CO_2 Emission Intensity	Nighttime Light Index
2011	0.5462	0.6783 *	0.8114 ***	0.3445	0.4871	0.5905 *
2012	0.5971	0.5469	0.7917 ***	0.4643	0.4311	0.6311 *
2013	0.6675	0.6047 *	0.7987 ***	0.5753	0.4453	0.5402
2014	0.6883 *	0.5779	0.7004 **	0.5039	0.4709	0.5194
2015	0.6877 *	0.5105	0.5514 **	0.4613	0.4997	0.4834
2016	0.8278 **	0.3628	0.5095 *	0.6213	0.5739	0.2894
2017	0.6798 *	0.4996	0.6122	0.3668	0.5920	0.3580
2018	0.6337	0.5026	0.6000	0.3461	0.6357	0.3078
2019	0.6112	0.5299	0.5573	0.5595	0.5812	0.3413
2020	0.6121	0.4083	0.3463	0.5978	0.6258	0.4173

4.3.2. Correlation Analysis

A correlation analysis was conducted using Origin software 2021 to examine the relationship between the system's coupling coordination degree and the driving factors with significant influence, as shown in Supplementary Figure S3. The results indicate that the coupling coordination degree is significantly positively correlated with GDP, urbanization rate, and greening coverage rate, while it is significantly negatively correlated with the nighttime light index. Among these factors, the greening coverage rate has the most substantial impact on the system's coupling coordination, with an average correlation coefficient of 0.703. The urbanization rate ranks second, with an average correlation coefficient of 0.56. GDP and the nighttime light index have opposing correlations with coupling coordination, with similar absolute values of correlation coefficients, 0.371 and −0.372, respectively.

5. Discussion

5.1. Discussion on the Spatiotemporal Evolution

The integration of the digital economy, green ecology, and energy utilization is accelerating, effectively driving the digital–green transformations of industries and lifestyles. This trend also provides robust support for the modernization of governance systems and the construction of an ecological civilization. The coupling coordination development exhibited a year-on-year increase from 2011 to 2021, indicating that the relationship among the four improved owing to benign intersystem development and the guidance of relevant policies. This is consistent with previous studies [49,50].

However, one key innovation in our study is the focus on the fluctuations and anomalies during specific periods, particularly the slowdown between 2016 and 2017, which has been largely overlooked in prior research that emphasized long-term trends. This slowdown can be attributed to several factors. Firstly, the period marked China's transition to a "new normal" economy, emphasizing supply-side structural reforms and reducing reliance on traditional industries [51]. This shift likely caused temporary disruptions in the coupling coordination development. Secondly, global economic uncertainties during this time may have impacted regional stability, affecting overall coupling growth [52,53]. Lastly, the digital economy, while a key driver of coordination, may have faced infrastructural and technological bottlenecks in 2016–2017, limiting its contribution to overall development [54,55].

In particular, the disparities in digital infrastructure development across provinces played a significant role in shaping these fluctuations. Provinces such as Zhejiang, Jiangsu, and Guangdong benefited from advanced digital infrastructure, high internet penetration rates, and strong policy support, which accelerated the integration of digital technologies into traditional industries, promoting sustainable growth and higher levels of coupling coordination. Conversely, provinces like Sichuan, Yunnan, and Guizhou faced significant digital infrastructure deficiencies, including slower internet speeds, less-developed cloud computing infrastructure, and insufficient investment in digital industries. This disparity limited their ability to fully harness the digital economy, slowing the overall progress of coupling coordination in these regions during critical periods such as 2016–2017.

Moreover, our analysis introduces an innovative focus on spatial disparities in the coordination of the four subsystems. As for the spatial characteristics of the coupling coordination development, the eastern region remains in the lead, prominently represented by Guangdong, which has reached a good level of coordination [49,56]. Next are Zhejiang, Fujian, and Jiangsu Provinces. For an explanation, see Weng, Lian, and Qin (2022) [57], who demonstrated that in the digital era, energy development in the eastern coastal region can adapt and self-heal rapidly because of robust infrastructure and economic systems. Additionally, they showed that ecological factors influence the application of digital technologies, further enhancing the economy's resilience. In contrast, the central and western provinces continue to lag behind due to a lack of sufficient digital infrastructure and investment, which hampers their ability to leverage digital technologies for economic growth and energy efficiency. These disparities underline the need for targeted digital infrastructure investments and policy support to bridge the digital divide and promote more balanced regional development. Low-value regions like Sichuan, Yunnan, and Guizhou rely heavily on their resource endowments but face significant challenges in science and technology innovation, weak digital infrastructure, and less coordinated growth. Therefore, it is necessary to leverage regional strengths to digitally empower industrial development, and comprehensive measures should be implemented to enhance development coordination and avoid falling into the "development imbalance trap".

Moreover, our study identifies a significant increase in Moran's I index from 0.327 to 0.370, indicating a propensity for spatial correlation to be geographically localized over time [58]. This contrasts with earlier studies that may not have captured such dynamics, likely due to differences in methodology, data granularity, or the period studied. All provinces can generate coupling coordination development feedback linkage with other provinces; however, the spatial correlation of coupling coordination development aligns with the principle of distance decay [59], showing a stronger tendency toward proximity correlation. In addition, The spatiotemporal analysis from 2011 to 2020 shows clear regional disparities in the coupling coordination degree within the Yangtze River Basin. Eastern coastal provinces, such as Zhejiang, Fujian, and Guangdong, consistently form high–high clusters due to their long-standing economic openness, well-developed infrastructure, and the impact of policies like the Special Economic Zones [60,61]. These findings align with previous studies that highlight the eastern region's advantages in digital economy growth and industrial diversification [62]. In contrast, the central and western provinces (e.g., Sichuan, Yunnan, and Tibet) form persistent low–low clusters, reflecting slower economic development, fewer infrastructure investments, and geographic barriers, despite efforts like the Western Development Strategy [63]. The appearance of high–low outliers in provinces such as Chongqing can be attributed to localized urban development policies, while low–high outliers in provinces like Guizhou stem from limited resource access and lagging infrastructure, even as neighboring areas advance. These spatial disparities underscore the uneven effects of national policies and suggest that more targeted, region-specific interventions are needed to address ongoing imbalances [64,65].

5.2. Discussion of the External Driving Factors

In this study, the effects of driving factors on coupling coordination development were also examined. Green coverage, GDP, urbanization rate, and nighttime light index had a significant impact on the coupling coordinated development of the Yangtze River Basin.

From 2011 to 2016, the increase in green coverage became a critical driving force, aligning with previous findings [66,67]. However, our study goes beyond earlier research by demonstrating how the integration of green technologies with the digital economy further amplified this effect, resulting in enhanced environmental outcomes and population inflow. This finding underscores the role of green technologies in facilitating the digital-green transition, which is largely underexplored in existing studies. Our research identifies the role of policy interventions during this period, such as ecological protection initiatives and green development programs, which played a pivotal role in reinforcing this trend [68].

In addition to these internal factors, global and national external factors also play a crucial role in shaping the coupling coordination of the Yangtze River Basin. The volatility of international trade and energy prices can exert significant pressure on the energy subsystem and the broader economic landscape, influencing both regional development and energy consumption patterns. For instance, fluctuations in global oil prices or trade tensions could disrupt energy markets and impact the cost of industrial operations, leading to uneven growth across regions. Similarly, global environmental policies, such as the Paris Agreement, and international efforts to mitigate climate change, have influenced national and regional environmental policies. These policies have accelerated the transition to cleaner energy sources and stricter environmental regulations, thereby affecting the coordination between ecological protection and economic growth.

Moreover, China's Belt and Road Initiative (BRI), an important international economic strategy, has influenced infrastructure development and trade flows, particularly in the western regions of the Yangtze River Basin. This initiative has provided opportunities for greater connectivity and economic development in otherwise underdeveloped areas, potentially impacting the long-term spatial dynamics of the region's coupling coordination development. However, these external forces also introduce new challenges, as regional economies must adapt to changing international market conditions and global policy shifts.

GDP exerted a significant influence during the 2014–2017 period, reflecting the role of economic growth in driving technological innovation and industrial upgrading. Since 2017, the foundation for digital transformation has been better, and cross-industry and cross-regional development integration is gradually being achieved. Rising incomes and urbanization led to increased demand for energy, digital services, and ecological goods, which further fueled GDP growth and reinforced the interconnected development of these sectors [69,70]. Liu et al. (2021) also confirmed the dynamic effects of GDP on the coupling coordination development of energy–economy–ecology in the Yangtze River Basin [71]. Their study made more detailed distinctions among the regions in which the driving factors are effective, and the results showed that the four provinces in the central region had lagging economies. Urbanization, another key driver, was identified as a powerful catalyst for coupling coordination development. While urbanization has been widely studied in the context of economic growth, our study introduces an innovative perspective by focusing on how urbanization interacts with the digital economy to drive the emergence of smart cities and new-type urbanization. The integration of digital technologies into urban planning has not only reduced economic pressure but also promoted sustainable urban environments, providing an important insight into the future trajectory of urban development in China [72]. Finally, the nighttime light index, often used as a proxy for economic activity, showed fluctuations corresponding with regional economic changes. While previous studies have recognized the index's value in economic analysis, our study innovatively links it to spatial–temporal coupling coordination, showing how variations in light intensity reflect broader trends in regional development. This adds a new dimension to the use of nighttime light data, allowing for more comprehensive spatial–temporal analyses of regional dynamics.

In summary, both internal factors, such as green coverage and urbanization, and external factors, including global trade, energy price fluctuations, and international environmental policies, significantly shape the dynamics of coupling coordination in the Yangtze River Basin. To achieve more balanced and sustainable regional development, policymakers need to consider not only domestic factors but also how global external forces might interact with regional systems and create new development opportunities or challenges.

5.3. Research Limitations

This study expanded on the literature in terms of indicator selection and research methodology, but several limitations remain. First, the temporal scope of the study, focused on the period from 2011 to 2020, may not capture longer-term trends and fluctuations; future research could extend this temporal range to better understand the evolving dynamics of the region. Although the data from 2011 to 2020 offers a comprehensive view of recent developments, this timeframe limits the ability to analyze longer-term trends or future projections. While projections beyond 2020 could provide valuable insights into potential future challenges, such analysis was not included because of the complexity and data limitations associated with long-term forecasting. Future studies could extend the timeline or incorporate projection models to explore the evolution of coupling coordination beyond 2020.

Second, although spatial patterns were analyzed at the Yangtze River basin level, finer spatial resolutions, such as city or county levels, could reveal more localized variations in coupling coordination, offering more precise insights for regional policy making. Third, influencing factors should not be restricted to current categories and should include granular data and additional natural factors. Studying the factors influencing coupling interactions on different spatial–temporal scales would also provide valuable references for economic growth and ecological civilization. Addressing these limitations in future research will enhance the depth and applicability of the findings. Moreover, while this study applied Global Moran's I to analyze the spatial autocorrelation and detect clustering patterns across the Yangtze River Basin, we acknowledge that other methods, such as Geary's C, could provide additional insights into local spatial dissimilarities. However, given the global perspective of this study and the focus on overall regional clustering, Global Moran's I was deemed the most suitable method. Future studies could benefit from incorporating Geary's C or other local spatial statistics to further investigate localized variations in coupling coordination development. This would provide a more detailed understanding of local-level spatial dynamics that complement the global patterns identified in this study.

6. Conclusions

This study explored the spatiotemporal evolution and driving factors influencing the coupling coordination of PEED in the Yangtze River Basin from 2011 to 2020. Our findings offer several key insights into the dynamics of regional development, spatial clustering, and the interplay between economic growth and environmental sustainability. Unlike previous research, this study emphasizes the temporal fluctuations and spatial disparities in coupling coordination, as well as the driving forces behind these dynamics.

(1) Temporal Evolution: Over the study period, the coupling coordination degree in the Yangtze River Basin exhibited a consistent upward trend, suggesting that the relationships among the four subsystems are becoming more integrated. However, our analysis identifies a notable slowdown in growth between 2016 and 2017, a period that corresponds with China's shift toward a "new normal" economy and supply-side reforms. This finding, largely overlooked in previous studies, highlights the impact of macroeconomic shifts on regional coordination dynamics. Spatially, the coupling coordination degree shows a distinct east-west divide, with higher levels in the eastern coastal regions and lower levels in the central and western provinces. This spatial differentiation reflects underlying economic, infrastructural, and policy disparities, which have led to uneven development patterns across the basin.

(2) Spatial Clustering and Autocorrelation: The study reveals significant spatial autocorrelation in the coupling coordination degree, with Moran's I increasing from 0.327 to 0.370 over the study period. This trend indicates that similar provinces are clustering together in terms of their development trajectories, particularly in the eastern region where high–high clusters are formed in provinces like Zhejiang, Fujian, and Guangdong. In contrast, the central and western regions, including Sichuan, Yunnan, and Tibet, consistently show low–low clusters. These findings suggest that regional

development policies, such as the Western Development Strategy, have not fully mitigated spatial imbalances, and more targeted interventions are needed to enhance coordination and close regional gaps.

(3) Driving Factors of Coupling Coordination: The dominant influencing factors of coupling coordination development vary across time and space. Greening coverage and GDP emerged as the most significant drivers, particularly in the earlier part of the decade, reflecting the role of environmental policies and economic growth in driving regional coordination. By contrast, urbanization rate and nighttime light index exhibited period-specific impacts, with the former playing a critical role in fostering urban growth and digital economy development. Interestingly, energy consumption and CO_2 emissions, while often emphasized in environmental studies, did not show significant effects on the coupling coordination degree in this study, pointing to a possible disconnect between energy policies and regional development dynamics.

(4) Policy Implications and Practical Contributions: Based on these findings, our study provides several targeted policy recommendations to promote more effective coupling coordination across the Yangtze River Basin, as follows: 1. Strengthen Ecological Protection and Industrial Upgrading. Focus on expanding green coverage and restoring degraded ecosystems in central and western regions, while offering incentives for clean industries and technological innovation in energy-efficient sectors. These measures will help balance ecological sustainability with economic development. For example, the successful model in Guangdong, which combines ecological restoration with industrial upgrading, can be adapted for regions like Sichuan and Yunnan. 2. Leverage Fiscal and Industrial Policies for Regional Integration. Use fiscal transfers and targeted investments in infrastructure to close the development gap between the eastern and western regions. Regional coordination can be improved through joint development projects, similar to initiatives seen in the Yangtze River Economic Belt, which promote collaboration in the digital economy and infrastructure sharing. 3. Promote Clean Energy Transition and Digital Economy Growth. Accelerate the clean energy transition by investing in renewable energy infrastructure and offering tax incentives for low-carbon technologies. Combine this with the development of the digital economy, using successful examples from provinces like Zhejiang and Guangdong, where digital innovations have been key drivers of economic competitiveness. These recommendations aim to foster balanced and sustainable development across the Yangtze River Basin, offering practical steps to enhance regional collaboration and long-term economic and environmental stability.

This study contributes to the literature on regional development by providing a spatial–temporal analysis of coupling coordination dynamics and identifying the key drivers of regional disparities. By combining insights from spatial econometrics and policy analysis, it fills an important gap in the understanding of how macroeconomic transitions and environmental policies shape regional coordination. Furthermore, the study offers a quantitative decision-making tool for policymakers to design targeted interventions that promote balanced growth and sustainable development. These findings not only have implications for the Yangtze River Basin but also offer a reference for other regions globally facing similar challenges in achieving economic-environmental balance.

Supplementary Materials: The following supporting information can be downloaded at: https://www.mdpi.com/article/10.3390/land13101726/s1.

Author Contributions: Conceptualization, T.L. and Q.Z.; methodology, T.L. and Q.Z.; software, T.L. and Q.Z.; validation, T.L. and Q.Z.; formal analysis, T.L.; investigation, T.L.; resources, T.L.; data curation, T.L.; writing—original draft preparation, T.L., Y.L. and Q.Z.; writing—review and editing, T.L., Y.L. and Q.Z.; visualization, T.L.; supervision, T.L. and Q.Z.; project administration, T.L. and Q.Z.; funding acquisition, T.L. and Q.Z. All authors have read and agreed to the published version of the manuscript.

Funding: This research was funded by the 2024 Hunan Province Social Science Achievement Review Committee Project: Research on the Construction Path of Resilient Cities on Both Sides of the Xiangjiang River Based on "Hunan Characteristics" [XSP24YBC069]; Fundamental Research Funds for the Central Universities of Central South University, grant number [2024ZZTS340].

Data Availability Statement: The data presented in this study are available on request from the corresponding author.

Conflicts of Interest: The authors declare no conflicts of interest.

References

1. Chen, L.; Zhong, Q.; Li, Z. Analysis of spatial characteristics and influence mechanism of human settlement suitability in traditional villages based on multi-scale geographically weighted regression model: A case study of Hunan province. *Ecol. Indic.* **2023**, *154*, 110828. [CrossRef]
2. Zhong, Q.; Fu, H.; Yan, J.; Li, Z. How does energy utilization affect rural sustainability development in traditional villages? Re-examination from the coupling coordination degree of atmosphere-ecology-socioeconomics system. *Build. Environ.* **2024**, *257*, 111541. [CrossRef]
3. Zhong, Q.; Li, Z. Long-term trends of vegetation greenness under different urban development intensities in 889 global cities. *Sustain. Cities Soc.* **2024**, *106*, 105406. [CrossRef]
4. Nawab, A.; Liu, G.; Meng, F.; Hao, Y.; Zhang, Y. Urban energy-water nexus: Spatial and inter-sectoral analysis in a multi-scale economy. *Ecol. Model.* **2019**, *403*, 44–56. [CrossRef]
5. Zhao, D.; Lin, J.; Bashir, M.A. Analyze the effect of energy efficiency, natural resources, and the digital economy on ecological footprint in OCED countries: The mediating role of renewable energy. *Resour. Policy* **2024**, *95*, 105198. [CrossRef]
6. Qin, P.; Xu, H.; Liu, M.; Du, L.; Xiao, C.; Liu, L.; Tarroja, B. Climate change impacts on Three Gorges Reservoir impoundment and hydropower generation. *J. Hydrol.* **2020**, *580*, 123922. [CrossRef]
7. Chang, K.; Zhang, H.; Li, B. The impact of digital economy and industrial agglomeration on the changes of industrial structure in the Yangtze River Delta. *J. Knowl. Econ.* **2024**, *15*, 9207–9227. [CrossRef]
8. Wang, J.Y.; Wang, S.J.; Li, S.J.; Feng, K.S. Coupling analysis of urbanization and energy-environment efficiency: Evidence from Guangdong province. *Appl. Energy* **2019**, *254*, 113650. [CrossRef]
9. Fu, L.; Liu, B.; Zhu, Z.; Cao, J.; Sun, C.; Yu, Z. Analysis of the coupling coordinated development and evolutionary trend of digital economy and ecological environment. *Front. Environ. Sci.* **2022**, *10*, 1006354. [CrossRef]
10. Ariken, M.; Zhang, F.; Liu, K.; Fang, C.; Kung, H.-T. Coupling coordination analysis of urbanization and eco-environment in Yanqi Basin based on multi-source remote sensing data. *Ecol. Indic.* **2020**, *114*, 106331. [CrossRef]
11. Ariken, M.; Zhang, F.; Chan, N.W.; Kung, H.T. Coupling coordination analysis and spatio-temporal heterogeneity between urbanization and eco-environment along the Silk Road Economic Belt in China. *Ecol. Indic.* **2021**, *121*, 107014. [CrossRef]
12. Zheng, H.; Khan, Y.A.; Abbas, S.Z. Exploration on the coordinated development of urbanization and the eco-environmental system in central China. *Environ. Res.* **2022**, *204*, 112097. [CrossRef] [PubMed]
13. Xiao, Y.; Zhong, J.L.; Zhang, Q.F.; Xiang, X.; Huang, H. Exploring the coupling coordination and key factors between urbanization and land use efficiency in ecologically sensitive areas: A case study of the Loess Plateau, China. *Sustain. Cities Soc.* **2022**, *86*, 104148. [CrossRef]
14. Chen, M.; Sun, Z.; Wang, Y.; Guo, S. Evaluation of coupling coordination among the urban physical environment, economy, and population: A case study of 36 main cities in China. *Adv. Civ. Eng.* **2019**, *2019*, 1576292. [CrossRef]
15. Zameer, H.; Yasmeen, H.; Wang, R.; Tao, J.; Malik, M.N. An empirical investigation of the coordinated development of natural resources, financial development and ecological efficiency in China. *Resour. Policy* **2020**, *65*, 101580. [CrossRef]
16. Dong, L.; Longwu, L.; Zhenbo, W.; Liangkan, C.; Faming, Z. Exploration of coupling effects in the Economy–Society–Environment system in urban areas: Case study of the Yangtze River Delta Urban Agglomeration. *Ecol. Indic.* **2021**, *128*, 107858. [CrossRef]
17. Liu, Y.; Liu, W.; Yan, Y.; Liu, C. A perspective of ecological civilization: Research on the spatial coupling and coordination of the energy-economy-environment system in the Yangtze River Economic Belt. *Environ. Monit. Assess.* **2022**, *194*, 403. [CrossRef]
18. Sun, C.; Yan, X.; Zhao, L. Coupling efficiency measurement and spatial correlation characteristic of water–energy–food nexus in China. *Resour. Conserv. Recycl.* **2021**, *164*, 105151. [CrossRef]
19. Han, D.; Yu, D.; Cao, Q. Assessment on the features of coupling interaction of the food-energy-water nexus in China. *J. Clean. Prod.* **2020**, *249*, 119379. [CrossRef]
20. Xing, L.; Xue, M.G.; Hu, M.S. Dynamic simulation and assessment of the coupling coordination degree of the economy-resource-environment system: Case of Wuhan City in China. *J. Environ. Manag.* **2019**, *230*, 474–487. [CrossRef]
21. Sun, Y.; Cui, Y. Analyzing the coupling coordination among economic, social, and environmental benefits of urban infrastructure: Case study of four Chinese autonomous municipalities. *Math. Probl. Eng.* **2018**, *2018*, 8280328. [CrossRef]
22. Wu, Q.; Cao, Y.; Fang, X.; Wang, J.; Li, G. A systematic coupling analysis framework and multi-stage interaction mechanism between urban land use efficiency and ecological carrying capacity. *Sci. Total Environ.* **2022**, *853*, 158444. [CrossRef]
23. Wang, M.; Zhao, X.; Gong, Q.; Ji, Z. Measurement of regional green economy sustainable development ability based on entropy weight-topsis-coupling coordination degree—A case study in Shandong Province, China. *Sustainability* **2019**, *11*, 280. [CrossRef]

24. Bahtebay, J.; Zhang, F.; Ariken, M.; Chan, N.W.; Tan, M.L. Evaluation of the coordinated development of urbanization-resources-environment from the incremental perspective of Xinjiang, China. *J. Clean. Prod.* **2021**, *325*, 129309. [CrossRef]
25. Yang, L.; Lin, Y.; Zhu, J.; Yang, K. Dynamic coupling coordination and spatial–temporal analysis of digital economy and carbon environment governance from provinces in China. *Ecol. Indic.* **2023**, *156*, 111091. [CrossRef]
26. Tomal, M. Evaluation of coupling coordination degree and convergence behaviour of local development: A spatiotemporal analysis of all Polish municipalities over the period 2003–2019. *Sustain. Cities Soc.* **2021**, *71*, 102992. [CrossRef]
27. Li, Q.; Zhao, Y.; Li, S.L.; Zhang, L.L. Spatial-temporal characteristics of the coupling coordination of social security and economic development in China during 2002–2018. *Reg. Sustain.* **2021**, *2*, 116–129. [CrossRef]
28. Abulimiti, M.; Simayi, Z.; Yang, S.; Chai, Z.; Yan, Y. Study of coordinated development of county urbanization in arid areas of China: The case of Xinjiang. *PLoS ONE* **2022**, *17*, e0276235. [CrossRef]
29. Feng, Y.J.; Zhu, A.K.; Liu, P.; Liu, Z.L. Coupling and coordinated relationship of water utilization, industrial development and ecological welfare in the Yellow River Basin, China. *J. Clean. Prod.* **2022**, *379*, 134824. [CrossRef]
30. Zhong, Q.; Li, Z.; He, Y. Coupling evaluation and spatial–temporal evolution of land ecosystem services and economic–social development in a city group: The case study of the Chengdu–Chongqing City Group. *Int. J. Environ. Res. Public Health* **2023**, *20*, 5095. [CrossRef] [PubMed]
31. Cui, D.; Chen, X.; Xue, Y.; Li, R.; Zeng, W. An integrated approach to investigate the relationship of coupling coordination between social economy and water environment on urban scale-A case study of Kunming. *J. Environ. Manag.* **2019**, *234*, 189–199. [CrossRef] [PubMed]
32. Liu, M.L.; Lu, M.Y.; Li, Z.Y. Coupling coordination analysis on digital economy-tourism development-ecological environment. *J. Clean. Prod.* **2024**, *470*, 143320. [CrossRef]
33. Zou, C.; Zhu, J.W.; Lou, K.L.; Yang, L. Coupling coordination and spatiotemporal heterogeneity between urbanization and ecological environment in Shaanxi Province, China. *Ecol. Indic.* **2022**, *141*, 109152. [CrossRef]
34. Liu, W.J.; Jiao, F.C.; Ren, L.J.; Xu, X.G.; Wang, J.C.; Wang, X. Coupling coordination relationship between urbanization and atmospheric environment security in Jinan City. *J. Clean. Prod.* **2018**, *204*, 1–11. [CrossRef]
35. Shan, L.; Jiang, Y.H.; Liu, C.C.; Zhang, J.; Zhang, G.H.; Cui, X.F. Conflict or Coordination? Spatiotemporal Coupling of Urban Population-Land Spatial Patterns and Ecological Efficiency. *Front. Public Health* **2022**, *10*, 890175. [CrossRef]
36. Liu, J.; Jin, X.B.; Xu, W.Y.; Gu, Z.M.; Yang, X.H.; Ren, J.; Fan, Y.T.; Zhou, Y.K. A new framework of land use efficiency for the coordination among food, economy and ecology in regional development. *Sci. Total Environ.* **2020**, *710*, 135670. [CrossRef]
37. Gan, L.; Shi, H.; Hu, Y.; Lev, B.; Lan, H.X. Coupling coordination degree for urbanization city-industry integration level: Sichuan case. *Sustain. Cities Soc.* **2020**, *58*, 102136. [CrossRef]
38. Qu, W.; Lian, H.; Wang, Y.; Ma, Y. Spatiotemporal Evolution of the Coupling Coordination Relationship of "Population-Environment" Development in the Xi'an Metropolitan Area. *Sustainability* **2023**, *15*, 4533. [CrossRef]
39. Kong, Q.S.; Kong, H.Y.; Miao, S.L.; Zhang, Q.; Shi, J.G. Spatial Coupling Coordination Evaluation between Population Growth, Land Use and Housing Supply of Urban Agglomeration in China. *Land* **2022**, *11*, 1396. [CrossRef]
40. Zhu, C.; Fang, C.L.; Zhang, L.F. Analysis of the coupling coordinated development of the Population-Water-Ecology-Economy system in urban agglomerations and obstacle factors discrimination: A case study of the Tianshan North Slope Urban Agglomeration, China. *Sustain. Cities Soc.* **2023**, *90*, 104359. [CrossRef]
41. Li, S.R.; Zhang, L.H.; Su, L.; Nie, Q.Y. Exploring the coupling coordination relationship between eco-environment and renewable energy development in rural areas: A case of China. *Sci. Total Environ.* **2023**, *880*, 163229. [CrossRef] [PubMed]
42. Liu, S.C.; Wu, P.J. Coupling coordination analysis of urbanization and energy eco-efficiency: A case study on the Yangtze River Delta Urban Agglomeration. *Environ. Sci. Pollut. Res.* **2023**, *30*, 63975–63990. [CrossRef] [PubMed]
43. Zhu, Y.Q.; Shen, S.W.; Du, L.Y.; Fu, J.; Zou, J.; Peng, L.A.; Ding, R. Spatial and Temporal Interaction Coupling of Digital Economy, New-Type Urbanization and Land Ecology and Spatial Effects Identification: A Study of the Yangtze River Delta. *Land* **2023**, *12*, 677. [CrossRef]
44. Xia, R.C.; Wei, D.Q.; Jiang, H.T.; Ding, Y.; Luo, X.Y.; Zhang, B.; Yin, J. Study on the coupling coordination development of China's multidimensional digital economy and industrial carbon emission efficiency. *Environ. Sci. Pollut. Res.* **2023**, *30*, 114201–114221. [CrossRef] [PubMed]
45. Li, J.S.; Sun, W.; Li, M.Y.; Meng, L.L. Coupling coordination degree of production, living and ecological spaces and its influencing factors in the Yellow River Basin. *J. Clean. Prod.* **2021**, *298*, 126803. [CrossRef]
46. Liu, Z.; He, C.; Zhang, Q.; Huang, Q.; Yang, Y. Extracting the dynamics of urban expansion in China using DMSP-OLS nighttime light data from 1992 to 2008. *Landsc. Urban Plan.* **2012**, *106*, 62–72. [CrossRef]
47. Yang, Z.; Fang, C.; Mu, X.; Li, G.; Xu, G. Urban green space quality in China: Quality measurement, spatial heterogeneity pattern and influencing factor. *Urban For. Urban Green.* **2021**, *66*, 127381. [CrossRef]
48. Acock, A.C. Working with missing values. *J. Marriage Fam.* **2005**, *67*, 1012–1028. [CrossRef]
49. Xiang, F.; He, W.; Li, Y.; Luo, L.; An, M.; Ramsey, T.S. Coupling coordination of the digital-energy system and its impact on energy efficiency: Evidence from the "2+ 26" cities of Beijing-Tianjin-Hebei, China. *Energy Strategy Rev.* **2024**, *54*, 101441. [CrossRef]
50. Bhatti, U.A.; Tang, H.; Khan, A.; Ghadi, Y.Y.; Bhatti, M.A.; Khan, K.A. Investigating the nexus between energy, socio-economic factors and environmental pollution: A geo-spatial multi regression approach. *Gondwana Res.* **2024**, *130*, 308–325. [CrossRef]
51. Tung, R.L. Opportunities and challenges ahead of China's "New Normal". *Long Range Plan.* **2016**, *49*, 632–640. [CrossRef]

52. Bonciani, D.; Ricci, M. The international effects of global financial uncertainty shocks. *J. Int. Money Financ.* **2020**, *109*, 102236. [CrossRef]
53. Zhang, D.; Lei, L.; Ji, Q.; Kutan, A.M. Economic policy uncertainty in the US and China and their impact on the global markets. *Econ. Model.* **2019**, *79*, 47–56. [CrossRef]
54. Luo, S.; Yimamu, N.; Li, Y.; Wu, H.; Irfan, M.; Hao, Y. Digitalization and sustainable development: How could digital economy development improve green innovation in China? *Bus. Strategy Environ.* **2023**, *32*, 1847–1871. [CrossRef]
55. Wang, J.; Wang, B.; Dong, K.; Dong, X. How does the digital economy improve high-quality energy development? The case of China. *Technol. Forecast. Soc. Chang.* **2022**, *184*, 121960. [CrossRef]
56. Han, X.; Wang, Y.; Yu, W.; Xia, X. Coupling and coordination between green finance and agricultural green development: Evidence from China. *Financ. Res. Lett.* **2023**, *58*, 104221. [CrossRef]
57. Weng, Q.; Lian, H.; Qin, Q. Spatial disparities of the coupling coordinated development among the economy, environment and society across China's regions. *Ecol. Indic.* **2022**, *143*, 109364. [CrossRef]
58. Ma, M.; Tang, J. Interactive coercive relationship and spatio-temporal coupling coordination degree between tourism urbanization and eco-environment: A case study in Western China. *Ecol. Indic.* **2022**, *142*, 109149. [CrossRef]
59. Miller, H.J. Tobler's first law and spatial analysis. *Ann. Assoc. Am. Geogr.* **2004**, *94*, 284–289. [CrossRef]
60. Dong, J.; Sheng, Y.; Shao, Q.; Jing, H.; Xia, M.; Li, D. The Impact of Digital Economy on the Coordinated Development of Urban and Rural Areas in the Context of Common Prosperity. *Int. J. Soc. Sci. Public Adm.* **2024**, *2*, 205–211. [CrossRef]
61. Liang, L.; Chen, M.; Luo, X.; Xian, Y. Changes pattern in the population and economic gravity centers since the Reform and Opening up in China: The widening gaps between the South and North. *J. Clean. Prod.* **2021**, *310*, 127379. [CrossRef]
62. Guo, J.; Yuan, X.; Song, W. Driving forces on the development of China's marine economy: Efficiency and spatial perspective. *Ocean Coast. Manag.* **2022**, *224*, 106192. [CrossRef]
63. Sun, L. Coordinated development of ecological environment in Yangtze River economic belt. In *Development of the Yangtze River Economic Belt in China*; Springer: Singapore, 2021; pp. 107–146.
64. Guo, B.; Wang, Y.; Zhang, H.; Liang, C.; Feng, Y.; Hu, F. Impact of the digital economy on high-quality urban economic development: Evidence from Chinese cities. *Econ. Model.* **2023**, *120*, 106194. [CrossRef]
65. Gu, S. Progress in and Prospects of the Reform of the System for Developing an Ecological Civilization. In *Beautiful China: 70 Years Since 1949 and 70 People's Views on Eco-civilization Construction*; Springer: Singapore, 2021; pp. 621–630.
66. Lee, H.; Lim, H.; Park, S. Quantitative assessment of green coverage changes under the human-biometeorological perspective: A simulation case study in Jeju, Republic of Korea. *Sustain. Cities Soc.* **2023**, *97*, 104734. [CrossRef]
67. Kanniah, K.D. Quantifying green cover change for sustainable urban planning: A case of Kuala Lumpur, Malaysia. *Urban For. Urban Green.* **2017**, *27*, 287–304. [CrossRef]
68. Sténs, A.; Roberge, J.-M.; Löfmarck, E.; Beland Lindahl, K.; Felton, A.; Widmark, C.; Rist, L.; Johansson, J.; Nordin, A.; Nilsson, U. From ecological knowledge to conservation policy: A case study on green tree retention and continuous-cover forestry in Sweden. *Biodivers. Conserv.* **2019**, *28*, 3547–3574. [CrossRef]
69. Jedwab, R.; Vollrath, D. Urbanization without growth in historical perspective. *Explor. Econ. Hist.* **2015**, *58*, 1–21. [CrossRef]
70. Al-Mulali, U.; Fereidouni, H.G.; Lee, J.Y.; Sab, C.N.B.C. Examining the bi-directional long run relationship between renewable energy consumption and GDP growth. *Renew. Sustain. Energy Rev.* **2013**, *22*, 209–222. [CrossRef]
71. Liu, J.; Tian, Y.; Huang, K.; Yi, T. Spatial-temporal differentiation of the coupling coordinated development of regional energy-economy-ecology system: A case study of the Yangtze River Economic Belt. *Ecol. Indic.* **2021**, *124*, 107394. [CrossRef]
72. Tian, Y.; Zhou, D.; Jiang, G. Conflict or Coordination? Multiscale assessment of the spatio-temporal coupling relationship between urbanization and ecosystem services: The case of the Jingjinji Region, China. *Ecol. Indic.* **2020**, *117*, 106543. [CrossRef]

land

MDPI

Article

A Multi-Scale Evaluation Model of Sustainable Development Goal 11.7 for Problem-Solving at Different Levels

Kan Wang, Xing Dang and Jianjun Bai *

School of Geography and Tourism, Shaanxi Normal University, Xi'an 710119, China;
wangkan@snnu.edu.cn (K.W.); d-xing@snnu.edu.cn (X.D.)
* Correspondence: bjj@snnu.edu.cn

Abstract: Sustainable Development Goal 11.7 (SDG 11.7) aims to promote the improvement of urban public spaces. However, the localization process of SDG 11.7 mainly relies on a bottom-up problem-solving approach, which fails to fully encompass the connotation of SDG 11.7. Additionally, existing evaluations primarily focus on a single scale, neglecting the impact of scale issues. These limitations can lead to imbalanced development or misallocation of responsibilities when guiding governments at different levels in promoting the sustainable development of public spaces. Therefore, this article introduces a multi-scale assessment model of SDG 11.7. It employs a top-down problem-solving approach to construct a sustainable development indicator framework, setting appropriate sustainable development indicators for various levels of government based on the connotation of SDG 11.7, and generates city-scale results by integrating three scales: apartment complexes, street blocks, and counties. Testing this model in Xi'an, China, revealed that it adequately captures four key aspects of SDG 11.7—safety, inclusiveness, accessibility, and greenness—through 11 indicators. The evaluation outcomes at the apartment complex, street block, and county levels effectively guide future development directions for various levels of government. Ultimately, the synthesis of these scales reveals the spatial pattern of SDG 11.7 at the city scale and identifies focal areas for development. Overall, this exploratory model demonstrates high accuracy and robustness, providing a comprehensive understanding of the essence of SDG 11.7. It also alleviates challenges posed by scale issues, offering decision support for monitoring SDG 11.7 across different levels of government in Chinese cities and promoting the process of sustainable development.

Keywords: SDG 11.7; top-down problem-solving approach; urban sustainable development indicator framework; multi-scale model; Xi'an city

1. Introduction

Cities serve as the primary nexus of human activity and play a pivotal role in advancing the United Nations' agenda for sustainable development [1]. Public spaces represent a main facet of implementing sustainable development strategies in cities, intricately linked to the urban environment and residents' well-being [2]. The United Nations Sustainable Development Goal 11.7 (SDG 11.7) particularly emphasizes the assessment of sustainability in urban public spaces, aiming to "provide universal access to safe, inclusive and accessible, green and public spaces, in particular for women and children, older persons and persons with disabilities by 2030" [3].

The connotation of SDG 11.7 mainly encompasses four aspects: safety, inclusiveness, accessibility, and greenness. Within the United Nations' global indicator framework, SDG 11.7.1 and SDG 11.7.2 are two indicators used to monitor SDG 11.7. These indicators have been extensively employed in assessments of SDG 11.7 [4,5]. As research deepens and concerns arise regarding the applicability of the United Nations' SDG indicator framework in cities with different contexts, scholars have begun to focus on the localization of SDG 11.7 indicators. Some studies have established a comprehensive set of indicators, including quality, proximity, and area, to evaluate the sustainability of parks in Denver, USA [6]. There are also studies that have utilized machine learning to extract Google Street View data, allowing for the assessment of accessible greenery for residents [7]. Furthermore, numerous studies have used accessibility indicators to evaluate residents' ability to access public spaces based on different modes of transportation or by distinguishing between different types of parks [8,9]. These efforts reflect the continuous improvement of SDG 11.7 indicators framework from various perspectives.

However, most assessment indicator sets have failed to comprehensively encompass the four key areas that SDG 11.7 aims to address. For instance, while spatial equity is directly involved in United Nations' SDG 11.7 framework, this aspect is not adequately reflected in the UN-SDG 11.7 indicators [10]. Existing research on the construction of the SDG 11.7 indicator framework similarly focuses on public space accessibility or the valuation of ecosystem services, neglecting issues such as inclusivity and safety [11], which results in biased evaluation outcomes. The reason for the inadequacy of sustainable development indicators in fully representing the connotation of SDG 11.7 is that the UN-SDGs framework represents a top-down system, while scholarly efforts to address sustainable development assessments often follow a bottom-up research paradigm, which makes even the UN-SDG 11.7 indicators unable to cover the connotation of SDG 11.7. While a bottom-up research paradigm allows for a more focused resolution of specific issues, the gaps in methods have resulted in an insufficient set of indicators, failing to capture the entire connotation of SDG 11.7 [12]. Neglecting the connotation of SDG 11.7 prevents stakeholders from focusing limited resources on the public space functions that are more important to urban residents, leading to imbalanced development [13]. Therefore, it is urgent to construct a top-down indicator framework to fully reflect the connotation of SDG 11.7.

Furthermore, there is a lack of attention to scale issues in the assessment of SDG 11.7. Scale refers to the spatial and temporal dimensions of the research object or process. In the realm of urban sustainable development, scale is invariably linked to administrative power and thus generally represented by administrative-level [14,15]. Numerous scholars have confirmed the scale-relatedness of sustainability [16]. Taking some SDG 11.7 indicators as examples, public space accessibility and greening rates vary significantly at different administrative levels [17]. However, these differences are related solely to scale rather than to actual objective changes. This highlights the need for research to consider the influence of scale issues on SDG 11.7 monitoring and to conduct multi-scale assessments. Unfortunately, very few studies have focused on the status of SDG 11.7 across different levels of administrative entities. Paying attention to the sustainable development status across different scales is crucial. On the one hand, evaluating SDG 11.7 solely at a singular scale might inadvertently limit considerations of other relevant scales, leading to a "scale trap" [18]. On the other hand, using single scale model for all administrative entities undoubtedly sets sustainable development goals beyond the scope of some governments' authority. These issues are all sources of the misalignment of responsibilities among administrative entities. Therefore, it is necessary to address the issue of responsibility mismatch between administrative entities by developing specific sustainable development indicators and evaluation methods for different levels of government, that is, to conduct multi-scale evaluations of SDG 11.7 at different levels of government [19].

Based on the above, this article constructs a comprehensive multi-scale SDG 11.7 indicator framework through a top-down problem-solving approach, which determines development indicators for apartment complexes, street blocks, and counties and proposes a multi-scale evaluation method to obtain SDG 11.7 results at the urban scale. This article mainly fills the gap of neglecting the connotation of SDG 11.7 and scale issues in related research. To illustrate the applicability of the model, Xi'an city is chosen as the case study area, providing a reference for monitoring SDG 11.7 in other cities in China.

2. Materials and Methods

2.1. Study Area, Data Source, and Processing

This article uses Xi'an city as a case study to validate the appropriateness of the multi-scale assessment model. Xi'an stands out as the sole megacity in northwest China and holds the distinction of being one of the first nine central cities approved by the State Council [20]. In recent years, Xi'an has undergone rapid development. However, this growth has come with challenges, primarily stemming from rapid urbanization and the encroachment upon farmland, resulting in serious environmental concerns. In response to the mounting awareness of sustainable development, the Xi'an government has escalated efforts in environmental stewardship. The "14th Five-Year Plan for Industrial Development of Xi'an" explicitly emphasizes the importance of public space allocation within land planning as a strategic measure to effectively address environmental issues. As such, this case study holds insights and references for other developing cities in China.

The study area (Figure 1), delineated by the urban boundaries, is established using the 2020 land use data of Xi'an and the maps of Xi'an's development zones and segmented management areas released by the Xi'an Civil Affairs Bureau in 2020. This study adopts the United Nations' definition of urban boundaries, which is determined by the degree of urbanization. The specific methodology involves extracting impermeable surfaces from the land use data and delineating the approximate grid range of the main urban areas by comparing it with satellite imagery. Subsequently, the smallest administrative divisions encompassing the main urban areas are chosen as the study area for this research. This approach aims to effectively reduce the fragmentation of the study area caused by directly extracting impermeable surface ranges, thereby mitigating the impact on spatial analysis related to SDG 11.7. Simultaneously, it ensures the integrity of administrative divisions to enhance the accuracy of multi-scale analysis.

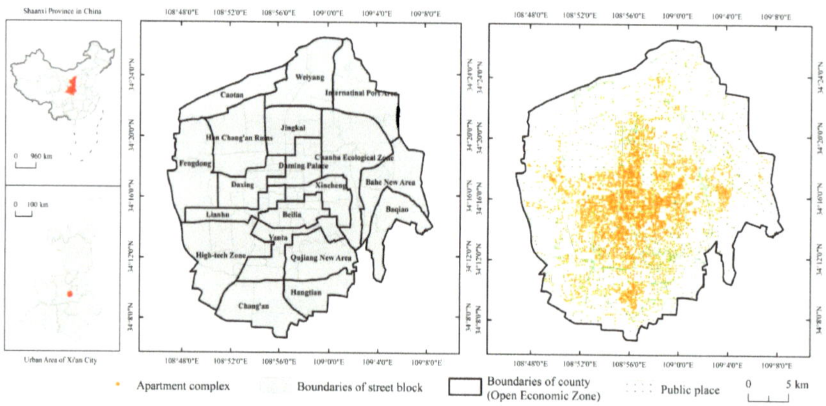

Figure 1. The study area within Xi'an city.

To effectively monitor and report on SDG 11.7, the United Nations defines public space as encompassing all open areas accessible to the public, including open public spaces and streets. Open public spaces are categorized based on their intended function, such as green public areas. The street space category includes streets, boulevards, and similar components. Taking into consideration the specific context within China, we emphasize the importance of public spaces that foster interaction with residents. These spaces should not only offer services like air purification but should also cater to recreational activities and scenic appreciation. Consequently, we exclude certain types of public spaces, such as urban green belts that residents cannot access, and ultimately obtain a total of 642 public spaces. The public space data come from the satellite image data of Xi'an City in 2020 intercepted by the RiverMap application and obtained through visual interpretation and digitization. This article conducts observations on Xi'an City's satellite images during both winter and summer to enhance the precision of our public space extraction results.

In China, an "apartment complex" refers to a group of buildings constructed by real estate companies for the purpose of housing residents. These complexes are typically managed by neighborhood committees, which function similarly to government entities. Therefore, this article treats "apartment complexes" as political entities and uses residential point data for analysis at this scale. Street block and county (including open economic zone) fall within China's administrative hierarchy as first- and second-level administrative units, and they play essential roles in maintaining the proper functioning of cities. The data used in this article cover the following scales: (1) Apartment Complex Data: A total of 4929 apartment complexes are identified through visual interpretation of 2020 satellite imagery from the RiverMap application. Data on the house prices of apartment complexes are obtained from the Anjuke website (URL: https://xa.anjuke.com, accessed on 10 October 2021). (2) Street Block Data: Maps are sourced from the Bureau of Surveying's standard map service website, and 59 street blocks are extracted through vectorization. (3) County Data: A total of 19 counties and development zones are acquired through vectorization of Xi'an' s development zone and segmented management area maps. (4) Road Network Data: Various types of roads within Xi'an are extracted using the RiverMap application. (5) Grid Data: An ArcGIS 10.1 platform's fishnet tool is employed to partition Xi'an City into 795 grids, each measuring 1 km × 1 km.

2.2. A Top-Down Indicator Framework for Problem-Solving at Different Levels

Guided by the principles of simplicity and ease of calculation, this article refines a set of SDG 11.7 indicators using a top-down problem-solving approach. This comprehensive framework consists of 11 key indicators (Table 1), and it has undergone refinement through four rounds of expert meetings, which can fully reflect the connotation of SDG 11.7.

The connotation of SDG 11.7 includes four aspects: safety, inclusiveness, accessibility, and greenness. This article selects the indicator spatial emergency response capacity (Sa) for the safety connotation of SDG 11.7, indicator resource area per household (Per), dispersibility (NNI), and dominance (Do) to reflect the inclusiveness connotation; indicator proximity (Ne), richness (Ac), NNI, and comprehensive coverage (Co) to reflect the accessibility connotation; and finally, indicator quality (Qa), Ac, Per, diversity (Sh), ecosystem services value (Es), total area (Su), and Co to reflect the greenness connotation. It should be noted that some of the above indicators can reflect multiple aspects of the connotation of SDG 11.7. On the basis of classifying the above indicators through the connotation of SDG 11.7, this article further classifies them based on their applicable scales. The following text introduces the specific classification process and indicators.

As the administrative power gradually extends from the apartment complex scale to the county scale (S1 to S3), we observe that the micro scale (S1) places greater emphasis on access to public spaces. To capture the sustainable development of S1, we employ three indicators: Qa, Ne, and Ac. The indicator Qa is expressed by the area of high-quality public space within a 2 km radius of the apartment complex. It is important to note that the quality of public spaces is calculated based on the weighted area of these spaces. Referring to previous studies, we assign weights of 1.5, 1.2, and 1 to the areas of water bodies, forests, and grasslands within public spaces, respectively. This allows us to calculate the weighted areas of 642 public spaces in Xi'an. We define the top 10% of these weighted areas as high-quality public spaces [21]. Ne is calculated using the shortest distance model, which refers to the time or spatial cost from demand points to public resources. Considering that the time distance is affected by the social and economic levels of different demand points [22], this article uses spatial distance to represent public space proximity. The formula is:

$$Ne_i = Min\left(Cost_{ij}\right) \tag{1}$$

where i is the demand point, j is the public space, $Cost_{ij}$ is the space cost from the i to the public space j, and Ne_i represents the shortest spatial distance from the demand point i to all the public spaces.

The Ac is expressed by the gravity model, which represents the attractiveness of public space of different quality to residential points within its service threshold and the difficulty of residential points to obtain these public spaces [23]. The formula is:

$$Ac_i = \sum_{j \in (d_{ij} \leq d_0)} \frac{S_j f(d_{ij})}{P_i} \tag{2}$$

where i is the demand point, j is the public space point, and Ac_i represents the richness of the public space obtained by demand i. S_j represents the supply capacity of j, which is expressed in the area of public space. P_i is the demand of i, which is expressed by the number of households. d_{ij} is the spatial distance between i and j, and d_0 is the service threshold of j. We divide the public space into four grades based on their quality, accounting for 10%, 20%, 30%, and 40% of the total number and set its service threshold at 2 km, 1.5 km, 1 km, and 0.5 km. $f(d_{ij}) = d_{ij}^{-\beta}$ is the power function considering the issues of spatial friction, which assumes that even within the same spatial limits, residents are more inclined to obtain public space with a closer distance. The value of β is between 1 and 2, and we set it to 1 by referring to the research results of Mao et al. [24]. Since the public space is a polygon, in order to achieve higher accuracy for the public space with a larger area, the OD matrix uses the distance from the demand point to the gate of the public space when calculating Ne and Ac. In the case of fully open public spaces, we segment the space's perimeter at 100 m intervals and calculate the distance from the point of demand to the edge point (Figure 2).

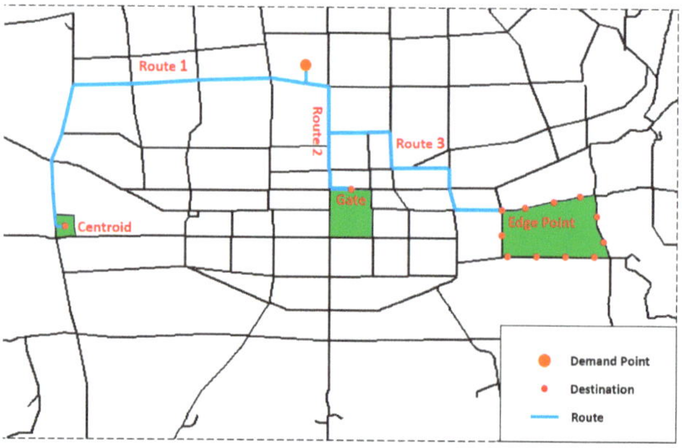

Figure 2. OD distance calculation scheme between apartment complex and public space.

As the most fundamental administrative level in China, S2 possesses specific administrative functions, and it places a heightened emphasis on the allocation of public resources. In this context, we establish four key indicators, namely Per, NNI, Do, and Sa, to comprehensively characterize the sustainable development of public spaces at the street block scale.

Per is expressed as the ratio of the total area of public space to the number of households within the administrative scope of the street block. NNI is used to describe the spatial distribution of public space and is analyzed by the Nearest Neighbor Index method [25]. The calculation method can be summarized as follows:

$$NNI = \frac{\overline{r_1}}{\overline{r_e}} = \frac{\overline{r_1}}{\frac{1}{2\sqrt{m/S}}} \tag{3}$$

where $\overline{r_1}$ represents the average actual nearest distance and $\overline{r_e}$ represents the theoretical nearest distance. m is the number of entrances and boundary points of the public space in the study area, and S is the area of the study area. When $NNI < 1$, it indicates that the public space is distributed in a concentrated manner, when $NNI = 1$, it indicates random distribution, and when $NNI > 1$, it indicates even distribution.

The indicator Do gauges the balance of services offered by public space resources within the region, specifically examining whether services like tourism and recreation are primarily controlled by one or a few significant public spaces. The calculation method is outlined below [26]:

$$Do = 1 - \sum_{k=1}^{n} P_k^2 \tag{4}$$

where n is the number of public spaces in the study area, P_k is the ratio of the area of public space k to the total area of public spaces in the area.

Given the limited occurrence of criminal incidents, such as robberies, in China, this article uses the spatial emergency response capacity of public space to expand the security connotation of SDG 11.7. Sa represents the ability of public spaces to facilitate the evacuation of people and mitigate disasters during emergencies, such as earthquakes. To assess this emergency capacity, the proportion of the population within the public space service threshold in comparison to the entire city's population is taken as the emergency capacity of the public space. The sum of the emergency capacities of all public spaces within the street block's boundaries is then computed to derive the Sa indicator.

At the county scale, we believe that greater emphasis should be placed on the overall quantity of public resources, owing to the substantial administrative influence wielded by counties. Consequently, we have established four indicators, with indicator *Sh* representing the diversity of various public space types within the region. In this study, we employ the Shannon diversity index to calculate *Sh* [27]. The calculation method is as follows:

$$Sh = -\sum_{t=1}^{T} H_t \ln H_t \tag{5}$$

where T represents the number of types of public spaces in the area. H_t represents the proportion of the area of the t-th type public space to the total area of the public space in the region.

The indicator Es signifies the ecosystem service value provided by public spaces within each county, considering four key aspects: environmental purification, climate regulation, temperature control, and tourism and leisure. The unit of measurement for Es is denominated in RMB (Chinese Yuan). For precise calculation, we reference the urban landscape ecosystem service evaluation index framework developed by Li and Zhou [28].

The Su indicator quantifies the total area of public spaces within the county, while Co reflects the extent of public space coverage throughout the county. In this article, we utilize the mean of both coverage and service coverage to establish a comprehensive coverage assessment. Coverage, in this context, denotes the ratio of public space area to the entire region's area, whereas service coverage refers to the proportion of the area that public spaces can effectively serve within their service thresholds in relation to the entire region. The calculation of service coverage involves overlapping analysis and the merging of overlapping areas.

Table 1. A top-down SDG 11.7 indicator framework.

Scale	Indicator	Description	Unit	SDG 11.7 Connotation				Notes [2]	References
				Safety	Inclusiveness	Accessibility	Greenness		
Apartment complex(S1)	Quality	Total area of available high-quality public space	m^2				√	Qa (+)	[6]
	Proximity	Proximity of the apartment complex to the public space	m			√		Ne (-)	[6]
	Richness	The richness of apartment complex's access to public space	m^2/per			√	√	Ac (+)	[29]
Street block(S2)	Resource area per household	Area of public space per household	m^2		√		√	Per (+)	[16]
	Dispersibility	Degree of dispersion of public space	/		√	√		NNI (+)	[25]
	Dominance	Advantage degree of public space area	/		√			Do (+)	[26]
	Spatial emergency response capacity	Emergency response capability of public space in streets in case of disasters	/	√				Sa (+)	[30]
	Equity [1]	Social equity of public space services	%		√			Eq (+)	[31]
County(S3)	Diversity	Diversity of public space	/				√	Sh (+)	[27]
	Ecosystem services value	Ecosystem service value of public space	yuan				√	Es (+)	[28]
	Total area	Total area of public space	m^2				√	Su (+)	[15]
	Comprehensive coverage	Overall coverage of public space within county	%			√	√	Co (+)	[32]

[1] This indicator is a supplementary indicator to supplement the social equity connotation of SDG 11.7. [2] The content in parentheses indicates that the indicators belong to positive or negative indicators.

2.3. Multi-Scale Evaluation Method for Urban SDG 11.7 Monitoring

After constructing a top-down SDG 11.7 indicator framework, this article further proposes a multi-scale evaluation method for urban SDG11.7 monitoring. To ensure comparability across various indicators, it is essential to establish boundaries for each indicator and normalize their values to a standardized range from 0 to 100.0 indicates the poorest performance, while 100 indicates the highest level of achievement. We utilize the term performance or score to characterize the degree of sustainable development, as opposed to solely focusing on indicator values, given the presence of certain negative indicators. Our approach draws from the methodology used to determine upper and lower bounds for data (decision tree method), as outlined in the 2018 SDG Index and Dashboard Report. This method effectively mitigates the potential impact of skewed data by mitigating the influence of extreme values on the data distribution, thereby enhancing the accuracy of standardized results [33].

The upper bounds of the indicators are determined with reference to the five-step decision tree method. If the conditions of the previous steps are met, the subsequent steps will not be carried out. (1) First, we use relevant absolute quantitative thresholds for the indicator framework, such as "absolute equality" to determine the boundary. (2) Secondly, we use the principle of "leaving no one behind" to determine the upper bounds of indicators related to coverage of public space. (3) We set the upper bounds of other indicators as the average of the top five performance indicators.

The determination of the lower bounds of indicators is similar to the upper bounds. (1) For indicators with clear absolute quantitative threshold meaning or completely contrary to the principle of "leaving no one behind", such as "absolute inequality" and "zero coverage", we use its worst possible performance to define the lower bounds. (2) For other indicators, the lower bound is defined as the value closest to the bottom 2.5% of the performance ranking of sustainable development indicators.

Following the establishment of upper and lower bounds for the indicators, we proceed to standardize their values, resulting in the final performance of the SDG indicators:

$$X = \frac{x - x_l}{x_u - x_l} \times 100 \tag{6}$$

where X represents the standardized score, i.e., standardized sustainable development performance; x represents the original value of the indicator; and x_u and x_l represent the upper and lower bounds of the indicator, respectively. The standardized score is set to 100 when the indicator value is greater than the upper bound and is set to 0 when the indicator value is less than the lower bound.

For the performance of SDG 11.7 at the apartment complex, street block, and county scales, this study calculates the weighted average of the performance of the indicators associated with each scale. Due to the complexity of hierarchical scales, multi-scale things in geography are interrelated and affected at different scales. Scale integration can span the interaction among scales to provide an overall assessment of the research area on a broader, more macroscopic level. Hence, building on the research findings of Wu et al. [34] and Yigitcanlar et al. [35], this article introduces a scale integration scheme (Figure 3). This scheme transfers the sustainability performance scores of S1, S2, and S3 to 1 km × 1 km grids, enabling city-scale analysis and allowing decision-makers to plan public spaces at the urban scale (S4). It is worth noting that, after experimenting with various grid sizes, this article has found that the one-kilometer grid effectively reflects the city's performance of SDG 11.7.

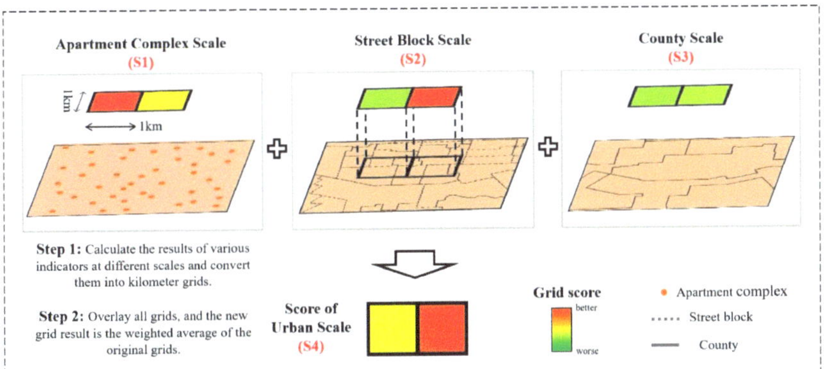

Figure 3. Scale integration scheme for multi-scale assessment.

To transform the points representing apartment complexes into a grid format, we employ interpolation to assign scores from a range of sustainable development indicators to

100 × 100 m resolution grids. By applying an overlapping analysis with the one-kilometer grid, we compute the average scores within each kilometer grid, resulting in the spatial distribution of each sustainable development indicator.

At both the street block and county scales, this article employs the area weighting method to superimpose the administrative boundaries onto the one-kilometer grid. There are two approaches for evaluating and analyzing sustainable development on a city scale (S4): (1) The first method involves directly calculating a weighted average of the one-kilometer grid data for the 11 indicators. (2) The second method entails first determining sustainability scores for the three different scales and subsequently averaging these three scale scores. It is worth noting that when the scores of the three scales are averaged, the impact of certain indicators can be either amplified or diminished due to the varying number of indicators at different scales. To mitigate this issue, we opt for the first scheme to transition between scales, ultimately generating the spatial pattern of SDG 11.7 at the S4 scale. The workflow of this study is shown in Figure 4.

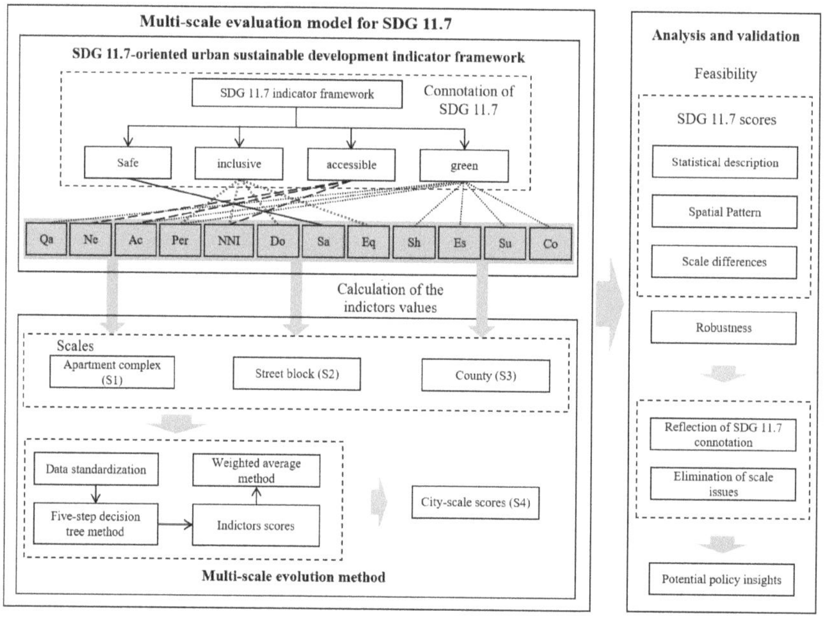

Figure 4. Workflow of this study.

3. Results

3.1. Multi Scale Monitoring of SDG 11.7 in Xi'an City

To obtain the spatiotemporal distribution of SDG 11.7 in Xi'an, we first calculate benchmarks for various sustainable development indicators (Table 2).

The evaluation results for Xi'an's progress in achieving SDG 11.7 are depicted in Figure 5. At the S1 scale, areas exhibiting higher levels of sustainable development are predominantly situated in proximity to Daming Palace, which is close to the urban center. This outcome underscores that the compact urban core area facilitates the residents' accessibility to public spaces in these locales. At the S2 scale, the regions boasting high scores in sustainable development are primarily concentrated within the heart of the High-tech Zone, Qujiang New Area, Hangtian, and Chanba Ecological Area. Evidently, these newer urban districts excel in the equitable distribution of public resources compared to the urban center, a conclusion that holds true at the S3 scale as well. Notably, the high-scoring areas for sustainable development at the S3 scale form a belt-shaped core, highlighting the adequate quantity of public spaces in these regions.

Table 2. Sustainable development indicators benchmarks.

Scale	Indicator	Upper Bound	Lower Bound	Units
S1	Quality	7,303,097	0	m^2
	Proximity	300	2515	m
	Richness	13,616.76	0	m^2/per
S2	Resource area per household	179	0	m^2
	Dispersibility	9.30	0	/
	Dominance	0.95	0	/
	Spatial emergency response capacity	0.32	0	/
S3	Diversity	1.53	0	/
	Ecosystem services value	6,191,608	0	yuan
	Total area	9,351,280	0	m^2
	Comprehensive coverage	57.57	0	%

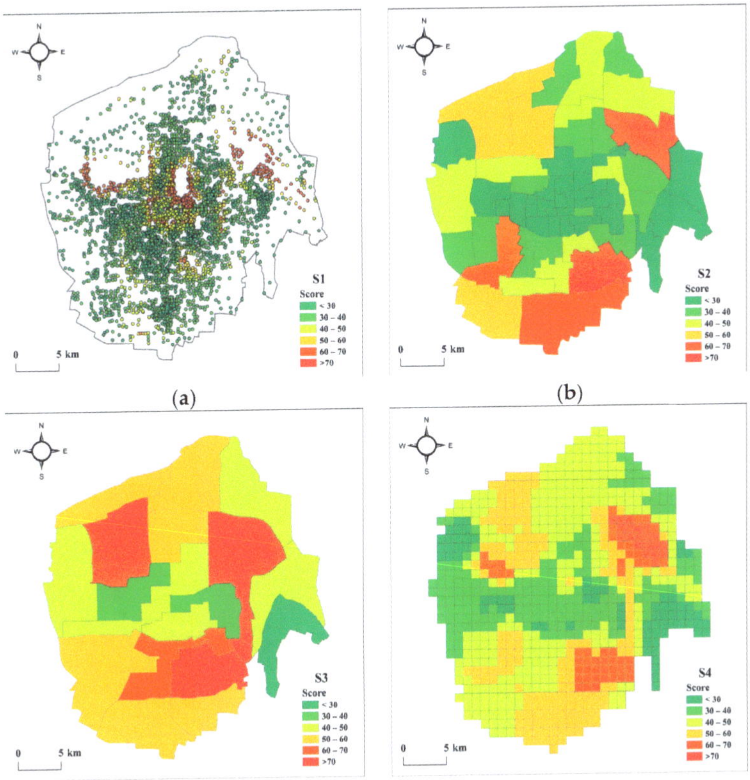

(a) (b) (c) (d)

Figure 5. Multi-scale spatial pattern of SDG 11.7 level in Xi'an. (a) S1 scale performance; (b) S2 scale performance; (c) S3 scale performance; (d) S4 scale performance.

The spatial pattern of SDG 11.7 observed at the S4 scale confirms the significant impact of rapid urbanization on sustainable development in the study area. Owing to the scarcity of land resources in the central urban region, public resources are primarily channeled into the new urban area, culminating in an "anti-core-edge" structure. The central urban area may exhibit superior performance in certain indicators due to historical factors (indicators of S1, for example), but in the broader context, the new urban area predominantly drives the city's sustainable development. It is important to note that although the overall performance of SDG 11.7 in newly developed urban areas is relatively good, there are low-value cores at both the eastern and western ends of the city. This is primarily due to the eastern part of Xi'an being a mountainous region, while the western area, having been developed later, remains in a rural–urban dual structure, resulting in inadequate levels of SDG 11.7.

As illustrated in Figure 6, the average scores for various sustainable development indicators are depicted. On the whole, SDG 11.7 in Xi'an is at a medium level, with ample room for enhancement. Among all the indicators, Ne and Co indicators demonstrate commendable performance, whereas Ac falls short. Given the relatively low sustainable development level of the new area at the S1 scale, urban managers should pay more attention to the acquisition of public space in future planning to enhance the performance of SDG 11.7.

Moreover, the spatiotemporal pattern of SDG 11.7 in this article differs significantly from those of other similar studies. Huang et al.'s assessment of public space accessibility in Xi'an reveals a spatial pattern of sustainability that decreases outward from the core of public spaces [36], while the results of this study do not support that conclusion. This discrepancy is particularly evident in the northwestern regions of Xi'an, where only a few areas near public spaces show relatively good levels of SDG 11.7. The reason for this divergence lies in the fact that the indicators used in other studies typically rely solely on accessibility to reflect certain aspects of SDG 11.7, while these studies are conducted at a single scale. The above further validates the necessity of constructing a multi-scale assessment model for SDG 11.7.

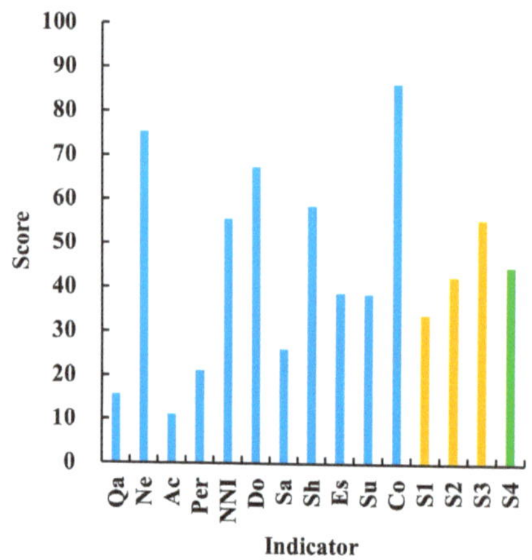

Figure 6. Performance of sustainable development indicators and different scales. To facilitate differentiation, the blue bars in the bar chart represent the individual indicators, the yellow bars represent the three scales of apartment complex, street block, and county, and the green bar represent the final results at the city scale. The same applies to the following figures.

3.2. Validation of the Model Robustness

The robustness of the model encompasses both uncertainty and sensitivity. We observe that a smaller number of indicators at each scale magnifies the influence of individual indicators on the outcomes during the analysis. When only a single indicator is used, the SDGs result becomes solely dependent on that particular indicator. To investigate the uncertainty arising from the variation in the number of SDG indicators, this study conducts an uncertainty analysis. Across the three distinct scales, the study explores various combinations of all sustainable development indicators within each scale. The outcomes yield a distribution of sustainable development performance, as depicted in Figure 7.

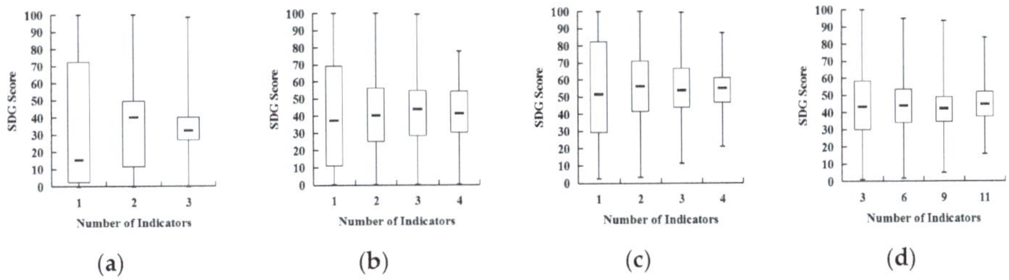

Figure 7. Uncertainty analysis for sustainable development performance. (**a**) S1; (**b**) S2; (**c**) S3; (**d**) S4.

The study observes that as the number of indicators increases, the uncertainty associated with the SDG results decreases, and the range of the overall sustainable development score tends to narrow. This trend is especially pronounced when examining the coefficient of variation (CV) of the indicators, as depicted in Figure 8. The coefficient of variation is notably high for individual indicators, while the CV for S4, which integrates all indicators, remains at a lower level.

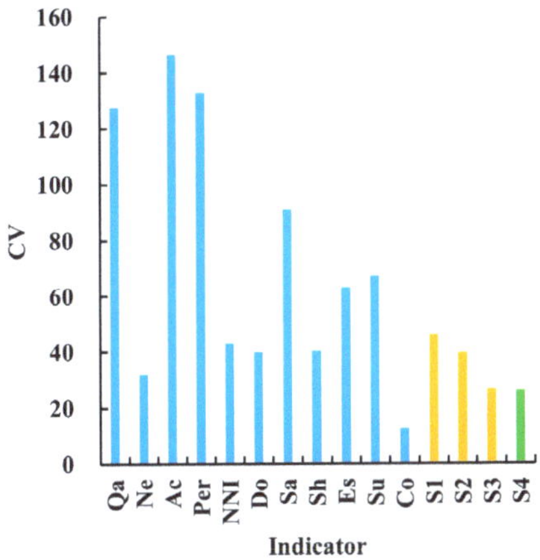

Figure 8. CV of SDG performance.

Furthermore, this article conducts a sensitivity analysis to assess how variations in individual indicator performance impact the assessment of sustainable development. This involves increasing and decreasing the value of each SDG indicator by 10%, followed by a recalibration of the sustainable development level to determine the sensitivity index, as depicted in Figure 9. Notably, the results reveal that alterations in a single indicator have a relatively minor impact on the overall outcomes, with the overall range staying within a 2% margin. In summary, the multi-scale sustainable development evaluation model outlined in this article exhibits a high degree of robustness.

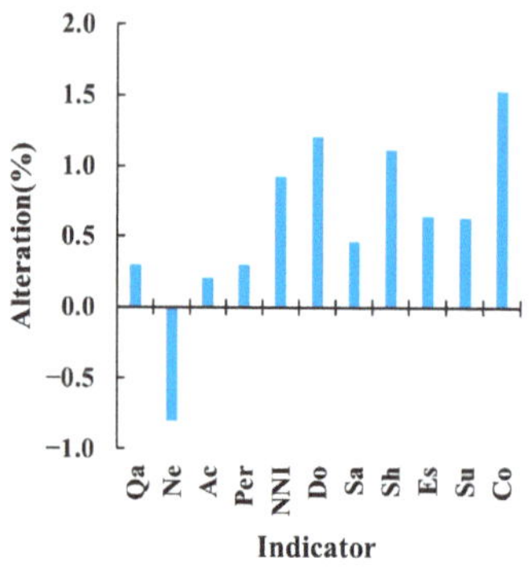

(a)

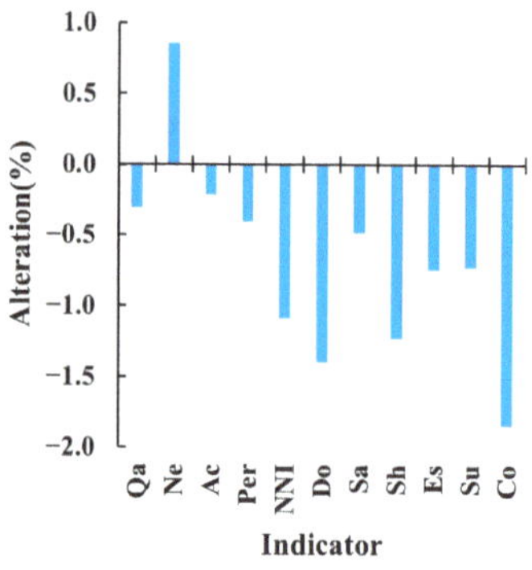

(b)

Figure 9. Sensitivity analysis of SDG performance. (**a**) Indicator's origin value increased by 10%. (**b**) Indicator's origin value decreased by 10%.

The robustness analysis of the model establishes the reliability of the framework developed in this study. It should be noted that the SDG11.7 framework, which includes multiple indicators, effectively reduces evaluation errors caused by a single indicator [33]. In addition, the calculation of indicators mainly relies on GIS spatial analysis, which greatly reduces the costs associated with data acquisition and computation [9]. The results of the spatial pattern of this case are consistent with Peng et al. [37], as they demonstrate the changing trend of urban sustainability from the core to the periphery. Overall, this evaluation model is both efficient and accurate.

4. Discussion

4.1. The Embodiment of Connotation of SDG 11.7 in the Indicator Framework

The advantage of a top-down problem-solving approach lies in its early consideration of the connotation of SDG 11.7 when establishing the indicator framework. Some previous comprehensive assessments of sustainable development goals have adopted similar approaches [17]. However, due to data limitations, some indicators are overlooked in early research, which also results in an incomplete reflection of the connotation of sustainable development goals.

In response to the connotation of SDG 11.7, this study establishes a total of 11 indicators that encapsulate various aspects of SDG 11.7. "Safety", represented by the Sa indicator, is previously described using crime rate data [38]. However, to overcome the limitations of statistical data, this article devises an index computed through spatial analysis methods. "Inclusiveness" is portrayed by the Per, NNI, Do, and Eq indicators, drawing inspiration from human geography indicators related to the allocation of public resources. These indicators reflect inclusiveness from different perspectives. For example, the indicator Per primarily reflects inclusiveness by assessing whether the configuration of public spaces meets the needs of all residents, while the indicators Do, NNI, and Eq, reflect inclusiveness by evaluating the spatial equity of public space distribution and its ability to cover vulnerable groups.

"Accessibility" is manifested through Ne, Ac, and NNI, with the core Ac indicator assessing residents' capacity to access public spaces after considering competition. The aforementioned indicators have been widely used in studies involving resources such as healthcare and green spaces [22], but they only reflect a portion of the connotation of SDG 11.7. This is also the reason why this study integrated indicators from different sources to reflect the entire connotation of SDG 11.7. "Greenness" is reflected by Qa, Ac, Per, Sh, and Es, drawing primarily from definitions of relevant indicators in landscape ecology to characterize a city's sustainable development level [27]. It should be noted that indicators related to landscape ecology are rarely used in the study of SDG11.7, resulting in a lack of relevant research reflecting its "greenness" connotation. The improvements presented in this article provide a reference for future research.

In view of the inclusion connotation of SDG 11.7, that is, the issue of equity, this study only reflects the meaning of spatial equity but does not provide relevant indicators of social equity [39]. This exclusion is due to the unavailability of comprehensive socio-economic data for residents in China. To ensure the model's applicability, the SDG 11.7 indicator framework does not include indicators related to social equity.

However, the robustness analysis has shown that introducing elements that reflect some aspects of social equity within the constraints of available data can enhance the accuracy of the assessment without unduly affecting the overall results. Therefore, we have employed apartment complex house prices as a proxy for certain socio-economic characteristics of residents. Through correlation analysis between house prices and SDG performance, the study has identified a substantial positive correlation between the level of sustainable development and residents' economic strength. This finding is further substantiated by both Mood's Median and the Jonckheere–Terpstra test. Mood's Median test reveals significant variations in sustainable development levels among regions with differing economic strengths, while the Jonckheere–Terpstra test confirms that sustainable development levels rise concomitant with increased economic strength. Hence, it is justifiable for us to introduce an indicator Eq (Table 1) that partially encapsulates the connotation of social equity and make revisions to the outcomes of the multi-scale evaluation model of SDG 11.7 so as to verify the hypothesis stated earlier.

This study defines residents of apartment complexes in the lowest 10% of housing prices as low-income groups. Subsequently, service thresholds (2 km, 1.5 km, 1 km, and 0.5 km) based on public space quality are used to create buffers and to count the number of low-income individuals within these buffers. Finally, the equity indicator Eq is calculated as the ratio of the number of low-income individuals with access to public space resources to the total number of low-income individuals. Notably, areas exhibiting higher equity levels in Xi'an are predominantly concentrated in the central urban area and the Qujiang New Area (Figure 10). The introduction of this equity indicator significantly reduces uncertainty at the street block scale as well. Broadly speaking, this additional indicator has a minor impact on the overall spatial pattern of sustainable development at the S4 scale. The most notable changes occur within the High-tech Zone and the Chanba Ecological Zone. Upon introducing the Eq indicator, the sustainable development performance of the old urban area in the High-tech Zone decreases, while the performance of the newly developed area in the High-tech Zone improves. Conversely, the performance of the old urban area in the Chanba Ecological Zone improves, while the performance of the newly developed area declines. Given the slower development pace of the old urban area, these results suggest that without considering the connotation of social equity, there may be a risk of underestimating the determination of economically driven urban new areas to achieve SDGs and overestimating the focus of ecologically oriented urban new areas on social equity.

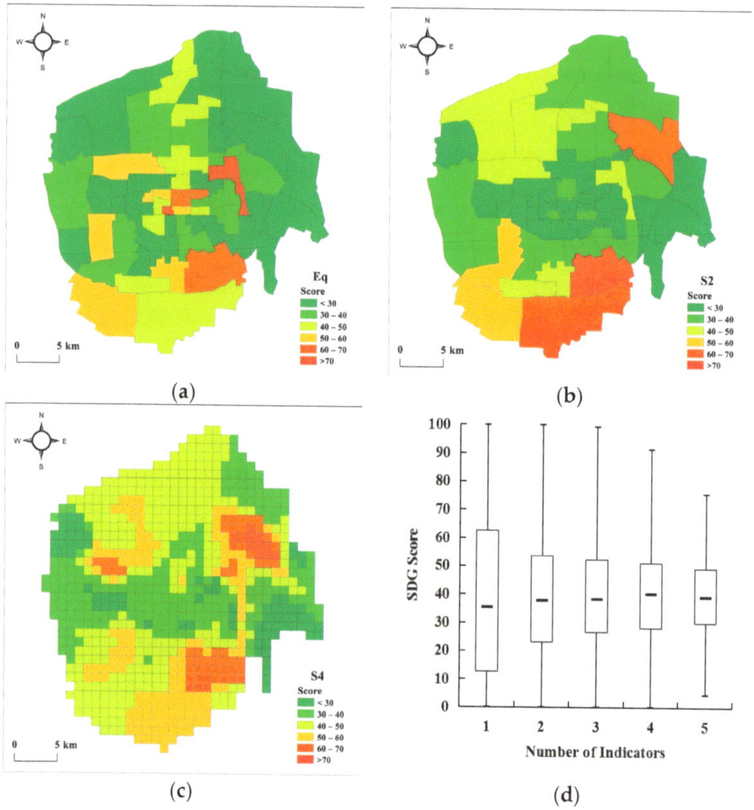

Figure 10. Revised spatial pattern of SDG 11.7 status of Xi'an. (**a**) Result of indicator Eq. (**b**) Revised S2 performance. (**c**) Revised S4 performance. (**d**) Uncertainty after the addition of the indicator Eq.

4.2. Weakening the Impact of Scale-Related Issues to Promote Governance Capacity at All Levels of Government

In the context of urban sustainable development, the absence of a multi-scale perspective might introduce bias into the assessment process [40]. Some studies have demonstrated differences in sustainable development assessment outcomes between multi-scale and single-scale approaches [35]. Consequently, some studies have highlighted the significance and complexity of selecting the optimal scale [41]. The model introduced in this article establishes development goals based on different scales, aiming to reduce the impact of scale issues as much as possible, thereby achieving a more precise simulation of the SDG 11.7 patterns and guiding policy implementation.

Currently, two scale-related issues warrant discussion: the modifiable areal unit problem (MAUP) and ecological fallacy [42,43]. The MAUP comprises scale effects and zoning problem—where scale effects refer to variations in statistical results caused by aggregating data into smaller or larger geographical units, while zoning problems signify result variations due to different partitioning schemes at a given scale. The ecological fallacy refers to biases or even completely contradictory outcomes when conclusions drawn at a specific scale are extrapolated to other scales [44].

The scale effect has a significant impact on the results of this study. When the scale gradually shrinks, some statistical indicators undergo dramatic shifts; that is, the negative externality is aggravated [45] when one of two adjacent spatial units achieves a high score in sustainable development due to the presence of public spaces, while the other languishes at a very low level of sustainable development due to a lack of such spaces. Although this scale effect is inevitable, we must point out that the sustainable development of local region cannot be disregarded even if the public space of its adjacent areas is sufficient. This discovery contradicts the notion that smaller scales yield more precise indicators. Although spatial indicators may enhance accuracy at the micro-level, the statistical indicators become less stable due to intensified edge effects. Given that edge effects cannot be entirely resolved or mitigated, our recourse is to exercise greater caution in indicator selection. This means reflecting the essence of SDGs through the simplification of indicators and the reduction of indicator redundancy.

The zoning problem is primarily evident at the county scale. Owing to the variations in the shapes and sizes of administrative divisions, certain areas with low sustainable development scores at the micro-level are influenced by their neighboring regions within the same county, resulting in what can be termed the "illusion of prosperity." Consequently, total quantity-based indicators may not accurately portray the sustainability of these areas. In recognition of the administrative unity within the county, we have made efforts to employ indicators that underscore the principle of equality to mitigate this impact. For instance, this article incorporates the use of indicators Es and Su to represent the SDG 11.7 level at the S3 scale. Additionally, we utilize the coverage indicator Co to alleviate the influence of zoning effects. Nonetheless, some scholars argue that data integration in statistical issues related to ecological fallacy might mask the needs of the most marginalized and vulnerable groups, as they merely reflect an overall improvement at the macro-level [10].

Overall, this article solves the problem of administrative responsibility misalignment caused by the mismatch between scale and development goals from a multi-scale perspective. However, it is crucial to acknowledge that the inherent scale-related disparities in assessment outcomes cannot be entirely eliminated. We strive to minimize these effects by employing the mentioned methodology and carefully selecting appropriate indicators to enhance the overall accuracy of the results.

4.3. Contributions and Limitations

SDG 11.7 is currently a focal point in urban sustainability assessments, with numerous studies analyzing the spatiotemporal evolution of SDG 11.7 patterns across different regions. However, the neglect of SDG 11.7′s connotation and scale issues in these studies is concerning. This article addresses these gaps, with key contributions, including (1) the development of a top-down sustainable development indicator framework that reflects the four aspects of SDG 11.7; (2) the establishment of a multi-scale assessment method for SDG 11.7 by setting development goals for different levels of administrative entities; and (3) the validation of the indicator framework and assessment method using Xi'an as a case study. Together, these elements constitute the uniqueness of this article.

Although the model constructed in this paper is set against the backdrop of Chinese cities, it has potential relevance for urban planning and sustainability assessments in other countries or regions. First, the connotation of SDG 11.7 is collectively agreed upon by UN member states, confirming the recognition of its four aspects—safety, inclusiveness, accessibility, and greenness—by most countries. Second, when establishing development goals for different administrative levels, this article emphasizes the progressive relationship between government levels and their development priorities (access, allocation, and overall quantity of public spaces), which is consistent with the administrative logic of most countries. Although different cities may have variations in the reasonable scale for selecting allocation and total quantity indicators, the use of indicators that reflect access issues at the micro-scale has been extensively validated. Of course, the indicator framework constructed in this article to assist governments at different levels in addressing issues of access, allocation, and total quantity of public spaces may face localization challenges. However, the indicators, such as accessibility and greening rates, developed in this article have been widely used in SDG 11.7 research across various regions, thereby providing guidance for policy formulation by governments at corresponding levels in those areas.

Inevitably, this study is not without its limitations. Firstly, data acquisition heavily relies on visual interpretation, especially the digitization of residential areas and public spaces, which is a time-consuming process. With the advancements in machine learning, some scholars have successfully employed algorithms to recognize and extract spatial features [46]. It is anticipated that in the future, urban sustainable development assessment can be accomplished entirely by computer-based methods. Secondly, the SDG 11.7 indicator framework exhibits shortcomings in adequately reflecting the connotation of social equity, necessitating supplementation with statistical data. Thirdly, while the article highlights that the choice of the number of indicators and fluctuations in indicator scores contribute to result uncertainty and sensitivity, it does not provide a specific number of indicators or sensitivity thresholds that could significantly reduce these biases. Future research could involve applying this model to other cities and conducting comparative analyses to determine more appropriate numbers of indicators and sensitivity thresholds, thereby further optimizing the indicator framework. Lastly, concerning the types of public spaces, this article does not encompass agricultural landscapes such as cultivated land on the outskirts of urban areas within the assessment. However, some studies have underscored that these agricultural landscapes contribute to ecological protection for cities and offer recreational and scenic services to urban residents, particularly those residing at the urban periphery [47]. This omission may lead to an underestimation of the sustainable development level in urban fringe areas.

5. Conclusions and Outlook

This article proposes a multi-scale assessment model for SDG 11.7 using a top-down problem-solving approach. The model integrates 11 indicators across three scales to comprehensively reflect the connotation of SDG 11.7, which includes safety, inclusiveness, accessibility, and greenness. By setting development goals for different levels of government and fully capturing the connotation of SDG 11.7, this study primarily fills the gap of neglecting the connotation of SDG 11.7 and scale issues in related research. In testing the model in Xi'an city, the model accurately assesses the sustainable development level of public spaces at the scales of apartment complex, street block, and county. Through scale transformation, we obtain the spatial pattern of SDG 11.7 at the urban level and identify areas requiring focused attention. Additionally, the model demonstrates strong robustness.

We believe that this model can be applied to monitor SDG 11.7 in other cities. First, by reflecting the connotation of SDG 11.7, the government can identify which aspects of public space need improvement rather than engaging in blind development. Second, by setting reasonable development goals for different levels of administrative entities, the model clarifies governmental responsibilities, significantly enhancing management efficiency. Finally, by providing the spatial pattern of SDG 11.7, managers can identify areas requiring focused attention. These advantages can help administrative entities effectively formulate policies and actions that balance environmental and development issues, thereby addressing the development imbalances and responsibility misalignments among various levels of government in promoting the sustainable development of public spaces. Future research can continue to improve the model's ability to reflect social equity connotations, such as by utilizing big data to develop spatial indicators instead of relying solely on statistical data, further enhancing the model's applicability and extending it to sustainability assessments of different social resources like housing, healthcare, and transportation.

Author Contributions: Conceptualization, K.W. and J.B.; methodology, X.D.; software, K.W.; validation, X.D.; formal analysis, K.W. and X.D.; investigation, K.W.; resources, K.W.; data curation, K.W.; writing—original draft preparation, K.W.; writing—review and editing, K.W. and J.B.; visualization, X.D.; supervision, J.B.; project administration, K.W. and J.B.; funding acquisition, J.B. All authors have read and agreed to the published version of the manuscript.

Funding: This research was funded by National Natural Science Foundation of China, grant number 42271289.

Data Availability Statement: Data will be made available on request.

Conflicts of Interest: The authors declare no conflicts of interest.

References

1. Giles-Corti, B.; Lowe, M.; Arundel, J. Achieving the SDGs: Evaluating indicators to be used to benchmark and monitor progress towards creating healthy and sustainable cities. *Health Policy* **2020**, *124*, 581–590. [CrossRef] [PubMed]
2. Rutt, R.L.; Gulsrud, N.M. Green justice in the city: A new agenda for urban green space research in Europe. *Urban For. Urban Green.* **2016**, *19*, 123–127. [CrossRef]
3. Chen, Q.; Du, M.; Cheng, Q.; Jing, C. Quantitative Evaluation of Spatial Differentiation for Public Open Spaces in Urban Built-Up Areas by Assessing SDG 11.7: A Case of Deqing County. *ISPRS Int. J. Geo-Inf.* **2020**, *9*, 575. [CrossRef]
4. Aguilar, R.; Kuffer, M. Cloud Computation Using High-Resolution Images for Improving the SDG Indicator on Open Spaces. *Remote Sens.* **2020**, *12*, 1144. [CrossRef]
5. Xu, M.; Xin, J.; Su, S.; Weng, M.; Cai, Z. Social inequalities of park accessibility in Shenzhen, China: The role of park quality, transport modes, and hierarchical socioeconomic characteristics. *J. Transp. Geogr.* **2017**, *62*, 38–50. [CrossRef]
6. Rigolon, A. Parks and young people: An environmental justice study of park proximity, acreage, and quality in Denver, Colorado. *Landsc. Urban Plan.* **2017**, *165*, 73–83. [CrossRef]
7. Ye, Y.; Richards, D.; Lu, Y.; Song, X.; Zhuang, Y.; Zeng, W.; Zhong, T. Measuring daily accessed street greenery: A human-scale approach for informing better urban planning practices. *Landsc. Urban Plan.* **2019**, *191*, 103434. [CrossRef]
8. Li, Z.; Fan, Z.; Song, Y.; Chai, Y. Assessing equity in park accessibility using a travel behavior-based G2SFCA method in Nanjing, China. *J. Transp. Geogr.* **2021**, *96*, 103179. [CrossRef]
9. Gupta, K.; Roy, A.; Luthra, K.; Maithani, S. GIS based analysis for assessing the accessibility at hierarchical levels of urban green spaces. *Urban For. Urban Green.* **2016**, *18*, 198–211. [CrossRef]
10. Ulbrich, P.; Porto de Albuquerque, J.; Coaffee, J. The Impact of Urban Inequalities on Monitoring Progress towards the Sustainable Development Goals: Methodological Considerations. *ISPRS Int. J. Geo-Inf.* **2018**, *8*, 6. [CrossRef]
11. Akuraju, V.; Pradhan, P.; Haase, D.; Kropp, J.P.; Rybski, D. Relating SDG11 indicators and urban scaling—An exploratory study. *Sustain. Cities Soc.* **2020**, *52*, 101853. [CrossRef]
12. Gazzeh, K.; Abubakar, I.R. Regional disparity in access to basic public services in Saudi Arabia: A sustainability challenge. *Util. Policy* **2018**, *52*, 70–80. [CrossRef]
13. Klopp, J.M.; Petretta, D.L. The urban sustainable development goal: Indicators, complexity and the politics of measuring cities. *Cities* **2017**, *63*, 92–97. [CrossRef]
14. Zheng, W.; Shen, G.Q.; Wang, H.; Hong, J.; Li, Z. Decision support for sustainable urban renewal: A multi-scale model. *Land Use Policy* **2017**, *69*, 361–371. [CrossRef]
15. Tan, P.Y.; Samsudin, R. Effects of spatial scale on assessment of spatial equity of urban park provision. *Landsc. Urban Plan.* **2017**, *158*, 139–154. [CrossRef]
16. Li, J.; Huang, X.; Chuai, X.; Yang, H. The impact of land urbanization on carbon dioxide emissions in the Yangtze River Delta, China: A multiscale perspective. *Cities* **2021**, *116*, 103275. [CrossRef]
17. Wang, Q.; Liu, C.; Hou, Y.; Xin, F.; Mao, Z.; Xue, X. Study of the spatio-temporal variation of environmental sustainability at national and provincial levels in China. *Sci. Total Environ.* **2022**, *807*, 150830. [CrossRef]
18. Russell, B. Beyond the Local Trap: New Municipalism and the Rise of the Fearless Cities. *Antipode* **2019**, *51*, 989–1010. [CrossRef]
19. Purvis, B.; Mao, Y.; Robinson, D. A multi-scale integrated assessment model to support urban sustainability. *Sustain. Sci.* **2021**, *17*, 151–169. [CrossRef]
20. Wang, L.; Cao, X.; Li, T.; Gao, X. Accessibility Comparison and Spatial Differentiation of Xi'an Scenic Spots with Different Modes Based on Baidu Real-time Travel. *Chin. Geogr. Sci.* **2019**, *29*, 848–860. [CrossRef]
21. Zhou, Z.; Li, M. Spatial-temporal change in urban agricultural land use efficiency from the perspective of agricultural multi-functionality: A case study of the Xi'an metropolitan zone. *J. Geogr. Sci.* **2017**, *27*, 1499–1520. [CrossRef]
22. Wang, K.; Bai, J.; Dang, X. Spatial Difference and Equity Analysis for Accessibility to Three-Level Medical Services Based on Actual Medical Behavior in Shaanxi, China. *Int. J. Environ. Res. Public Health* **2020**, *18*, 112. [CrossRef]
23. Xing, L.; Liu, Y.; Liu, X. Measuring spatial disparity in accessibility with a multi-mode method based on park green spaces classification in Wuhan, China. *Appl. Geogr.* **2018**, *94*, 251–261. [CrossRef]
24. Mao, K.; Chen, Y.; Wu, G.; Huang, J.; Yang, W.; Xia, Z. Measuring Spatial Accessibility of Urban Fire Services Using Historical Fire Incidents in Nanjing, China. *ISPRS Int. J. Geo-Inf.* **2020**, *9*, 585. [CrossRef]
25. Tang, C.; Guo, X.; Zhou, G.; Wu, J.; Chen, W. Spatial distribution and influencing factors of innovation platforms in urban agglomerations of the middle reaches of the Yangtze River Basin. *Prog. Geogr.* **2020**, *39*, 531–541. [CrossRef]
26. Lucas, K.; van Wee, B.; Maat, K. A method to evaluate equitable accessibility: Combining ethical theories and accessibility-based approaches. *Transportation* **2015**, *43*, 473–490. [CrossRef]
27. Strong, W.L. Biased richness and evenness relationships within Shannon–Wiener index values. *Ecol. Indic.* **2016**, *67*, 703–713. [CrossRef]

28. Li, M.; Zhou, Z. Positive and negative ecosystem services evaluation and its spatial pattern analysis on urban landscape: A case study of Xi'an City. *Acta Geogr. Sin.* **2016**, *71*, 1215–1230.
29. Li, L.; Du, Q.; Ren, F.; Ma, X. Assessing Spatial Accessibility to Hierarchical Urban Parks by Multi-Types of Travel Distance in Shenzhen, China. *Int. J. Environ. Res. Public Health* **2019**, *16*, 1038. [CrossRef]
30. Yin, J.; Xu, S.; Jing, Y.; Yin, Z.e.; Liao, B. Evaluating the impact of fluvial flooding on emergency responses accessibility for a mega-city's public services: A case study of emergency medical service. *Acta Geogr. Sin.* **2018**, *73*, 1737–1747.
31. Liu, R.; Tai-chee, W.; Liu, S. The informal housing market in Beijing's rural areas: Its formation and operating mechanism amidst the process of urbanization. *Geogr. Res.* **2010**, *29*, 1355–1358.
32. Zhu, Z.; He, Q.; Qin, W. Spatial Livability of Residential Areas in Changsha City. *Sci. Geogr. Sin.* **2020**, *40*, 1859–1867.
33. Xu, Z.; Chau, S.N.; Chen, X.; Zhang, J.; Li, Y.; Dietz, T.; Wang, J.; Winkler, J.A.; Fan, F.; Huang, B.; et al. Assessing progress towards sustainable development over space and time. *Nature* **2020**, *577*, 74–78. [CrossRef] [PubMed]
34. Wu, H.; Jiang, Z.; Lin, A.; Zhu, W.; Wang, W. Analyzing spatial characteristics of urban resource and environment carrying capacity based on Covert-Resilient-Overt: A case study of Wuhan city. *Acta Geogr. Sin.* **2021**, *76*, 2439–2457.
35. Yigitcanlar, T.; Dur, F.; Dizdaroglu, D. Towards prosperous sustainable cities: A multiscalar urban sustainability assessment approach. *Habitat Int.* **2015**, *45*, 36–46. [CrossRef]
36. Huang, X.; Li, H.; Zhang, Y.; Xie, L. Research and Evaluation of Accessibility of Xian City Park Green Space Based on Population Discretization and Transportation Network Analysis. *Remote Sens. Inf.* **2014**, *29*, 98–102.
37. Peng, L.; Zhang, L.; Li, X.; Wang, P.; Zhao, W.; Wang, Z.; Jiao, L.; Wang, H. Spatio-Temporal Patterns of Ecosystem Services Provided by Urban Green Spaces and Their Equity along Urban–Rural Gradients in the Xi'an Metropolitan Area, China. *Remote Sens.* **2022**, *14*, 4299. [CrossRef]
38. Venter, Z.S.; Shackleton, C.; Faull, A.; Lancaster, L.; Breetzke, G.; Edelstein, I. Is green space associated with reduced crime? A national-scale study from the Global South. *Sci. Total Environ.* **2022**, *825*, 154005. [CrossRef]
39. Boisjoly, G.; Serra, B.; Oliveira, G.T.; El-Geneidy, A. Accessibility measurements in São Paulo, Rio de Janeiro, Curitiba and Recife, Brazil. *J. Transp. Geogr.* **2020**, *82*, 102551. [CrossRef]
40. Wen, X.; Deng, X.; Zhang, F. Scale effects of vegetation restoration on soil and water conservation in a semi-arid region in China: Resources conservation and sustainable management. *Resour. Conserv. Recycl.* **2019**, *151*, 104474. [CrossRef]
41. Pang, B.; Zhao, J.; Zhang, J.; Yang, L. Calculating optimal scale of urban green space in Xi'an, China. *Ecol. Indic.* **2023**, *147*, 110003. [CrossRef]
42. Fernández, I.C.; Wu, J. Assessing environmental inequalities in the city of Santiago (Chile) with a hierarchical multiscale approach. *Appl. Geogr.* **2016**, *74*, 160–169. [CrossRef]
43. Lee, S.-I.; Lee, M.; Chun, Y.; Griffith, D.A. Uncertainty in the effects of the modifiable areal unit problem under different levels of spatial autocorrelation: A simulation study. *Int. J. Geogr. Inf. Sci.* **2018**, *33*, 1135–1154. [CrossRef]
44. Buyantuyev, A.; Wu, J.; Gries, C. Multiscale analysis of the urbanization pattern of the Phoenix metropolitan landscape of USA: Time, space and thematic resolution. *Landsc. Urban Plan.* **2010**, *94*, 206–217. [CrossRef]
45. Nicholls, S. Measuring the accessibility and equity of public parks: A case study using GIS. *Manag. Leis.* **2001**, *6*, 201–219. [CrossRef]
46. Zhou, Y.; Smith, S.J.; Elvidge, C.D.; Zhao, K.; Thomson, A.; Imhoff, M. A cluster-based method to map urban area from DMSP/OLS nightlights. *Remote Sens. Environ.* **2014**, *147*, 173–185. [CrossRef]
47. Song, B.; Robinson, G.M.; Zhou, Z. Agricultural transformation and ecosystem services: A case study from Shaanxi Province, China. *Habitat Int.* **2017**, *69*, 114–125. [CrossRef]

land

Article

Land Stewardship and Development Behaviors Under an Ecological-Impact-Weighted Land Value Tax Scheme: A Proof-of-Concept Agent-Based Model

Dakota B. Walker [1,2,3,*], Alican Mertan [4], Joshua Farley [1,2,3], Donna Rizzo [2,5] and Travis Reynolds [1,2]

1 Department of Community Development & Applied Economics, University of Vermont, 146 University Place, Burlington, VT 05405, USA
2 Gund Institute for Environment, University of Vermont, 210 Colchester Ave, Burlington, VT 05405, USA
3 Leadership for the Ecozoic, 146 University Place, Burlington, VT 05405, USA
4 Department of Computer Science, University of Vermont, 33 Colchester Ave, Burlington, VT 05405, USA
5 Department of Civil & Environmental Engineering, University of Vermont, 33 Colchester Ave, Burlington, VT 05405, USA
* Correspondence: dakota.walker@uvm.edu

Abstract: Sprawling land development patterns have exacerbated ecological degradation, social fragmentation, and public health problems. Perverse incentives arise from the ability to privatize collectively created value in land rents and socialize ecological costs. Land value taxation (LVT) has been shown to encourage urban infill development by reducing or eliminating rent-seeking behavior in land markets. However, despite its purported benefits, this tax reform is value monistic in its definition of optimal land use and, therefore, does little to address the lack of non-market information to inform land use decisions. We propose an ecological-impact-weighted land value taxation policy (ELVT) which incorporates the ecological footprint of land use into one's land value tax burden. We test both proposed policies (LVT and ELVT) relative to a "status quo" (SQ) property tax scheme, utilizing a conceptual spatially explicit agent-based model of land use behaviors and housing development. Our findings suggest that both tax interventions can increase the capital intensity and decrease the land intensity of housing development. Furthermore, both tax interventions can lead to a net profit loss for speculators and a decrease in the average housing unit price. The ELVT scheme is shown to significantly increase urban nature provisions and dampen the loss of ecological value across a region.

Keywords: suburban sprawl; housing; land value tax; ecological deficit tax; agent-based model; private conservation; pro-environmental practices

1. Introduction

Suburban sprawl has become a defining feature of the modern landscape in the U.S. and other countries [1,2]. It is characterized by dispersed, low-density residential buildings reliant on automobile travel for access to amenities and commercial areas [3]. Today, most U.S. citizens reside in suburban neighborhoods [4], places which account for over 90% of new impervious development [1]. Such rates of development continue to outpace population growth, suggesting an increase in per capita land consumption [1].

The negative effects of sprawling land development span the social [5,6], public health [7,8], economic [9,10], and ecological domains [11,12]. Concerns about sprawl have been raised as far back as 1958, particularly regarding its impact on open-space amenities and public expenditures [13]. More recently, research has demonstrated the economic inefficiencies of sprawl, which can increase public expenditures, as it necessitates an extension of the boundary of public services and infrastructure [14,15]. Suburbs have also contributed to the United States' oversized greenhouse gas emissions due to their associated higher residential energy burden [16] and longer commuting distances [17]. Jones and Kammen [18] found that suburban households have 25% higher carbon footprints than urban households and comprise around 50% of the total U.S. household carbon footprint.

Of particular relevance to this paper are the environmental consequences of sprawling land development—arguably the most central concern about sprawl today [19]. Suburban development remains a leading cause of the loss of forested rural lands, along with the essential ecosystem services they provide. At the current development rates, an estimated 18 million acres of U.S. agricultural land could be lost, fragmented, or degraded by 2040, with 12 million acres projected to be converted for low-density residential use [20]. However, much of this research simplifies ecological footprint according to land use categories, overlooking the variability in the impacts that different suburban developments can have. For example, turf grass lawns, which cover three times more area than any other irrigated crop in the U.S. [21], are a major contributor to habitat and biodiversity loss. Despite their ubiquity, lawns are not essential to suburban life, but rather a mere landscaping choice shaped by cultural norms. Reforming such practices could significantly reduce the ecological impact of suburban development, even if the trend toward suburbanization continues.

Despite the contribution of each new development to these broader social dilemmas, individual actors remain unresponsive to the collective costs they pose—an evident example of a market failure. In response, governmental agencies at the local, county, state, and in some cases, national level have implemented urban containment policies. These efforts include comprehensive planning, public land acquisition, and various regulations and incentives to establish clear boundaries between urban and rural areas, promoting urban infill and preserving rural land [19]. Common policies include zoning restrictions, development impact fees, property tax adjustments, density bonuses, brownfields redevelopment subsidies, and the public conservation of peripheral lands, among other things [19]. Some research has found these policies to have at least some effects in promoting compact development [22,23], particularly when enforced at the state level [24] or enacted as a suite of regulations, incentives, and public investment policies [19]. Others note the counterproductive effects of these policies, such as increased housing costs [25], which lead to spillover or leapfrog sprawl in surrounding municipalities [26,27]. Despite their best efforts, urban planners and governmental bodies at various scales have not been able to redirect the inertia of development away from suburban forms. Exploring more effective urban containment policies remains paramount and has been explicitly called for by local and state governments [19].

Many argue that prior policy interventions have ignored a crucial driver of suburban sprawl—speculation, e.g., [27,28]. Land, being fixed in supply, derives its value from the demand it garners, its desirability for use or financial returns, and the number of demanders [29]. Increases in a parcel's desirability and value may result from municipal development or rising consumer interest, neither of which stems from actions made by the parcel's owner. Thus, it is considered as economic rent [29]. Speculative investment in land prioritizes economic rent capture over capital investment, resulting in "leapfrog development". Strategies typically involve purchasing large swaths of land outside the city periphery and holding that land out of use until its value appreciates due to the extension of city services, at which point, speculators may sell and purchase land beyond the new periphery [27].

The rent-generating potential of land creates perverse incentives for land use decisions, encouraging land hoarding based on speculative value rather than the land's immediate use for housing or commercial purposes. In other words, land prices are increasingly set according to their potential to appreciate rather than their use value, which can embed boom–bust cycles in land markets and force many to seek housing on the more affordable city outskirts. It also undermines collective welfare through private appropriation of the public and nature-derived values of place [30]. Policies and public investment, which aim to make city centers more livable and amenable to development, can instead lead to displacement and growing wealth inequality, so long as publicly created value remains available for private appropriation [31]. As just one example, the benefits of urban greenspace have been shown to capitalize on property prices and result in the displacement of existing communities and market-based exclusivity of the subsequent benefits of greenspace [32,33][1]. Thus, addressing the speculative dynamics in land markets is essential for crafting more effective urban containment policies.

Land value tax (LVT) is a policy tool for curbing speculation-driven suburbanization by redistributing the benefits of public improvement and local land stewardship toward the community at large, rather than to neighboring parcels alone via price appreciation [30,35]. Originally proposed in the late 19th century by Henry George [30], land value taxation consists of a tax levied solely on the unimproved value of land. As an alternative to the conventional property tax, LVT shifts taxes off capital improvements (which stem from private investment) and on to land rents (which stem from public investment). Consequently, it removes the deadweight loss associated with taxes on land improvements, and removes an incentive to speculate by redirecting the fruits of public service investments back to the municipality. This policy has sustained interest among researchers and policymakers, and has been implemented in over 30 countries, including many places in the U.S. [36]. Researchers have demonstrated the effectiveness of LVT policies in promoting urban infill over urban sprawl [37–39], preserving exurban land for nature [40,41], and reducing housing costs [42,43].

However, despite their improvements both to equity and land use efficiency by buffering the exploitation of land rents, conventional land value taxation schemes remain reliant on markets to define the "optimal use" for land and, therefore, lack the ability to relay non-market benefits and costs to landowners in order to, for example, safeguard ecological thresholds [44]. Thus, this sustained reliance on markets and their associated value monism presents a shortcoming regarding the applicability of land value taxation to ecologically sound land stewardship.

Moreover, conventional land value taxation falls short of directly incentivizing private urban or suburban land stewardship due to the non-discriminate distribution of the social welfare created. While a landowner would bear the full cost of land regeneration (including the direct cost and potentially increased tax burden from land value appreciation), they would receive a negligible portion of any monetary benefits (via the broad distribution of new tax revenue).

Although a slew of public action has focused on conservation or incentivizing private land stewardship, such endeavors typically engage in large-scale conversions in rural regions [45]. Urban and suburban areas, where the majority of Americans live, present an underutilized opportunity for the broader adoption of ecological stewardship [45–47]. Addressing this gap requires a paradigm shift away from human settlements as sacrifice zones countervailed by far-off conserved nature and towards a view that recognizes the potential for human–nature relationships to be mutually enhancing [48,49]. Urban ecological stewardship also directs the benefits of ecosystem services to within urban and suburban centers, with co-benefits including food access [50], cultural transformation via crowding-in pro-environmental behavior [51], urban resilience [52], and the variety of other benefits associated with proximity to greenspace [53].

1.1. Policy Proposal

This research seeks to introduce and justify the theoretical expansion of a land value taxation scheme aimed at encouraging the broader adoption of land stewardship practices within and outside urban centers. We propose an ecological-impact-weighted land value tax scheme (ELVT) which penalizes land degradation and subsidizes ecological stewardship in high-ecological-importance areas. Under the ELVT scheme, a conventional LVT rate is first applied and then scaled up or down according to non-market information and landowner decision making. The ecological weighting considers the following two factors: the value that a particular parcel would have to collective ecological health and service provisions (deemed "ecological potential"), and the degree to which the parcel, in its current state, is meeting that potential (deemed "ecological value"). For example, a parking lot or industrial site directly abutting unfragmented natural land may have a high ecological potential and low ecological value. So too might a parking lot that is far from any greenspace. Ultimately, such designations of ecological potential and ecological value would be at the discretion of the enacting municipality.

Under ELVT, ecological value would have an inverse relationship with tax burden, the magnitude of which is mediated by ecological potential. Those who live in a location with high ecological potential and have converted their lawn into a garden or native habitat (for example) would face a lower tax rate, which is subsidized by the higher tax rate of those who maintain ecologically degrading land use practices but still benefit from the ES provisions around them. For parcels with a low ecological potential, the tax would mirror that of a conventional LVT.

ELVT provides a means to align individual action with collective value creation. It coheres with the original justification for land value taxation, namely re-socializing collective values, but allows for a broader definition of such values. This policy also aligns with broader policy suggestions such as common asset trusts, which builds an institutional capacity to manage common goods toward collective benefit [54], and has received some recent attention in political spheres (see [55]). Finally, ELVT could provide a passive means of agglomeration incentives, in which land stewardship increases the land value and benefits of restoration among neighboring parcels, thus incentivizing their restoration and bolstering the contiguity of intact land.

While the effects of LVT have a long history of examination, to the authors' best knowledge, researchers have yet to explore the outcomes associated with the particular mechanisms of the proposed ELVT scheme. Nevertheless, limited research on similar proposals supports further examination. Lafuite et al. [41] showed that a natural land depletion tax has the effect of internalizing the value of biodiversity, resulting in more diversity being preserved at equilibrium through more labor-intensive agricultural practices that reduce natural land conversion. The tax was also predicted to mitigate the vulnerability of the economy to overshoot and collapse given the time-delayed externalities of biodiversity loss [41]. Xiong and Li [56] proposed an ecological deficit tax that compensates the difference between land's ecological carrying capacity and its current ecological footprint. Using a computable general equilibrium model, the authors found that the ecological deficit tax had the effect of reducing the region's ecological footprint in the long run. Similarly, Muellbauer [57] made a case for a "Green Land Value Tax". Though not focused on ecological land stewardship, it entails a conventional land value tax scheme, except that property owners would receive a discount for measures which lower energy use.

1.2. Study Overview

In this study, we utilize a spatially explicit agent-based model of land use behaviors and housing development as a proof-of-concept exploration of the emergent outcomes associated with LVT and ELVT policies. We employ boundedly rational decision making among homeowners, developers, and speculators operating within a regional land and housing market. Our analysis focuses on four experiments in which the model is tested under different parameter settings. In experiment 1, we model the three tax schemes (with approximate revenue neutrality) under three different configurations of homeowner preference, which influence the resulting development patterns toward suburban sprawl, compact urban infill, and something in between. In experiment 2, we perform a sweep of the percentage of speculator agents in the model to explore their role in mediating spatial and socio-economic outcomes. In experiment 3, we perform a sweep of the magnitude of divergence between the LVT and ELVT tax rates to explore the effect of ELVT under increasingly extreme operationalizations. In experiment 4, we relax the assumption of revenue neutrality and perform a sweep of the LVT rates to assess the spatial and economic effects of more aggressive rent socialization tax policies. Our key finding is that both LVT and ELVT tax interventions increase housing density and decrease land-intensive housing, while ELVT further densifies the urban core by moderating the loss of natural beauty, which can otherwise drive prospective homeowners away from the urban core.

2. Methods

Urban development is an inherently complex process; emergence results from non-linear interactions between strongly coupled social, environmental, and economic systems, featuring spatial and multi-scale interactions and context-adaptive decision making amongst a heterogenous population of land use decision makers [58]. Such complexity precludes the use of traditional economic or statistical models [59]. Instead, researchers have increasingly turned to agent-based models (ABMs) [60]. ABMs take a micro-lens of analysis, controlling only individual decision making rules and system contexts, therefore facilitating the examination of how complex emergent macrophenomena can result from individual action, as well as the outcomes associated with intervening at the individual level [61].

This research takes inspiration from a robust foundation of academic study using spatially explicit ABMs to understand residential locational choices and urban sprawl. Magliocca et al. [62] incorporated endogenous, spatially heterogenous land and housing price formation to demonstrate the occurrence of urban sprawl and leapfrog development driven by consumer preferences and various economic influences on the profitability of farmland conversion. Brown and Robinson [63] utilized empirical residential locational preferences and showed how more realistic heterogeneity in preferences led to greater urban sprawl in location choice. Jackson et al. [64] demonstrated the emergence of gentrification as mediated by residential migration and resulting changes in land rents. Still, others have shown the important dynamics that exist between land use activity, ecological services, and resilience, e.g., [65,66].

Researchers have also utilized ABMs to explore the effectiveness of tax interventions associated with land value on urban sprawl. Hosseinali et al. [67] found that a policy that increased the land cost by 50% (presumably by way of a tax) in areas planned for growth was the most successful tested policy in encouraging denser development patterns. Furthermore, Chen [68] demonstrated the effect of differential bargaining power on the effect of land taxes. He found that increasing exurban land costs through a development tax was ineffective in a "thinly traded" land market, because it increased buyer bargaining power such that they could capture a greater surplus from the exchange and were, thus, encouraged to purchase exurban land.

Our ABM simulates a regional housing market and resulting land use patterns to the extent necessary to explore the effect of property taxes on developer, speculator, and homeowner decision making. We built the model from scratch using Python (Version 3.10.8). The complete code and instructions to reproduce our results are publicly accessible in our code repository (https://github.com/DakotaBWalker/Speculation-and-Land-Tax-ABM) (accessed on 30August 2024). A cellular automata structure is used to represent a hypothetical geographic region as a grid of cells (parcels) with land use states and underlying environmental characteristics. Spatially heterogenous cell characteristics provide information from which agents choose actions that fulfill their motivations and maximize their utility, constrained by imperfect information and bounded rationality.

2.1. Environment

The hypothetical geographic region is represented by a 100 × 100 grid of cells, with each cell representing a plot of land with underlying locational, economic, and environmental characteristics. These cells take one of the following four discrete land use states: conserved land that is unavailable for agent purchase or development, undeveloped land that is also un-owned, land that is undeveloped but owned by an agent, and land that has been developed with housing. All cells falling into the lattermost state also contain values which describe the number of housing units in the cell, as well as the number of those units currently available for purchase. The environment also characterizes ecological value (EV)

as discrete states where conserved or undisturbed forest land is equal to 1, disturbed but undeveloped land is equal to 0, and developed land is equal to −1, unless restoration has occurred, in which case, *EV* is equal to 0.5.

Four environmental characteristics form the foundation of agent preferences. Based on the hedonic price principle, we assume that a particular property can be understood as a bundle of characteristics from which consumers derive utility based on their particular preferences and importance. Thus, variation in property prices can be used to reveal preferences for property attributes and associated willingness to pay. In the case of residential choice, price is understood as a function of physical, neighborhood, and locational attributes, e.g., [69,70]. For example, empirical research has demonstrated a strong association between real estate prices and surrounding forest amenities or natural beauty [71–73], neighborhood population or building density [74,75], proximity to a central business district [70,76], and property size [75,77,78].

The natural beauty (*NB*) matrix represents the hedonic value of a plot for environmental, aesthetic, or recreational purposes. Pre-specified forested cells are given the highest NB value of 1. A distance decay function is then used to gradually lower the NB value with increasing distance to a forested cell. *NB* values are also reduced by one sixth of the cell's density value (however, with a fixed *NB* min value of 0.3) to represent the typical loss of natural beauty in high-density urban locations. The proximity to the central business district (*CBD*) matrix represents each cell's proximity to the central cell, normalized to between 0 and 1. The housing density (*density*) matrix represents the development intensity of the surrounding cells, which is calculated with the sum of housing units in the surrounding 7×7 cell grid divided by the maximum value of surrounding development, assumed to be an average of two units per cell. The lot size (*lot_size*) matrix describes the acreage of each cell, which is assumed to increase with distance to the central cell (*dist*) as follows:

$$lot\_size = 0.075 \times (dist + 1) + 0.25 + \varepsilon \tag{1}$$

$$\varepsilon = N\left(o, \frac{lot\_size}{4}\right) \tag{2}$$

Initial lot sizes are specified, then given random variance (ε), with the standard deviation being proportional to the initial lot size. As such, we assume that lot sizes fall between approximately 0.325 acres at the center and approximately 6 acres at the outermost cell. *lot_size* affects agent preferences in the following two ways: first, it is used to calculate the total land tax burden, which directly affects agent willingness to pay; second, it is used to calculate the lot size per housing unit once development occurs, which affects homeowner housing preference.

Other environmental characteristics speak to the costs and profitability of land or housing ownership. Per acre land values are specified for each cell, representing the cost per acre of purchasing a particular plot of land devoid of improvements. Because our model does not endogenize price formation in the land market, we approximate land values based on homeowner preferences, neighborhood housing density, and total regional housing as follows:

$$LVPA = land\_pref \times \left(nbr\_housing + 1\right)^{1.4} \times total\_housing^{0.9} + LVmin \tag{3}$$

where *land_pref* is the average homeowner land preference (0,1) for each cell across 50 randomly sampled homeowners, *nbr_housing* is the sum of occupied housing units in the surrounding 13×13 neighborhood, *total_housing* is the sum of all occupied housing units across all cells, and *LVmin* is the specified minimum land value per acre. Upon calibrating the LV equation with the specified exponents, the approximation reproduces a characteristic power law distribution of land values across a region, as well as the land price appreciation that results from nearby or aggregate development (See Supplementary Materials; [79]).

The environment also includes separate tax rate matrices for improvement value (*IVT_rt*) and land value (*LVT_rt*). For the SQ tax scheme, both tax rate matrices are equal and uniform across space. For the LVT tax scheme, *LVT_rt* is spatially uniform, while

IVT_rt is homogenously equal to zero. This also applies to the ELVT scheme, however, with LVT_rt varying across space according to EV and EP. The formula for calculating LVT_rt under an ELVT regime is as follows:

$$LVT\_rt = \tau + E_{burden} \tag{4}$$

$$E_{burden} = \frac{EP(-EV)}{eco\_burden\_denom} \tag{5}$$

where τ is the baseline LVT_rt and *eco_burden_denom* is a scalar used to arbitrarily control the magnitude of divergence from the base rate at EP and EV extremes. Due to the tax rate variation that results from land use changes under ELVT, each cell also includes hypothetical tax rate values if each parcel is restored or developed to allow agents to anticipate the taxes associated with their actions.

The EP matrix was specified to represent one of the many ways a municipality might communicate the locations in which land stewardship is most important. We used a distance decay function to create a gradient of values from 0 to 3 as a function of distance to any conserved land. In practice, ecological potential could be a far more sophisticated description of ecological integrity, essential habitats, etc., or a more arbitrary description of, for example, neighborhoods in need of better stormwater management.

2.2. *Agents*

Agents represent land users of various types who can buy, sell, develop, and restore parcels of land. Agents spend money buying land, developing land, restoring land, and paying taxes. Agents gain money by way of an assumed fixed exogenous income (for homeowner agents) and land or housing sales. Agents are heterogeneous in their preferences, wealth, and altruism. They make decisions based on the expected benefits of different actions relative to their own motivations, constrained by their wealth. Decision making is implemented under imperfect information and bounded rationality, in which agents are given a randomly generated subset of all possible options and choose randomly from a smaller subset of those options that satisfy at least a given threshold of relative benefit. Agents interact indirectly, through sensing and reacting to the environment, which is impacted by other agents, as well as through a housing marketplace where purchase price is directed from the buyer to the seller.

"Homeowner" agents seek to buy housing that meets at least a given threshold of utility, constrained by their wealth. To model agent utility with respect to a housing unit, we use a modified multiplicative Cobb–Douglas function [80], which has been used extensively to model consumer choices by way of preferences and product characteristics [81] and is commonplace within the urban economic modeling of residential choices [63,82–85]. The function takes the form shown in Equation (6), where the utility that a particular homeowner (r) gains from a particular housing unit (h) is a function of three vector variables, each with a length of four (i.e., $m = 4$). One vector comprises objective cell characteristics (γ_h); one comprises the agent's preferred values for each characteristic (β_r); and the third vector comprises the preference weight (α_i) that the agent assigns to each cell characteristic.

$$U_{r,h} = \prod_{i=1}^{m} \left(1 - \left|\beta_{r,i} - \gamma_{h,i}\right|\right)^{\alpha_{r,i}} \tag{6}$$

Homeowner agents draw each preference weight ($\alpha_{r,i}$) by randomly sampling within a given range from a uniform probability distribution. Agent preference weights are then normalized such that they sum to 1, which allows for utility to be compared across agents. The objective cell characteristics and preferred values (also sampled at random from a given range) fall between 0 and 1. Thus, agent utilities are constrained to between 0 and 1, where a utility of 1 would occur only through a perfect match between the preferred and actual values for all four cell characteristics. For natural beauty (NB) and CBD proximity (CBD) cell characteristics, the idea value (β) is assumed to be the max value (i.e., 1) for all agents, which allows the equation to be simplified to the objective cell characteristic raised to the preference weight. For density and lot size cell characteristics, agents are given unique preferred values (*ideal_density* and *ideal_lot_size*), such that the same lot size may fall close to the preferred lot size of one agent, but may be far from ideal for another agent. Another way of representing the operationalized utility function is shown in Equations (7)–(9). This particular implementation of the Cobb–Douglas utility function most resembles that of Brown et al. [84].

$$U_{i,j} = NB_h{}^{\alpha_{NB,r}} \times d\_match_{r,h}{}^{\alpha_{density,r}} \times CBD_h{}^{\alpha_{CBD,r}} \times lot\_match_{r,h}{}^{\alpha_{lot\_size,r}} \quad (7)$$

$$d\_match_{r,h} = (1 - |ideal\_density_r - density_h|) \quad (8)$$

$$lot\_match_{r,h} = (1 - |ideal\_lot\_size_r - lot\_size_h|) \quad (9)$$

The utility of agent r for a particular housing unit h is then multiplied with their wealth (assumed to represent the maximum amount of money they would be willing to spend on a home) to determine their willingness to pay ($WTP_{r,h}$) for the unit. WTP is also affected by the present value of the unit's property taxes, given a homeowner future discount rate (*future_disc_rt_homeowner*), as well as the neighborhood vacancy rate, as follows:

$$WTP_{r,\text{h}} = U\_hu_{r,\text{h}} \times wealth_r \times (1 + vacancy\_multiplier) - present\_value\_tax \quad (10)$$

$$vacancy\_multiplier = (0.25 - vacancy\_rt_h) * 0.25 \quad (11)$$

Because the vacancy rate is fixed at between 0 and 0.5, *vacancy_multiplier* can reduce or increase WTP by as much as 6.25%.

Under the ELVT scheme, homeowners that already own a home have the option to restore their land. Restoring land refers to the act of investing a pre-determined sum (*restoration_cost*) per acre to change the land use attributes of a cell such that it then provides greater ecosystem services (EV changes from -1 to 0.5). The probability of choosing to restore (*p_restore*) for a homeowner (r) and owned parcel (h) is as follows:

$$P\_restore_{r,h} = altruism_r \times economic\ benefit_{r,h} \times neighborhood\ adoption_h \quad (12)$$

In which *altruism* (0,1) reflects the agent's own devotion to providing collective goods, *economic benefit* is the expected present value of savings on tax payments divided by income, and *neighborhood adoption* represents the percentage of plots in the surrounding 7×7 cell grid that have restored their land. The homeowner chooses to restore if their *P_restore* value is greater than a random number between 0 and 1 drawn from a uniform distribution.

Homeowners also have the option to sell their current home if they become overburdened by the cost of ownership or if the associated utility of other housing units on the market at or below their willingness to pay exceeds the utility that they derive from their current home by at least a given threshold (*pref_dif_to_sell*).

"Developer" agents seek to supply homeowners with housing by buying and developing land. They make decisions about where to buy land for development based on expected profit, represented by the percentage difference between the 75th percentile of the aggregate willingness to pay across a random sample of homeowners and the expected cost of developing. Costs include the present value of the tax burden after development, given developers' future discount rate (*future_disc_rt_developer*). Potential development sites are constrained to a random subset of plots (*rnd_off_limits*) to invoke imperfect information, as well as to locations in which the cost of development would not exceed their wealth and where the vacancy rate is less than 0.25. Upon calculating the expected profit for each potential development site, developers then choose randomly between the top x% (*rnd_choice_pct*) of plots with regard to their associated expected profit, invoking bounded rationality.

Upon buying land, developers determine how many units to build. Developers are each given unique values for square footage per unit, building costs per square foot, and the desired improvement value to land value ratio by drawing randomly with uniform probability from given parameter ranges (*unit_sf_range*, *build_cost_psf*_range, and *desired_IV_LV_ratio_range*, respectively). The agent (*d*) can then calculate how many units to build (*N_units*) on a plot (*h*) as follows:

$$N\_units_{i,j} = \frac{LC_{d,h}}{IC\_per\_unit_{d,h}} \times desired\_IV\_LV\_ratio_d \tag{13}$$

$$LC_{d,h} = LV_d \times \left(1 + \frac{LVT\_rt\_developed_h}{future\_disc\_rt_d}\right) \tag{14}$$

$$IC\ per\ unit_{d,h} = unit\_sf_d \times build\_cost\_psf_d \times \left(1 + \frac{IVT\_rt_h}{future\_disc\_rt_d}\right) \tag{15}$$

Under the ELVT tax scheme, developers may consider the profitability of building half the units they intend and restoring the other half of the site. They consider the half restoration if their personal altruism value is greater than a randomly drawn value between 0 and 1. If so, they estimate the cost per unit of developing half the units, given a reduced tax rate. If the cost per unit plus a profit premium of 15% remains lower than homeowner willingness to pay, they choose to conduct the half-restoration development.

Developers then estimate willingness to pay per unit by sampling 20 homeowner agents at random. If the cost per unit is less than or equal to the 75th percentile of homeowner willingness to pay, the development occurs. The price per unit is set somewhere between the cost per unit and the estimated willingness to pay, with the exact point between the two based on the neighborhood vacancy rate as follows:

$$price = cost\_per\_unit + \frac{WTP - cost\_per\_unit}{2} \times (1 + vacancy\_multiplier) \tag{16}$$

$$vacancy\_multiplier = (0.25 - vacancy\_rt) \times 4 \tag{17}$$

Therefore, we assume that a vacancy rate of zero would set the price at *WTP* and a vacancy rate of 0.5 would set the price at the *cost_per_unit*, which reflects how differential bargaining power between buyers and sellers affect price and how increasing supply can drive profits to zero [86]. Upon setting a price, the housing units are added to the market and made available for homeowners to buy.

"Speculator" agents seek to buy and sell land or housing units in such a way as to capitalize on land value appreciation that results from other agent actions. Their decision making is based on expected profit of buying, holding, then selling a plot, calculated as follows:

$$BC\_ratio = \frac{dLV}{LVT\_rt} \tag{18}$$

where *dLV* is the average percentage change in land value per year over the past 3 timesteps. If the maximum *BC_ratio* across all available plots is greater than 10%, the speculator chooses at random between the plots with the x% (*rnd_choice_pct*) highest *BC_ratio*. Speculators continue buying plots at each timestep, so long as the aforementioned profitability condition is met.

At each timestep, speculators sell one of their owned plots either randomly with a probability of 0.1 or if their wealth is less than or equal to zero. Under the ELVT tax scheme, speculators can also consider land restoration. For all owned unrestored land, they look at the profit potential of restoring the land (as a result of lower taxes), as follows:

$$P\_restore = \frac{\left(\frac{LV \times (LVT\_rt - LVT\_rt\_restored)}{future\_disc\_rt} - restoration\ cost\right)}{restoration\ cost} \tag{19}$$

2.3. Feedback

Several key feedback loops between agent actions and environment form the basis of recursivity in the model (Figure 1). With regard to preferences, development affects both density (positively) and natural beauty (negatively), which then affects the preferred locations at the next timestep. Newly restored cells also affect natural beauty (positively). Changes in development and locational preferences impact land values and the resulting cost of purchasing and developing, tax burdens, and returns for existing owners of land or housing. Fast-appreciating land value attracts speculators to the area, with whom developers must compete for access to desirable land and homeowners must compete for access to housing.

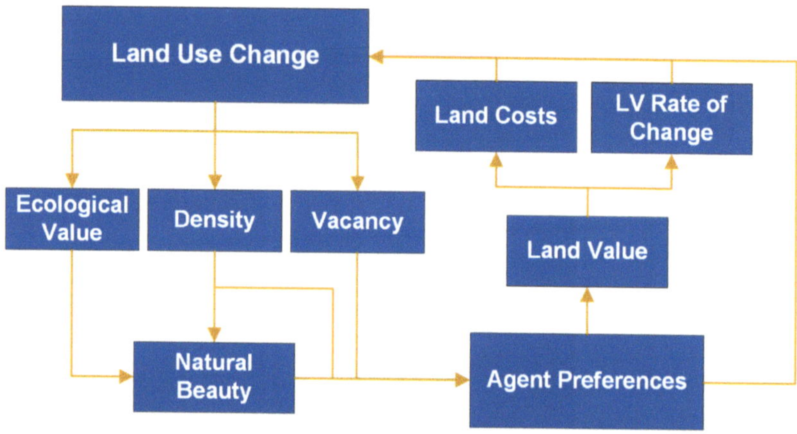

Figure 1. Feedback interactions among environmental and agent attributes.

2.4. Scheduling

The model is run by first initializing the starting number of agents, as well as the cell grid environment with all corresponding geographic characteristics. At each timestep, agents act by buying, selling, or restoring cells. After all agents act, the ancillary cell characteristics (e.g., density and vacancy) are then updated based on the changed land use or housing ownership. A pre-specified number of new agents are then added to the simulation before the timestep advances and the agents act again. The high-level process sequence is visualized in Figure 2.

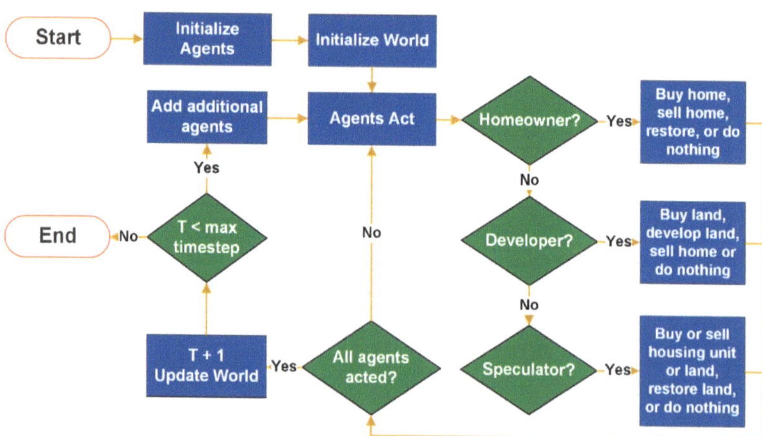

Figure 2. High-level model process sequence.

2.5. Parameters and Initialization

As a conceptual model primarily for the purpose of exploratory analysis, our model features limited empirical calibration. Like many other conceptual ABMs, e.g., [87], we hesitate to set some parameter values for empirical data without emulating other aspects of the context in which those empirical data are based. Rather, we seek to employ general rule-of-thumb values, where necessary, to form parameter ranges from which agents draw at random.

Table 1 summarizes the parameters and associated values or value ranges used for the model. The model begins by initializing a specified number of agents (*n_init_agents*), whose type is drawn at random from a weighted list (*ag_type_wt*), along with the agent's initial characteristics depending on the agent type. A notable assumption regards development costs and the preferred amount of capital investment. We assume a unit square footage range between 1000 and 2500 and a per square foot cost range from USD 100 to USD 200, which represents all project costs except for land costs [88–90]. Therefore, the average cost to build a unit (not including the land cost) is USD 225,000. We also assume that developers seek to build units such that the land cost represents between 12.25 and 20% of the total costs of development on average (*desired_IV_LV_ratio_range* = [4, 7]) [89,91–93].

We then initialize the environment's land use states and ecological values, which were randomly generated, but remained fixed for all simulation runs (Figure 3). From this, we derive ecological potential as a distance decaying value from conserved land, and natural beauty as a distance decaying value from any forested cell ($EV = 1$), with stronger values weighted toward those that are conserved. Finally, tax rates are chosen based on the total tax revenue of initial runs of the model, such that the tax revenues in the final timestep under the LVT and ELVT scenarios are roughly revenue neutral in relation to the status quo tax scenario.

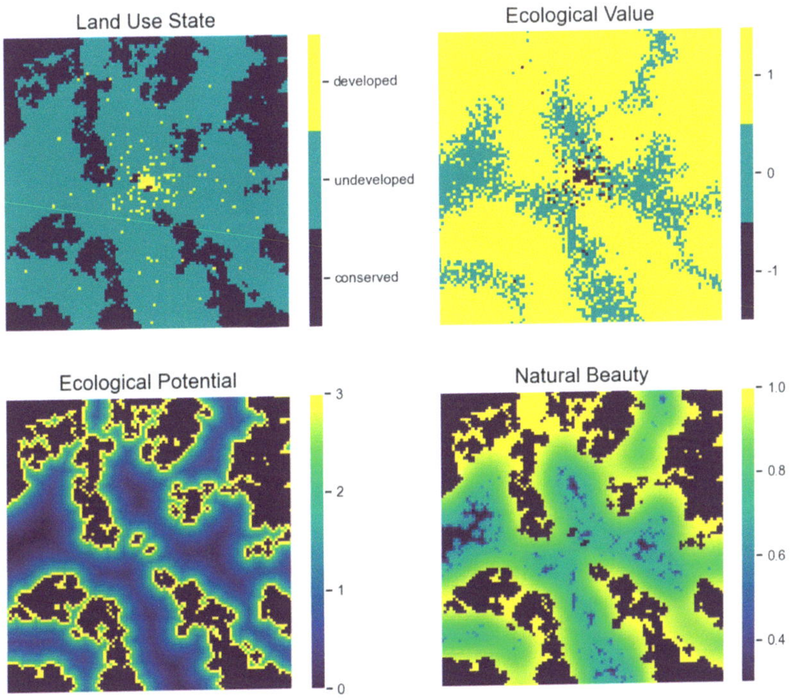

Figure 3. Initial spatial distribution of land use and ecological value, as well as resulting ecological potential and natural beauty.

Table 1. Model parameters under "baseline" configuration.

Parameter	Description	Baseline Configuration
max_timestep	Number of timesteps for which the model runs	20
P_initial_development	Used to scale the randomly generated initial housing development	0.2
N_init_agents	The number of agents initialized into the simulation at timestep 0	300
N_new_agents	The number of agents added to the simulation at each subsequent timestep	30
ag_type_wt	The weights used to condition randomly chosen agent type when each agent is initialized. Organized as [developer, homeowner, speculator]	[0.15, 0.80, 0.05]
ln(*starting_wealth_homeowner*)	The mean(std dev.) of the normally distributed values from which homeowner agents draw then exponentiate to derive starting wealth	13.25 (0.25)
ln(*starting_wealth_developer*)	The mean(std dev.) of the normally distributed values from which developer agents draw then exponentiate to derive starting wealth	16 (0.5)
ln(*starting_wealth_speculator*)	The mean(std dev.) of the normally distributed values from which speculator agents draw then exponentiate to derive starting wealth	15 (1)
rnd_off_limits	The probability from which each available land cell draws to determine whether it is not available for purchase (re-drawn at each timestep)	85%
rnd_choice_pct	The top percentage of cell location choices from which agents randomly choose	10%
altruism_homeowner_range	The range of uniform probability altruism values from which homeowners draw	[0.2, 1]
altruism_developer_range	The range of uniform probability altruism values from which developers draw	[0.1, 0.5]
future_disc_rt_homeowner	The rate at which homeowners discount potential income in future years	10%
future_disc_rt_developer	The rate at which developers discount potential income in future years	25%
future_disc_rt_speculator	The rate at which speculators discount potential income in future years	33%
pref_dif_to_sell	The minimum difference in utility between the most preferred on-market housing unit and the currently owned housing unit above which the homeowner sells their current property	0.15
NB_pref_range	The range of values from which homeowners sample uniformly to determine their natural beauty preference weighting	[0.01, 0.4]
CBD_pref_range	The range of values from which homeowners sample uniformly to determine their *CBD* preference weighting	[0.2, 0.6]
density_pref_range	The range of values from which homeowners sample uniformly to determine their *density_match* preference weighting	[0.2, 0.4]
lot_size_pref_range	The range of values from which homeowners sample uniformly to determine their *lot_size_match* preference weighting	[0.01, 0.2]
Ideal_density_range	The range of uniform probability *ideal_density* values from which developers draw	[0.1, 0.7]
Ideal_lot_size_range	The range of uniform probability *ideal_lot_size* values from which developers draw	[0.05, 1]
LVT_rt	The rate at which land value is taxed. Formatted as [SQ rate, LVT rate, ELVT rate] and chosen to approximate revenue neutrality	[0.035, 0.10, 0.10]

Table 1. *Cont.*

Parameter	Description	Baseline Configuration
IVT_rt	The rate at which improvement value is taxed. Formatted as [SQ rate, LVT rate, ELVT rate]	[0.035, 0, 0]
eco_burden_denom	The denominator of the eco burden calculation. Used to arbitrarily control the magnitude of divergence from the base rate at EP and EV extremes	50
unit_sf_range	The range of uniform probability *unit_sf* values from which developers draw to calculate their *cost_per_unit*	[1000, 2500]
build_cost_psf_range	The range of uniform probability *build_cost_psf* values from which developers draw to calculate their *cost_per_unit*	[100, 200]
desired_IV_LV_ratio_range	The range of uniform probability *desired_IV_LV_ratio* values from which developers draw to calculate how many units to build	[4, 7]
restoration_cost	The cost to restore one acre of land	USD 5000

2.6. Experiment Design

We perform four experiments to assess the relative outcomes under different parameter assumptions. Table 2 summarizes the parameters used in each experiment, with all other parameters being fixed, using the values specified in Table 1. In experiment 1, we model three land use scenarios via changes in homeowner preferences, as follows: a baseline scenario represents middle-of-the-road urban development; a sprawl scenario represents a stronger preference for larger lot sizes, low-density, and high-natural-beauty plots; and a high-density scenario represents a stronger preference for locations in close proximity to CBD and with a high density.

Table 2. Unique parameter values used in model experimentation.

Experiment 1: Land Use Patterns			
Parameter	**Baseline Scenario**	**Sprawl Scenario**	**High-Density Scenario**
NB_pref_range	[0.01, 0.4]	[0.3, 0.5]	[0.01, 0.2]
CBD_pref_range	[0.2, 0.6]	[0.1, 0.3]	[0.5, 0.7]
density_pref_range	[0.2, 0.4]	[0.2, 0.5]	[0.4, 0.6]
lot_size_pref_range	[0.01, 0.2]	[0.2, 0.4]	[0.05, 0.1]
Ideal_density_range	[0.2, 0.5]	[0.01, 0.3]	[0.3, 0.8]
Ideal_lot_size_range	[0.05, 0.75]	[0.25, 1.0]	[0.05, 0.15]
LVT_rt ([SQ, LVT, ELVT])	[0.035, 0.1, 0.1]	[0.035, 0.135, 0.135]	[0.035, 0.1, 0.1]
IVT_rt ([SQ, LVT, ELVT])	[0.035, 0, 0]	[0.035, 0, 0]	[0.035, 0, 0]
Experiment 2: Speculator Count Sensitivity Sweep			
Parameter	Min	Max	Increment
ag_type_wt_speculator	0.01	0.29	0.07
N_init_agents	300	420	30
N_new_agents	30	42	3
Experiment 3: ELVT Magnitude Sensitivity Sweep			
Parameter	Min	Max	Increment
eco_burden_denom	20	140	30
Experiment 4: LVT Rate Sensitivity Sweep			
Parameter	Min	Max	Increment
LVT_rt	0.05	0.25	0.05

Note. All experiments use the baseline parameter settings (Table 1) except where specified under each experiment.

In experiment 2, we assess the impact of a growing speculator population on sprawl and housing availability. We perform this via a sweep on the probability of choosing a speculator agent (*ag_type_wt*) when randomly assigning initial and new agents. We also increase the total number of agents so as to keep the homeowner and developer counts relatively constant for each iteration. In experiment 3, we assess the impact of ELVT under increasing divergence from LVT rates at the extremes (i.e., the magnitude of the subsidy and penalty) via a sweep of *eco_burden_denom* values. In experiment 4, we assess the system outcomes under increasing LVT rates.

Model stochasticity arises from the various randomly drawn elements of the model (e.g., homeowner preferences, available land at each timestep, and agent cell choice), as well as through the path dependence from the effect of prior agent actions on environmental conditions. To account for this, we utilized the Monte Carlo average and standard error over 20 runs of each unique initialization to converge on a robust estimate of each system metric. Therefore, for the multi land use tax comparison, a total of nine unique initializations yielded 180 runs. For each sensitivity analysis, a total of five unique initializations yielded 100 runs.

2.7. Analysis

Given the limited calibration and conceptual framing of the model, we adhere to an exploratory analytical approach, focusing on broad system dynamics and emergent patterns to build intuition and inspire future inquiry, rather than attempting to forecast specific regional land use outcomes or provide a means of decision support for policymakers [94]. More specifically, we direct attention to the relative changes among both tax interventions (as compared to status quo) pertaining to the following questions: (1) To what degree do changes in homeowner preferences lead to differing land development patterns? (2) Can differing tax schemes reinforce or redirect land use patterns and emergent outcomes for three common land use patterns as enacted by changes in homeowner preferences? (3) How does the magnitude of speculator interest affect such emergent patterns, and can LVT and ELVT dampen their effect? (4) How does ELVT differ from LVT under various tax scheme operationalizations? Table 3 summarizes the specific metrics used to investigate our research questions pertaining to urban densification, housing availability, ecological integrity, and land use incentives.

Table 3. Measured land use, economic, and ecological outcomes.

Category	Metric	Description
Urban Densification	Avg HU lot size (acres)	The average lot size per housing unit for housing which is occupied by homeowners
	High-density housing rate	The percentage of homes which occupy cells with a density above 0.4
	Largest patch size	The largest number of contiguous developed cells across the region
	Housing within urban boundary	The percent of housing units that are within 20 cells from the center
Housing Availability	Total housing	The sum of all housing units in the region
	Avg HU price	Average price for one housing unit
	Avg homeowner without home	Percent of homeowners who either have not found a home or have recently sold their home to move
	Avg vacancy rate	The average rate of vacancy across all cells and all timesteps
Ecological Integrity	Ecological value change	The percentage change in the sum of ecological value relative to initial ecological value
	Urban ecological value	The sum of regional ecological value

Table 3. *Cont.*

Category	Metric	Description
Land Use Incentives	Avg tax burden by home type	The average tax burden for homeowners with high-density housing (housing_unit_lotsize < 0.125), mid-density housing (0.125 < housing_unit_lotsize < 0.25), and low-density housing (housing_unit_lotsize > 0.25)
	Avg tax burden by agent type	The average tax burden of homeowner, developer, and speculator agents
	Avg change in wealth	The average change in wealth for each agent type from timestep 0 to the final timestep

Note. Metric is calculated for values of the final timestep unless otherwise stated in the description.

3. Results

3.1. Experiment 1

Under the conventional SQ tax scheme, our model broadly reproduced archetypal development patterns via changes in agent preference weightings (Figure 4). The "baseline" land use scenario demonstrated a central urban core with a downward-sloping gradient of housing density extending out to scattered suburban arms of development. The "sprawl" land use scenario demonstrated low-density decentralized housing clusters with minimal urban core housing. The "high-density" land use scenario demonstrated a tight core of high-density housing with relatively few suburban dwellings beyond it.

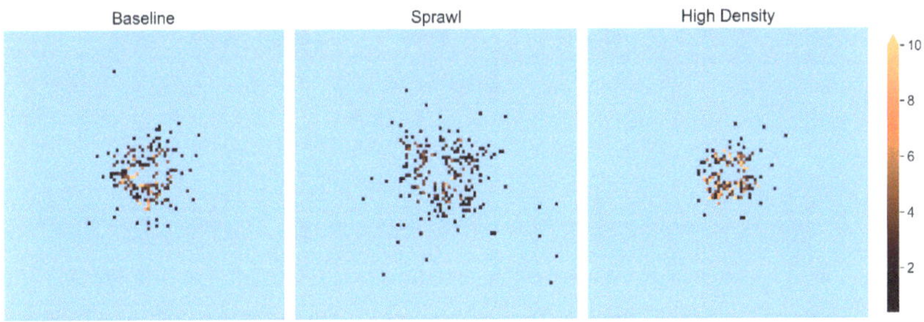

Figure 4. Housing units per acre for housing created by "developer" agents from single simulation run. *Note.* Pre-specified initial development (fixed across all land use scenarios) not shown in the figure creates a small urban core and scatter outer development for all simulations.

For the baseline scenario, the initial development tended toward the center. The broad ranges of homeowner preferences seemed to reinforce project costs as the primary driver of location. As initial development increased land value in the center, developers moved just outside to capture lower land costs with similar homeowner willingness to pay. However, as land values increased further, developers could increase the number of units built, and, therefore, profitability again favored the urban core. This dynamic continued until the loss of natural beauty or the increase in density lowered homeowner preference for the urban core, at which point, broader exurban development ensued.

In the spawl scenario, development locations heavily favored locations with a high natural beauty, initially in close proximity to the urban core. However, such development increased density and reduced natural beauty, such that subsequent development moved farther from the center toward higher-natural-beauty locations.

In the high-density scenario, homeowner preferences steeply declined with distance from initial development because of their high *ideal_density* preference. As the central core was further developed, homeowner preference for these locations increased, resulting in a positive feedback loop of tight urban development. Developments moved outward only when all more central land was unavailable.

Figure 5 visualizes the average and standard error for each model outcome under each land use scenario and tax scheme. Relative to the SQ scheme, both LVT and ELVT significantly increased the total housing count and were associated with higher-density dwellings on average. Furthermore, exurban housing development was less frequent under both tax interventions (Figure 6). Both tax interventions were also associated with a significant decrease in housing prices on average (Figure 5). In the baseline scenario, the average housing unit price decreased by 27.8% and 30.8% under LVT and ELVT, respectively. In the sprawl scenario, the average housing price decreased by 18.9% and 2.5% under LVT and ELVT, respectively. In the density scenario, the average housing price decreased by 36.8% and 37.0% under LVT and ELVT, respectively. We attribute this to the increase in development spurred by the tax interventions, which not only allowed developers to spread development costs over more units, but also acted to increase the vacancy rate and, therefore, push prices down. As a result, far more homeowners found houses in this model, with an average of 64% of homeowners unhoused in the status quo scenario as compared to averages of 27.5% and 28.1% under LVT and ELVT, respectively, averaged over all land use scenarios.

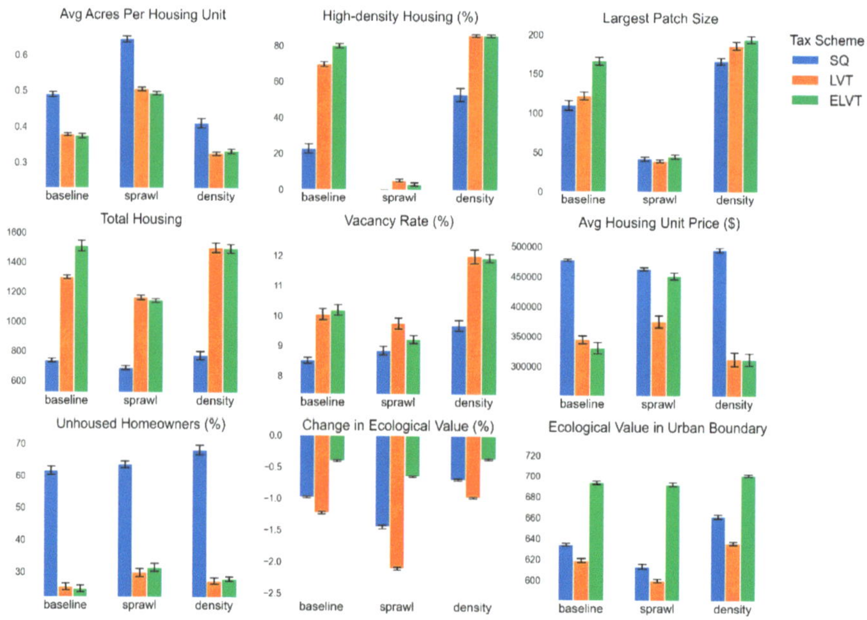

Figure 5. Urban densification, housing availability, and ecological integrity metrics for each land use scenario and tax scheme.

Both tax interventions also had the effect of greatly decreasing the profitability of speculative investment (Figure 7). Relative to SQ, the average tax burden of speculators increased by 512% and 580% under LVT and ELVT, respectively. As a result, the change in speculator wealth, on average, decreased from 6.7% under SQ to −1.8% and −1.0% under LVT and ELT, respectively. Figure 7 also shows a significant decrease in homeowner wealth, despite a small increase in tax burden. However, coupled with the finding that the average housing prices decreased precipitously, this may speak to a broader alignment of housing with use value and, therefore, becoming less of an investment asset amongst homeowners. Finally, despite a large increase in tax burden, developer's wealth, on average, increased under both tax interventions. This, too, may encourage developers to favor high-volume, capital-intensive investment in housing above leapfrog development strategies.

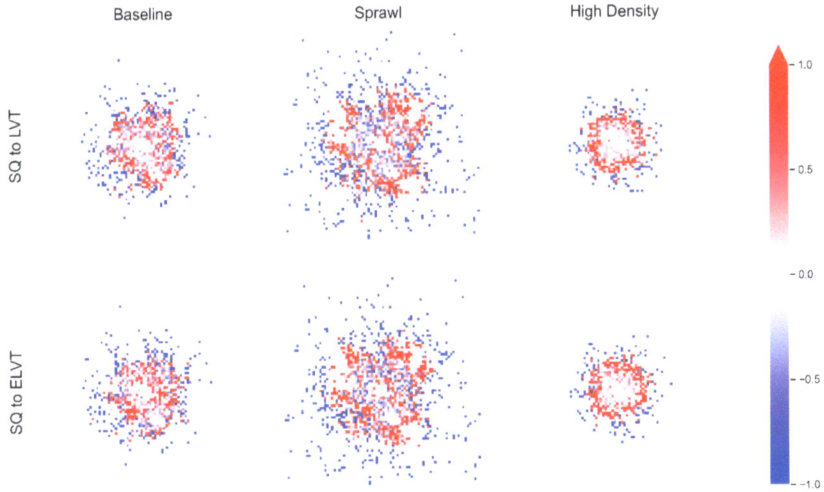

Figure 6. Percentage change in Monte Carlo average likelihood of development for LVT and ELVT relative to SQ tax scheme.

Figure 7. Changes in wealth and tax burden by agent and house type for the baseline land use scenario.

Some notable differences emerged when comparing the effects of the LVT and ELVT schemes. Broadly, ELVT acted to increase urban infill densities, likely as a result of higher urban natural beauty and subsequent homeowner preference. Also, where LVT tended to exacerbate regional ecological value loss relative to SQ, ELVT greatly decreased EV loss to almost neutral levels.

3.2. Experiment 2: Speculator Count Sensitivity Analysis

The results of experiment 2 demonstrate the significant sway speculator agents had in the model (Figure 8). As the number of speculators increased, they tended to decrease the availability of high-value land and housing, which forced broader, low-density exurban development. Moreover, given the increased competition for available housing, housing prices tended upward, as did the percentage of agents unable to find housing. Even though the LVT and ELVT tax interventions increased the average tax burden of speculators, those tax interventions also led to faster-appreciating land values as a result of more development. As such, speculators were still able to profit from the region's land and housing market.

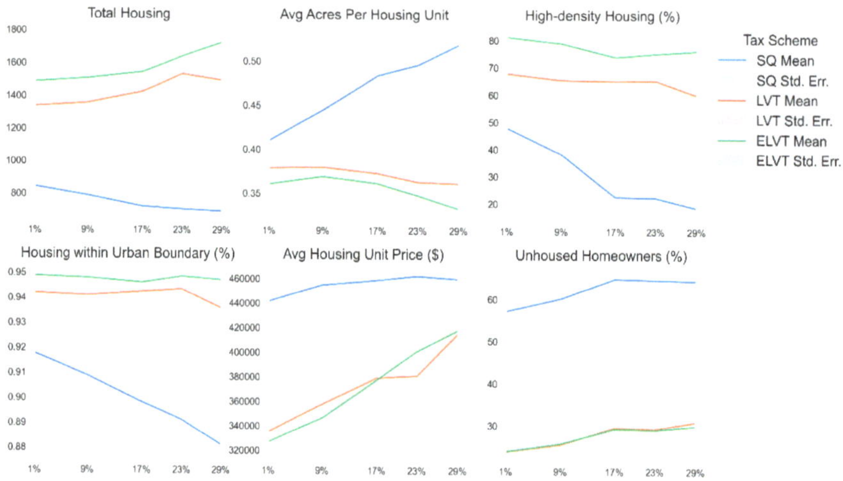

Figure 8. Select urban densification, housing availability, and ecological integrity metrics for sweep of percentage of agents of speculator type.

3.3. Experiment 3: ELVT Magnitude Sensitivity Sweep

The results of experiment 3 signify the need for greater attention directed toward the decision-making criteria of adopting restoration practices. A decrease in the *eco_burden_denom* value increased the divergence between *ELVT_rt* and *LVT_rt* (i.e., increased the magnitude of the increase or decrease in tax burden in high-ecological-potential areas), which should have led to a broader adoption of land restoration. Yet, Figure 9 shows a slight decrease in the number of restored parcels between the lowest and highest *eco_burden_denom* values. This may have been the result of a higher urban density rate, which meant less homeowners had the minimum lot size for which restoration was possible. Nonetheless, the higher tax rate on ecologically important land acted to decrease the average acres per housing unit and increase the percentage of housing with the urban boundary.

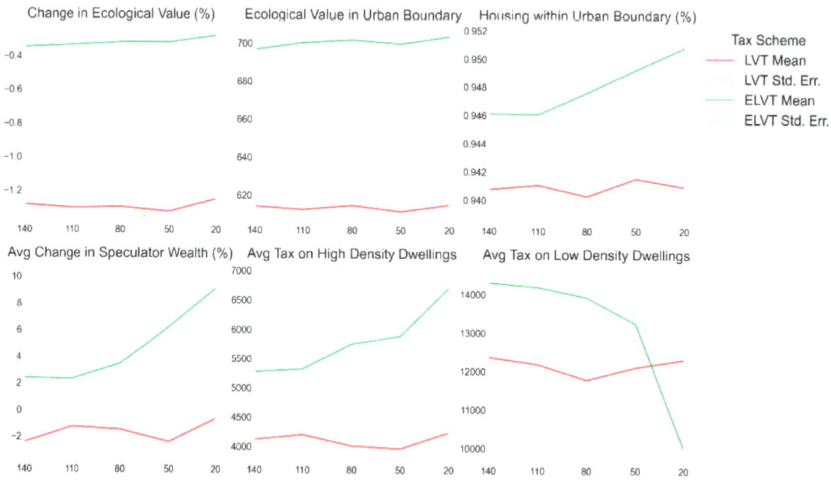

Figure 9. Select urban densification, housing availability, and ecological integrity metrics for sweep of *eco_burden_denom* values.

The results also show a significant decrease in the speculator tax burden that resulted from decreasing the *eco_burden_denom* values. Land-owning speculators were again able to capture a large portion of land rents so long as they restored the land parcel to receive a lower tax burden. Such restorations would increase the surrounding natural beauty and, therefore, increase the land value for the benefit of speculative investment. This, coupled with the finding that the average tax burden on low-density housing decreased and the average tax burden on high-density housing increased, suggests that, as the *ELVT*_rt diverges further from the *LVT_rt*, some of the beneficial effects of land value taxation may diminish.

3.4. Experiment 4: LVT Rate Sensitivity Sweep

In experiment 1, we set the LVT and ELVT tax rates such that resulting tax revenue in the final timestep was relatively close to revenue neutral, as compared to SQ. In experiment 4, we showed that land value taxation can further decrease urban sprawl and speculator interest as tax rates increase. We found that, as *LVT_rt* increased, many of the beneficial effects of the revenue-neutral tax intervention were magnified. As shown in Figure 10, the total housing and housing density continued to increase while housing prices continued to fall. Moreover, where the revenue-neutral tax intervention did not directly address the urban boundary size, a higher *LVT_rt* significantly decreased the largest patch size and increased the percentage of housing within the urban boundary (Figure 10). Taken together, these results suggest that, with a high enough rate, land value taxation can lead to dense, walkable cities that exert less pressure on surrounding exurban land.

Figure 10. Select urban densification, housing availability, and ecological integrity metrics for sweep of LVT_rt.

4. Discussion

In this paper, we presented a novel policy tool aimed at buffering the problematic trajectories imparted by rentier capitalism and resulting land use tendencies. The expanded land value tax scheme not only limits the financial viability of land speculation, but also provides a mechanism by which communities can price in the social costs of ecological degradation and subsidize land stewardship. With this conceptual agent-based model, we were able to conduct exploratory analyses in order to build our understanding of the broad changes in outcomes and land use patterns that result from the LVT and ELVT tax interventions.

The results of this study show general alignment with many of the theoretically and empirically demonstrated effects of land value taxation on land development, including higher housing development rates and urban housing capital density [42,95], decreased urban sprawl [37,42], and lower housing unit prices [42,43]. Under a revenue-neutral tax shift (experiment 1), both LVT and ELVT acted to increase the capital intensity of housing development, thereby increasing density and decreasing housing prices. Furthermore, speculation was taxed heavily and became a profit-losing endeavor. However, under the revenue-neutral scenario, the location of developments was still more strongly mediated by household locational preferences than costs. Given a preference for low or medium density, urban sprawl persisted. As we increased the rate at which land value was taxed (experiment 4), homeowners favored highly land-efficient developments in the urban core, more so than any exurban lands they may have favored at a lower tax rate.

We also showed that speculators exert significant sway in land patterns and housing prices (experiment 2). Under the SQ tax scheme, a higher speculator count decreased the total housing and the percentage of housing within the urban boundary, thereby increasing the average acres per housing unit. However, under both the LVT and ELVT schemes, all such metrics stayed relatively consistent as the speculator count increased. While at low speculator count, both tax interventions resulted in a negative wealth change for speculators, as the count increased, wealth change as well as housing prices trended upward.

Our findings also suggested many beneficial effects of ELVT. The proposed tax intervention broadly acted to increased urban housing densities, increase urban greenspace provisions, and greatly decrease the loss of ecological value across the hypothetical region. However, the decision-making criteria for the adoption of land restoration did not show much response to a growing subsidy for doing so. We partially attribute this to homeowners having less means for restoration as housing densifies. Finally, we found that the disincentive to speculate on land value under LVT was diminished under ELVT as the subsidy of land restoration increased; speculators were again able to profit off land rents, so long as they restored the land.

One cautionary result from the model was the faster appreciation of land values under both the LVT and ELVT tax scenarios due to the increased development and density they spurred. Despite the expectation that much of this appreciation would be taxed away and, therefore, attract less price appreciation due to speculative interest, a higher property tax burden can still displace existing residents [96]. This is particularly worrisome under the ELVT scheme given the well-documented evidence of green gentrification dynamics [32]. However, the degree to which gentrification would be dampened were speculation eliminated remains a question. Furthermore, the validity of this result remains questionable, given our proxy land value estimation; further examination is needed with the incorporation of the many drivers of upward or downward pressure on land values, including vacancy, commercial and rental interest, buyer and seller bargaining power, and speculator demand.

By establishing this proof-of-concept study, we hope to provide direction and inspiration for future empirically grounded inquiries. Such empirical data would aim to calibrate this model to a particular region according to local consumer preferences, developer decision-making behavior, land stewardship practices and adoption likelihood, and land values and development costs, as well as more detailed estimations of locally defined ecologically important land and the effects of small-scale land stewardship on ecosystem service provisions. Such local calibrations would also present an opportunity to test how an ELVT policy would interact with an already complex legal framework of land use and property rights.

Furthermore, future research could improve the model's realism by including a wider array of land use decision criteria and actor types. For example, the inclusion of existing landowners may help to shed light on willingness to sell, tax capitalization in land, and farmland conversion dynamics [62]. A primary feature of land value taxation is to increase the cost of holding urban land idle; therefore, this study's omission of pressure on existing vacant landowners missed an important aspect of the story. Furthermore, the assumption of one central business district was overly simplistic, and may have decreased sprawl potential as compared to when commercial areas and jobs follow suburbanites [84]. Likewise, the redevelopment of land was not incorporated into the model, which may have unnecessarily locked neighborhoods into a low-density form when redevelopment would have increased the housing capacity in desirable locations. Finally, a comparison of the impacts of the tax scheme vs. other parameter variables (or other policy approaches) can be utilized to shed light on the relative importance of the tax intervention. One example of this is Bell et al. [87], who used a Random Forest Classifier model to describe the relative importance of various parameters for particular outcomes.

While the implementation of the proposed tax reform is not recommended until it can be tested in the context of the locality, such local calibration requires detailed social, economic, and ecological accounting of the region, which can be the focus of local or state policy. As an important proxy for the capitalization of public investment in nearby parcels of land, assessing land value, separate from a parcel's improvement value, can shed light on the potential tax revenue, distributional, and equity impacts of a switch to land value taxation. Municipalities can instruct property tax assessors to list improvement value and land value separately, even if the tax structure remains the same. While accurate land value estimation is often noted as a hindrance to the implementation of a LVT, recently the use of mass appraisal techniques has drastically streamlined the process, e.g., [97,98]. Likewise, given the significant impact of speculative investment on sprawl, municipalities might consolidate public information on property purchases and local investment behaviors to assess the ubiquity of speculation-based land purchases. This, in tandem with land value data, can help to identify the localized effects of speculative purchases on housing price appreciation and gentrification. Finally, states can reform legislation to allow municipalities to implement a differential rate of taxation on improvement value and land value (a "split-rate tax") if they so choose. This has long been the case in Pennsylvania, which now has 20 jurisdictions using a split-rate tax model [36] and is currently being considered in Minnesota [99].

Supplementary Materials: The following supporting information can be downloaded at: https://www.mdpi.com/article/10.3390/land13111795/s1, Figure S1: All Spatial Variables for One Run under SQ Tax Scheme and "Baseline" LU configuration; Figure S2: All Spatial Variables for One Run under LVT Tax Scheme and "Baseline" LU configuration; Figure S3: All Spatial Variables for One Run under ELVT Tax Scheme and "Baseline" LU configuration.

Author Contributions: Conceptualization, D.B.W. and J.F.; methodology, D.B.W. and A.M.; software, D.B.W. and A.M.; validation, D.B.W. and A.M.; formal analysis, D.B.W. and A.M.; investigation, D.B.W. and A.M.; resources, D.B.W. and A.M.; data Curation, D.B.W. and A.M.; writing—original draft preparation, D.B.W.; writing—review and editing, D.B.W., J.F., D.R. and T.R.; visualization, D.B.W. and A.M.; supervision, J.F., D.R. and T.R.; project administration, D.B.W.; funding acquisition, J.F. All authors have read and agreed to the published version of the manuscript.

Funding: This research was supported by the Leadership for the Ecozoic initiative at McGill University and the University of Vermont.

Data Availability Statement: The Python source code for this agent-based model is publicly available in our GitHub code repository (https://github.com/DakotaBWalker/Speculation-and-Land-Tax-ABM) (accessed on 30August 2024).

Conflicts of Interest: The authors declare no conflicts of interest.

Notes

1 For a broader discussion on the unequal distribution of the benefits of urban greenspace, see Juntti and Ozsezer-Kurnuc [34].

References

1. Vogler, J.B.; Vukomanovic, J. Trends in United States Human Footprint Revealed by New Spatial Metrics of Urbanization and Per Capita Land Change. *Sustainability* **2021**, *13*, 12852. [CrossRef]
2. Theobald, D.M. Landscape Patterns of Exurban Growth in the USA from 1980 to 2020. *Ecology and Society* **2005**, *10*, 32. [CrossRef]
3. Rafferty, J. The Problem of Urban Sprawl. Saving Earth | Encyclopedia Britannica. 2019. Available online: https://www.britannica.com/explore/savingearth/urban-sprawl (accessed on 20 March 2023).
4. Bucholtz, S.; Molfino, E.; Kolko, J. *The Urbanization Perceptions Small Area Index: An Application of Machine Learning and Small Area Estimation to Household Survey Data*; US Department of Housing and Urban Development: Washington, DC, USA, 2020.
5. Mazumdar, S.; Learnihan, V.; Cochrane, T.; Davey, R. The Built Environment and Social Capital: A Systematic Review. *Environ. Behav.* **2018**, *50*, 119–158. [CrossRef]
6. Mouratidis, K. Is Compact City Livable? The Impact of Compact versus Sprawled Neighbourhoods on Neighbourhood Satisfaction. *Urban Stud.* **2018**, *55*, 2408–2430. [CrossRef]
7. Zhao, Z.; Kaestner, R. Effects of Urban Sprawl on Obesity. *J. Health Econ.* **2010**, *29*, 779–787. [CrossRef] [PubMed]
8. Hamidi, S.; Ewing, R.; Tatalovich, Z.; Grace, J.B.; Berrigan, D. Associations between Urban Sprawl and Life Expectancy in the United States. *Int. J. Environ. Res. Public Health* **2018**, *15*, 861. [CrossRef]
9. Gaffney, M. Containment Policies for Urban Sprawl. In *Approaches to the Study of Urbanization*; Government Research Center, University of Kansas: Lawrence, KS, USA, 1964; pp. 115–133.
10. Bento, A.M.; Franco, S.F.; Kaffine, D. Is There a Double-Dividend from Anti-Sprawl Policies? *J. Environ. Econ. Manag.* **2011**, *61*, 135–152. [CrossRef]
11. Simkin, R.D.; Seto, K.C.; McDonald, R.I.; Jetz, W. Biodiversity Impacts and Conservation Implications of Urban Land Expansion Projected to 2050. *Proc. Natl. Acad. Sci. USA* **2022**, *119*, e2117297119. [CrossRef]
12. Groffman, P.M.; Cavender-Bares, J.; Bettez, N.D.; Grove, J.M.; Hall, S.J.; Heffernan, J.B.; Hobbie, S.E.; Larson, K.L.; Morse, J.L.; Neill, C.; et al. Ecological Homogenization of Urban USA. *Front. Ecol. Environ.* **2014**, *12*, 74–81. [CrossRef]
13. Whyte, W.H. *The Exploding Metropolis*; University of California Press: Oakland, CA, USA, 1993; ISBN 978-0-520-08090-4.
14. Gielen, E.; Riutort-Mayol, G.; Miralles i Garcia, J.L.; Palencia Jiménez, J.S. Cost Assessment of Urban Sprawl on Municipal Services Using Hierarchical Regression. *Environ. Plan. B Urban Anal. City Sci.* **2021**, *48*, 280–297. [CrossRef]
15. Welch, P. *Development Patterns and Municipal Finances: An Analysis of Sprawl and Spending in 82 U.S. Cities*; University of North Carolina: Chapel Hill, NC, USA, 2017.
16. Ewing, R.; Rong, F. The Impact of Urban Form on U.S. Residential Energy Use. *Hous. Policy Debate* **2008**, *19*, 1–30. [CrossRef]
17. Zhang, L.; Hong, J.; Nasri, A.; Shen, Q. How Built Environment Affects Travel Behavior: A Comparative Analysis of the Connections between Land Use and Vehicle Miles Traveled in US Cities. *J. Transp. Land Use* **2012**, *5*, 40–52. [CrossRef]
18. Jones, C.; Kammen, D.M. Spatial Distribution of U.S. Household Carbon Footprints Reveals Suburbanization Undermines Greenhouse Gas Benefits of Urban Population Density. *Environ. Sci. Technol.* **2014**, *48*, 895–902. [CrossRef] [PubMed]
19. Bengston, D.N.; Fletcher, J.O.; Nelson, K.C. Public Policies for Managing Urban Growth and Protecting Open Space: Policy Instruments and Lessons Learned in the United States. *Landsc. Urban Plan.* **2004**, *69*, 271–286. [CrossRef]
20. Xie, Y.; Hunter, M.; Sorensen, A.; Nogeire-McRae, T.; Murphy, R.; Suraci, J.P.; Lischka, S.; Lark, T.J. U.S. Farmland under Threat of Urbanization: Future Development Scenarios to 2040. *Land* **2023**, *12*, 574. [CrossRef]
21. Milesi, C.; Running, S.W.; Elvidge, C.D.; Dietz, J.B.; Tuttle, B.T.; Nemani, R.R. Mapping and Modeling the Biogeochemical Cycling of Turf Grasses in the United States. *Environ. Manag.* **2005**, *36*, 426–438. [CrossRef]
22. Dempsey, J.A.; Plantinga, A.J. How Well Do Urban Growth Boundaries Contain Development? Results for Oregon Using a Difference-in-Difference Estimator. *Reg. Sci. Urban Econ.* **2013**, *43*, 996–1007. [CrossRef]
23. Woo, M.; Guldmann, J.-M. Urban Containment Policies and Urban Growth. *Int. J. Urban Sci.* **2014**, *18*, 309–326. [CrossRef]
24. Woo, M.; Guldmann, J.-M. Impacts of Urban Containment Policies on the Spatial Structure of US Metropolitan Areas. *Urban Stud.* **2011**, *48*, 3511–3536. [CrossRef]
25. Dawkins, C.J.; Nelson, A.C. Urban Containment Policies and Housing Prices: An International Comparison with Implications for Future Research. *Land Use Policy* **2002**, *19*, 1–12. [CrossRef]
26. Burge, G.S.; Trosper, T.L.; Nelson, A.C.; Juergensmeyer, J.C.; Nicholas, J.C. Can Development Impact Fees Help Mitigate Urban Sprawl? *J. Am. Plan. Assoc.* **2013**, *79*, 235–248. [CrossRef]
27. Foldvary, F.E.; Minola, L.A. The Taxation of Land Value as the Means towards Optimal Urban Development and the Extirpation of Excessive Economic Inequality. *Land Use Policy* **2017**, *69*, 331–337. [CrossRef]
28. Clawson, M. Urban Sprawl and Speculation in Suburban Land. *Land Econ.* **1962**, *38*, 99–111. [CrossRef]
29. Ricardo, D. *On the Principles of Political Economy and Taxation*, 1st ed.; Murray: London, UK, 1817.
30. George, H. *Progress and Poverty: An Inquiry into the Cause of Industrial Depressions, and of Increase of Want with Increase of Wealth: The Remedy*; D. Appleton: New York, NY, USA, 1879.
31. Stratford, B. The Threat of Rent Extraction in a Resource-Constrained Future. *Ecol. Econ.* **2020**, *169*, 106524. [CrossRef]
32. Bockarjova, M.; Botzen, W.J.W.; van Schie, M.H.; Koetse, M.J. Property Price Effects of Green Interventions in Cities: A Meta-Analysis and Implications for Gentrification. *Environ. Sci. Policy* **2020**, *112*, 293–304. [CrossRef]
33. Yazar, M.; Hestad, D.; Mangalagiu, D.; Saysel, A.K.; Ma, Y.; Thornton, T.F. From Urban Sustainability Transformations to Green Gentrification: Urban Renewal in Gaziosmanpaşa, Istanbul. *Clim. Chang.* **2020**, *160*, 637–653. [CrossRef]

34. Juntti, M.; Ozsezer-Kurnuc, S. Factors Influencing the Realisation of the Social Impact of Urban Nature in Inner-City Environments: A Systematic Review of Complex Evidence. *Ecol. Econ.* **2023**, *211*, 107872. [CrossRef]
35. Batt, H.W. Development and Wealth: A Georgist Perspective. *Am. J. Econ. Sociol.* **2012**, *71*, 1004–1046. [CrossRef]
36. Masterson, V. What Is Land Value Tax and Could It Fix the Housing Crisis? Available online: https://www.weforum.org/agenda/2022/03/land-value-tax-housing-crisis/ (accessed on 21 February 2023).
37. Cho, S.-H.; Gyu Kim, S.; Roberts, R.K. Measuring the Effects of a Land Value Tax on Land Development. *Appl. Spat. Anal.* **2011**, *4*, 45–64. [CrossRef]
38. Junge, J.R.; Levinson, D. Financing Transportation with Land Value Taxes: Effects on Development Intensity. *J. Transp. Land Use* **2012**, *5*, 49–63. [CrossRef]
39. Banzhaf, H.S.; Lavery, N. Can the Land Tax Help Curb Urban Sprawl? Evidence from Growth Patterns in Pennsylvania. *J. Urban Econ.* **2010**, *67*, 169–179. [CrossRef]
40. Kalkuhl, M.; Edenhofer, O. Ramsey Meets Thünen: The Impact of Land Taxes on Economic Development and Land Conservation. *Int. Tax Public Financ.* **2017**, *24*, 350–380. [CrossRef]
41. Lafuite, A.-S.; Denise, G.; Loreau, M. Sustainable Land-Use Management under Biodiversity Lag Effects. *Ecol. Econ.* **2018**, *154*, 272–281. [CrossRef] [PubMed]
42. Choi, K.-W.; Sjoquist, D.L. Economic and Spatial Effects of Land Value Taxation in an Urban Area: An Urban Computable General Equilibrium Approach. *Land Econ.* **2015**, *91*, 536–555. [CrossRef]
43. DiMasi, J.A. The Effects of Site Value Taxation in an Urban Area: A General Equilibrium Computational Approach. *Natl. Tax J.* **1987**, *40*, 577–590. [CrossRef]
44. Wyatt, P. *Property Valuation*; John Wiley & Sons: Hoboken, NJ, USA, 2022; ISBN 978-1-119-76741-1.
45. Miller, J.R.; Hobbs, R.J. Conservation Where People Live and Work. *Conserv. Biol.* **2002**, *16*, 330–337. [CrossRef]
46. Beatley, T.; Brown, J.D. The Half-Earth City. *William Mary Environ. Law Policy Rev.* **2021**, *45*, 775.
47. Tallamy, D.W. *Nature's Best Hope: A New Approach to Conservation That Starts in Your Yard*; Timber Press: Portland, OR, USA, 2020; ISBN 978-1-60469-900-5.
48. Berry, T. *The Dream of the Earth*; Sierra Club Books: San Francisco, CA, USA, 1988; ISBN 978-1-57805-135-9.
49. Leopold, A. A Land Ethic. In *A Sand County Almanac: With Other Essays on Conservation from Round River*; Oxford University Press: Oxford, UK, 1966; pp. 237–263. ISBN 978-0-345-32135-0.
50. Garcia, M.T.; Ribeiro, S.M.; Germani, A.C.C.G.; Bógus, C.M. The Impact of Urban Gardens on Adequate and Healthy Food: A Systematic Review. *Public Health Nutr.* **2018**, *21*, 416–425. [CrossRef]
51. Soga, M.; Gaston, K.J. Extinction of Experience: The Loss of Human–Nature Interactions. *Front. Ecol. Environ.* **2016**, *14*, 94–101. [CrossRef]
52. Staddon, C.; Ward, S.; De Vito, L.; Zuniga-Teran, A.; Gerlak, A.K.; Schoeman, Y.; Hart, A.; Booth, G. Contributions of Green Infrastructure to Enhancing Urban Resilience. *Environ Syst Decis.* **2018**, *38*, 330–338. [CrossRef]
53. Hunter, R.F.; Cleland, C.; Cleary, A.; Droomers, M.; Wheeler, B.W.; Sinnett, D.; Nieuwenhuijsen, M.J.; Braubach, M. Environmental, Health, Wellbeing, Social and Equity Effects of Urban Green Space Interventions: A Meta-Narrative Evidence Synthesis. *Environ. Int.* **2019**, *130*, 104923. [CrossRef] [PubMed]
54. Weston, B.H.; Bollier, D. *Green Governance: Ecological Survival, Human Rights, and the Law of the Commons*; Cambridge University Press: Cambridge, UK, 2014; ISBN 978-1-107-03436-5.
55. Farley, J.; Costanza, R.; Flomenhoft, G.; Kirk, D. The Vermont Common Assets Trust: An Institution for Sustainable, Just and Efficient Resource Allocation. *Ecol. Econ.* **2015**, *109*, 71–79. [CrossRef]
56. Xiong, Z.; Li, H. Ecological Deficit Tax: A Tax Design and Simulation of Compensation for Ecosystem Service Value Based on Ecological Footprint in China. *J. Clean. Prod.* **2019**, *230*, 1128–1137. [CrossRef]
57. Muellbauer, J. Why We Need a Green Land Value Tax and How to Design It. In *OECD Fiscal Federalism Studies*; Dougherty, S., Kim, H.-A., Eds.; OECD: Paris, France, 2023; ISBN 978-92-64-73536-1.
58. Heckbert, S.; Baynes, T.; Reeson, A. Agent-Based Modeling in Ecological Economics. *Ann. N. Y. Acad. Sci.* **2010**, *1185*, 39–53. [CrossRef] [PubMed]
59. Dosi, G.; Roventini, A. More Is Different... and Complex! The Case for Agent-Based Macroeconomics. *J. Evol. Econ.* **2019**, *29*, 1–37. [CrossRef]
60. Crooks, A.; Heppenstall, A.; Malleson, N.; Manley, E. Agent-Based Modeling and the City: A Gallery of Applications. In *Urban Informatics*; Shi, W., Goodchild, M.F., Batty, M., Kwan, M.-P., Zhang, A., Eds.; The Urban Book Series; Springer Singapore: Singapore, 2021; pp. 885–910. ISBN 9789811589829.
61. Bonabeau, E. Agent-Based Modeling: Methods and Techniques for Simulating Human Systems. *Proc. Natl. Acad. Sci.* **2002**, *99*, 7280–7287. [CrossRef]
62. Magliocca, N.; McConnell, V.; Walls, M. Exploring Sprawl: Results from an Economic Agent-Based Model of Land and Housing Markets. *Ecol. Econ.* **2015**, *113*, 114–125. [CrossRef]
63. Brown, D.G.; Robinson, D.T. Effects of Heterogeneity in Residential Preferences on an Agent-Based Model of Urban Sprawl. *Ecol. Soc.* **2006**, *11*, art46. [CrossRef]
64. Jackson, J.; Forest, B.; Sengupta, R. Agent-Based Simulation of Urban Residential Dynamics and Land Rent Change in a Gentrifying Area of Boston. *Trans. GIS* **2008**, *12*, 475–491. [CrossRef]

65. Guzy, M.R.; Smith, C.L.; Bolte, J.P.; Hulse, D.W.; Gregory, S.V. Policy Research Using Agent-Based Modeling to Assess Future Impacts of Urban Expansion into Farmlands and Forests. *Ecol. Soc.* **2008**, *13*, art37. [CrossRef]
66. Tsai, Y.; Zia, A.; Koliba, C.; Bucini, G.; Guilbert, J.; Beckage, B. An Interactive Land Use Transition Agent-Based Model (ILUTABM): Endogenizing Human-Environment Interactions in the Western Missisquoi Watershed. *Land Use Policy* **2015**, *49*, 161–176. [CrossRef]
67. Hosseinali, F.; Alesheikh, A.A.; Nourian, F. Agent-Based Modeling of Urban Land-Use Development, Case Study: Simulating Future Scenarios of Qazvin City. *Cities* **2013**, *31*, 105–113. [CrossRef]
68. Chen, Y. Effects of Development Tax on Leapfrog Sprawl in a Thinly Traded Land Market. *Land Use Policy* **2020**, *92*, 104420. [CrossRef]
69. Usman, H.; Lizam, M.; Burhan, B. A Review of Property Attributes Influence in Hedonic Pricing Model. In Proceedings of the 2nd African International Conference on Industrial Engineering and Operations Management, Harare, Zimbabwe, 7–10 December 2020.
70. Yang, L.; Wang, B.; Zhou, J.; Wang, X. Walking Accessibility and Property Prices. *Transp. Res. Part D Transp. Environ.* **2018**, *62*, 551–562. [CrossRef]
71. Daams, M.N.; Sijtsma, F.J.; Vlist, A.J. van der The Effect of Natural Space on Nearby Property Prices: Accounting for Perceived Attractiveness. *Land Econ.* **2016**, *92*, 389–410. [CrossRef]
72. Kim, Y.-S.; Johnson, R.L. The Impact of Forests and Forest Management on Neighboring Property Values. *Soc. Nat. Resour.* **2002**, *15*, 887–901. [CrossRef]
73. Nilsson, P. The Influence of Urban and Natural Amenities on Second Home Prices. *J. Hous. Built Environ.* **2015**, *30*, 427–450. [CrossRef]
74. Fesselmeyer, E.; Seah, K.Y.S. The Effect of Localized Density on Housing Prices in Singapore. *Reg. Sci. Urban Econ.* **2018**, *68*, 304–315. [CrossRef]
75. Cho, S.-H.; Clark, C.D.; Park, W.M.; Kim, S.G. Spatial and Temporal Variation in the Housing Market Values of Lot Size and Open Space. *Land Econ.* **2009**, *85*, 51–73. [CrossRef]
76. Lukavec, M.; Kadeřábková, B. How Much Does a Minute of Commuting Time Cost? An Examination of Property Prices in Relation to Distance to the City Center in Prague, Czech Republic. *Stavební Obz. Civ. Eng. J.* **2017**, *26*, 555–567. [CrossRef]
77. Kopits, E.; McConnell, V.; Miles, D. Lot Size, Zoning, and Household Preferences. *Hous. Policy Debate* **2012**, *22*, 153–174. [CrossRef]
78. McConnell, V.; Walls, M. *The Value of Open Space: Evidence from Studies of Nonmarket Benefits*; Resources for the Future: Washington, DC, USA, 2005.
79. Albouy, D.; Ehrlich, G.; Shin, M. Metropolitan Land Values. *Rev. Econ. Stat.* **2018**, *100*, 454–466. [CrossRef]
80. Cobb, C.W.; Douglas, P.H. A Theory of Production. *Am. Econ. Rev.* **1928**, *18*, 139–165.
81. Voorneveld, M. *From Preferences to Cobb-Douglas Utility. Stockholm School of Economics*; The Economic Research Institute (EFI): Stockholm, Sweden, 2008.
82. Huang, Q.; Parker, D.C.; Filatova, T.; Sun, S. A Review of Urban Residential Choice Models Using Agent-Based Modeling. *Environ. Plan. B Plan. Des.* **2014**, *41*, 661–689. [CrossRef]
83. Filatova, T.; Parker, D.; Veen, A. Agent-Based Urban Land Markets: Agent's Pricing Behavior, Land Prices and Urban Land Use Change. *J. Artif. Soc. Soc. Simul.* **2009**, *12*.
84. Brown, D.G.; Page, S.; Riolo, R.; Zellner, M.; Rand, W. Path Dependence and the Validation of Agent-based Spatial Models of Land Use. *Int. J. Geogr. Inf. Sci.* **2005**, *19*, 153–174. [CrossRef]
85. Magliocca, N.; Safirova, E.; McConnell, V.; Walls, M. An Economic Agent-Based Model of Coupled Housing and Land Markets (CHALMS). *Comput. Environ. Urban Syst.* **2011**, *35*, 183–191. [CrossRef]
86. Harding, J.P.; Knight, J.R.; Sirmans, C.F. Estimating Bargaining Effects in Hedonic Models: Evidence from the Housing Market. *Real Estate Econ.* **2003**, *31*, 601–622. [CrossRef]
87. Bell, A.; Parkhurst, G.; Droppelmann, K.; Benton, T.G. Scaling up Pro-Environmental Agricultural Practice Using Agglomeration Payments: Proof of Concept from an Agent-Based Model. *Ecol. Econ.* **2016**, *126*, 32–41. [CrossRef]
88. Abraham, R.; Tynan, C. How Much Does It Cost to Build a House in 2023? Available online: https://www.forbes.com/home-improvement/contractor/cost-to-build-a-house/ (accessed on 3 June 2023).
89. Lynch, E. *Cost of Constructing a Home: 2023*; Economics & Housing Policy; National Association of Home Builders (NAHB): Washington, DC, USA, 2023.
90. Theriault, M. How Much Does It Cost to Build a House? (2023 Edition). Available online: https://www.newhomesource.com/learn/cost-to-build-house-per-square-foot/ (accessed on 3 June 2023).
91. Foo, C.H. The Impact of Capital Contributions and Compliance Costs on Housing Affordability. *Build. Investig.* **2018**, 34–38.
92. Rabinowitz, A. *Land Investment and the Predevelopment Process: A Guide for Finance and Real Estate Professionals*; Quorum Books: Westport, CT, USA, 1988; ISBN 978-0-89930-326-0.
93. Hoyt, H.; Schuetz, J. *Making Apartments More Affordable Starts with Understanding the Costs of Building Them*; Brookings: Washington, DC, USA, 2020.
94. Matthews, R.B.; Gilbert, N.G.; Roach, A.; Polhill, J.G.; Gotts, N.M. Agent-Based Land-Use Models: A Review of Applications. *Landsc. Ecol.* **2007**, *22*, 1447–1459. [CrossRef]

95. Plassmann, F.; Tideman, T.N. A Markov Chain Monte Carlo Analysis of the Effect of Two-Rate Property Taxes on Construction. *J. Urban Econ.* **2000**, *47*, 216–247. [CrossRef]
96. Martin, I.W.; Beck, K. Gentrification, Property Tax Limitation, and Displacement. *Urban Aff. Rev.* **2018**, *54*, 33–73. [CrossRef]
97. Kolbe, J.; Schulz, R.; Wersing, M.; Werwatz, A. Land Value Appraisal Using Statistical Methods. *Z. Immobilienökonomie* **2019**, *5*, 131–154. [CrossRef]
98. Bencure, J.; Tripathi, N.; Miyazaki, H.; Ninsawat, S.; Kim, S. Development of an Innovative Land Valuation Model (iLVM) for Mass Appraisal Application in Sub-Urban Areas Using AHP: An Integration of Theoretical and Practical Approaches. *Sustainability* **2019**, *11*, 3731. [CrossRef]
99. Hubbard, R. Taxing Land vs. the Buildings on It at Core of House Bill—Session Daily—Minnesota House of Representatives. Available online: https://www.house.mn.gov/SessionDaily/Story/18250 (accessed on 29 September 2024).

land

MDPI

Article

Multi-Scenario Simulation of Urban–Rural Land Use Spatial Reconstruction in Highly Urbanized Areas: A Case Study from the Southern Jiangsu Region

Changjun Jiang [1] and Huiguang Chen [1,2,*]

[1] College of Public Administration, Nanjing Agricultural University, Nanjing 210095, China; jcjland@stu.njau.edu.cn

[2] China Resources & Environment and Development Academy, Nanjing Agricultural University, Nanjing 210095, China

* Correspondence: chenhuiguang@njau.edu.cn

Abstract: China's rural population flowing into highly urbanized areas has led to the spatial reconstruction of urban–rural land use. Exploring the laws and trends of urban–rural land use in highly urbanized areas is of great significance in promoting rural transformation. This paper takes the southern Jiangsu region as a research area and uses a system dynamics (SD) model to simulate the demand for different land types based on economic, social, policy, and environmental (ESPE) factors. Future land use simulation (FLUS) is used to simulate the spatial evolution trend of urban–rural land use based on point–axis elements. The results show that the agricultural production space is severely squeezed by the urban living space. Under the scenario of rapid expansion, the decrease in arable land quantity and the demand area for rural residential areas are the largest. Under the scenario of high-quality development, the decrease in arable land area and the demand for land in rural residential areas are lowest. Based on the spatial simulation, it is reported that the areas with more intense land use spatial reconstruction in the three scenarios are mainly concentrated in the region's urban–rural border areas. The future evolution of urban–rural land is summarized into three models: (1) single-center-driving expansion, (2) patchy expansion near the city center, and (3) multi-center-driving expansion. This paper proposes targeted policy recommendations to provide a scientific reference for solving the conflict between urban and rural land use.

Keywords: urban–rural land use; spatial reconstruction; multi-scenario simulation; SD-FLUS model; ESPE factors

1. Introduction

Over the past 40 years of China's reform and opening, urbanization and industrialization have entered a stage of rapid development. The development of urbanization has seriously changed the original development trajectory of rural areas [1,2]. By the end of 2022, China's urbanization rate had reached 65.22%, especially in the southern Jiangsu region, where the urbanization level has exceeded 80%. Under the backdrop of rapid urbanization, there are serious problems with land use, which has become an important factor affecting social stability in China. The leasing of construction land can realize great benefits in a short period of time. When faced with huge fiscal deficit pressure, local governments are more inclined to reverse the financial pressure in the short term [3]. Therefore, the rapid expansion of urban construction land in suburban areas has squeezed the development space of rural land. Examples include the non-agriculturalization of cultivated land and the deterioration of the ecological environment of cultivated land. Scientific predictions and the rational planning of land spatial development and utilization are effective means of implementing land resource management. Simulating the spatial patterns of urban–rural land use types in different scenarios can provide a reference basis for rural transformation and development.

Land use cover change (LUCC) is the foundation of resource, environmental, and ecological research and scientific management. Essentially, LUCC provides important background data for land resource planning [4–6]. The study of simulating land use changes plays a significant role in revealing the impact of human activities on regional ecological environment change. Such research can also provide support for land management decisions [7,8]. The most important content of LUCC research is land use spatial simulation. Through a relevant driving factor set, the spatial distribution of land use is obtained, which helps to analyze characteristics and hotspots of spatial change. Currently, there are few studies that examine land use spatial simulation in highly urbanized areas. There is also a scarcity of research on seeking policy recommendations for urban–rural transformation in highly urbanized areas. Scholars at home and abroad have achieved fruitful results in simulating and predicting land use change, but these studies have been used to serve the decision-making pertaining to national or regional socioeconomic development. The simulation types are mainly divided into three types: land quantity prediction, land spatial distribution prediction, and comprehensive prediction.

Land quantity simulation models can predict various types of future land use areas based on the current land use status and socioeconomic factors. The most typical prediction models currently include the grey prediction model GM (1,1), multiple regression model, Markov model, and SD simulation model [9–12]. These prediction models can be combined with man–land relationship theory to predict changes in land quantity in a certain region. The disadvantage of such models is that they can only predict the quantity of land types within specific boundaries, such as administrative boundaries (e.g., provinces, cities, and counties); they cannot predict the spatial distribution of land use. On the contrary, the land spatial distribution prediction model simulates the development of land types in the future. Such a model can simulate the spatial distribution of land use at the micro level; the simulation results also have good spatiotemporal dynamics. However, the land spatial distribution prediction model lacks macro constraints on regional socioeconomic and population development. Moreover, previous studies on land use spatial prediction were mostly based on one or two periods of historical land use spatial data for prediction. In addition, the selection of sample data in previous periods was subjective. The classic models include cellular automata simulation (CA), multi-agent simulation (MAS), and the CLUE-S model, PLUS model, and FLUS model [13–17]. The comprehensive prediction model adopts the method of combining the land quantity prediction model with the land spatial prediction model; the aim is to overcome the inherent limitations of a single model. The FLUS model demonstrates strong simulation capabilities in diverse fields such as urban development planning, land change simulation, and ecological risk prediction. This model has been widely applied at global and regional scales [18,19]. Currently, the SD model is an effective way of simulating future regional land demand, incorporating variables of socioeconomic-influencing factors. The FLUS model, with the advantages of simulating urban development planning, land use change, and ecological risk prediction, is considered to have excellent performance. Thus, the FLUS model can be combined with system dynamics (SD) models to carry out high-resolution, long-term, and high-precision land spatiotemporal change simulations.

The southern region of Jiangsu is located in a highly urbanized area in China. Here, the mechanism of the factors of social and economic transformation is more complex, as well as the spatial reconstruction of land use. Exploring the spatial evolution of land use in typical cities in the southern region of Jiangsu has practical significance. Firstly, this study predicts the land demand in 2035 based on a set of "economic-social-policy-environment" (ESPE) factors. According to different ESPE development situations, inertia development scenarios, high-quality development scenarios, and rapid expansion scenarios are set up. Secondly, combining land demand simulated by the SD, the spatial reconstruction characteristics of land use in 2035 are simulated using the "point–axis" factor. Finally, typical regions are selected for analysis, and targeted recommendations are provided. The contribution of this study mainly includes two aspects: (1) the research work is conducive to exploring mechanisms and methods to alleviate urban–rural land use conflicts in highly urbanized areas; (2) the use of a model combining SD and FLUS for analysis enriches the research methods of land use simulation. Additionally, one of the key issues addressed in the SD-FLUS model is the rational selection of multidimensional indicators and the construction of their relationship with changes in land quantity.

2. Theoretical Framework, Materials, and Methods

2.1. Theoretical Framework

Most of the research above on the factors influencing land use change focuses on the economy, society, policy, and environment. Based on this, a theoretical framework is constructed, as shown in Figure 1. The changes in land elements cannot be separated from the driving force of economic factors. Due to the development of the urban economy, the agglomeration of production scale has led to the expansion of urban boundaries. Moreover, the investment in fixed assets is becoming increasingly active, which increases the demand for construction land [20,21]. With the transfer of labor from the primary industry to the secondary and tertiary industries, more agricultural land has also been converted into non-agricultural construction land. For social factors, population change is the main cause of land use change. In the process of urbanization, people's various material needs for daily life form various construction land demands such as production space, public facility land, and infrastructure land. Land use changes will ultimately be based on policy systems. The relevant departments have implemented measures such as returning farmland to forests and renovating rural residential areas to meet regional planning goals. Climate change will affect vegetation growth and agricultural production, and natural factors determine the material basis of different land uses.

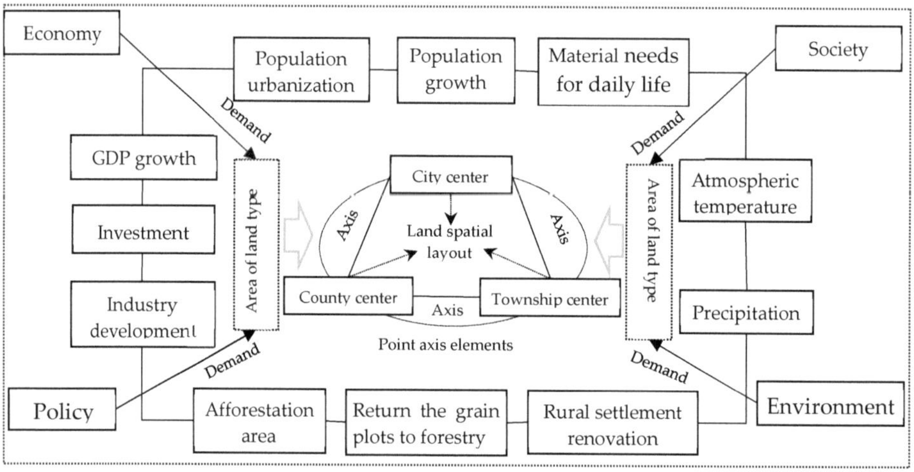

Figure 1. Theoretical framework.

Factors such as economy, society, policy, and environment can have a direct impact on the quantity of land use. However, the spatial changes in different land use types tend to be based more on point and axis elements. The point–axis development mode basically conforms to the objective laws of productivity spatial motion. The economic center is always first concentrated in a few areas with better conditions, such as the central areas of cities, counties, and townships. Due to the exchange of production factors requiring transportation routes and water supply lines, points are connected to form an axis. The development of a social economy is often arranged according to the point–axis development model, and land resources are an important carrier of the point–axis system. The axis elements in the southern Jiangsu region are dense, mainly including highways, railways, pilot roads, and rivers. Point–axis elements are important factors in the spatial layout of land use.

2.2. *Overview of the Study Area*

The southern region of Jiangsu Province includes five cities—Nanjing, Suzhou, Wuxi, Changzhou, and Zhenjiang—with a land area of 28,000 km^2 (Figure 2). Southern Jiangsu is one of the regions with the highest population and economic density in China and is also the birthplace of the "Sunan Model" [22–24]. At present, the urbanization rate in some cities in the southern Jiangsu region has reached 80%. For example, in Nanjing, the urbanization rate of the permanent resident population is 86.8%, which exceeds the average urbanization rate of developed countries. Meanwhile, the output value of the tertiary industry accounted for about 55% of GDP.

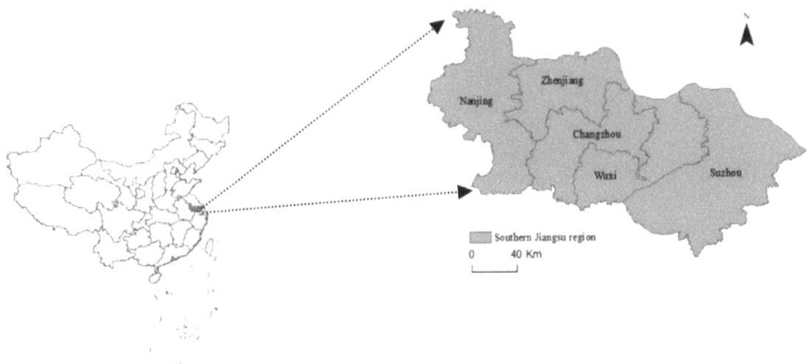

Figure 2. Study area.

2.3. *Model Logic and Data Sources*

The simulation process coupling socioeconomic factors and the mechanism of "point–axis" spatial factors is shown in Figure 3 below. This paper forms a cause-and-effect diagram and a flow–storage diagram of the socioeconomic factors related to the spatial reconstruction of land use. This is done by simulating the demand for different types of land and the evolution scenarios in the region. Then, using the land demand predicted by the SD model combined with the FLUS model, the future evolution of the spatial pattern of urban–rural land use in southern Jiangsu is simulated.

The SD model of the demand for spatial reconstruction of land use is constructed, with driving factors including "economic-social-policy-environment" dimensions (Table 1). Economic factors include fixed assets investment, GDP growth rate, agricultural investment proportion, secondary industry added value, and real estate investment proportion. Social factors include population growth rate, urbanization rate, total population, per capita demand for meat, per capita demand for aquatic products, and the driving force of agricultural mechanization. Policy factors include the afforestation area, returning farmland to forests, and rural settlement improvement. Environmental factors include temperature and precipitation. The four elements "economic-social-policy-environment" (ESPE) simultaneously act on the land use system. The SD is used to simulate the demand for different types of land use in the southern Jiangsu region under different scenarios. The spatial simulation of land use based on the FLUS model uses "point–axis" driving factors for simulation [25]. The "point–axis" factors mainly include the distance to prefecture-level cities, distance to railways, distance to county centers, distance to national highways, distance to towns, distance to county roads, and distance to rivers.

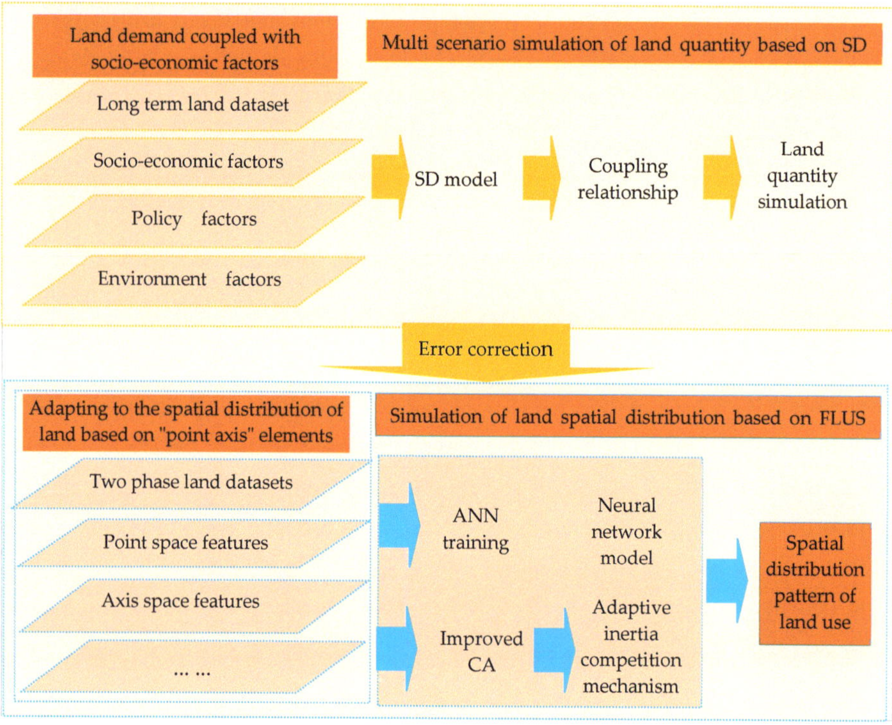

Figure 3. Simulation process based on the SD-FLUS model.

Table 1. Factors driving spatial reconstruction of urban–rural land use in southern Jiangsu.

Factor Layer	ESPE	Spatial Driving Factor	
		Point/km	Axis/km
Economy	Fixed assets investment (CNY 10,000) GDP growth rate (%) Agricultural investment ratio (%) Value added by secondary industry (CNY 100 million) Real estate investment ratio (%)	Distance to prefecture cities Distance to county center Distance to township	Distance to railway Distance to highway Distance to county road Distance to river
Society	Population growth rate (%) Population urbanization rate (%) Total population (1000 people) Per capita demand for meat (kg) Per capita demand for aquatic products (kg) Power of agricultural mechanization (10,000 kw)		
Policy	Afforestation area (km^2) Returning the grain plots to forestry (km^2) Rural settlement renovation (km^2)		
Environment	Atmospheric temperature (°C) Precipitation (mm)		

The data on ESPE factors are from the statistical yearbooks and official websites of various cities. The spatial driving factors are sourced from the OpenStreetMap website. Land use data comes from the resource and environmental science data center of the Chinese Academy of Sciences.

2.4. Model Settings

2.4.1. The SD Model

Land use system refers to an organic whole with a certain structure and function that is formed by the mutual influence of subsystems in a region, such as population, land resources, land environment, economy, society, and management regulation [26,27]. Therefore, the land use system is a comprehensive and extremely complex system. The SD method, as a simulation method that is based on feedback control theory, can precisely simulate the complexity of land use systems. Specifically, the SD method uses the analysis and research of information feedback to solve and understand system problems [28]. Due to the high complexity of land use systems with many state variables, constructing a corresponding SD model around the "land demand" is difficult. Therefore, there is a need to find the main relationship structure and grasp the key elements rather than try to reflect the entire system in detail. The spatial boundary of the SD model is set for the southern Jiangsu region. The simulation base year is 2010, with a time step of one year. After obtaining the simulated values of land use demand from 2010 to 2020, this study compares them with the actual values until the accuracy requirements are met (i.e., when the error between the actual and simulated values is within 15%) [29]. Then, the base period and the end of the system simulation are adjusted to obtain the quantity of land use demand in 2035. The ultimate aim is to explore the spatial change trend of cultivated land and rural settlements.

2.4.2. The FLUS Model

The FLUS model has made significant improvements based on traditional CA [19]. Based on the neural network algorithm (ANN), the probability of each land use type in each pixel of the research area is calculated [30,31]. This paper extracts land use data for southern Jiangsu in 2010 and 2020 and applies the PLUS model CA module. The land use data and the "point–axis" factor set for southern Jiangsu in 2010 are inputted. This study input the following into the self-adaptive inertia and competition mechanism CA module: the 2010 land use status map of the southern Jiangsu region, the 2010 land use development suitability atlas, the restricted conversion area, the actual values of the various land types in 2020, the conversion rules, and the neighborhood weights. The FLUS model simulation is based on the GeoSOS-FLUS V1.0 Beta platform [32,33], and the Kappa coefficient and Figure of Merit index are used to verify the simulation's accuracy. The Kappa coefficient of this simulation is 0.89, and the overall simulation accuracy is 0.92, far exceeding the basic value of 0.75 for the model's operational accuracy.

3. Results

3.1. Spatial and Temporal Evolution Characteristics of Land Use

In Figure 4, the types of land use included cultivated land (CL), rural residential areas (RS), urban construction land (UCL), woodland (WL), grassland (GL), water areas (WTA), and unused land (UL). It shows the spatial distribution of land use in southern Jiangsu in 2000, 2010, and 2020. In 2000, the change of construction land was mainly distributed throughout the urban areas of Nanjing, Wuxi, and Suzhou. Meanwhile, the urban areas of Changzhou and Zhenjiang had relatively small areas of construction land. In 2010, the construction land in the southern Jiangsu region increased significantly, mainly around the urban expansion of Nanjing, Wuxi, Suzhou, and Changzhou. However, the construction land in Zhenjiang did not increase significantly. Compared with the expansion of urban construction land from 2000 to 2010, the expansion speed of urban construction land in southern Jiangsu decreased from 2010 to 2020. However, the expansion of construction land is still obvious near the Taihu Lake area. During the period from 2010 to 2020, a series of measures, e.g., permanent basic farmland, urban development boundaries, and ecological red lines, were gradually introduced, thereby defining the scope of permanent agricultural land and ecological areas. During the study period, the ecological space in southern Jiangsu, including inland waters, forest land, and grassland, did not change significantly.

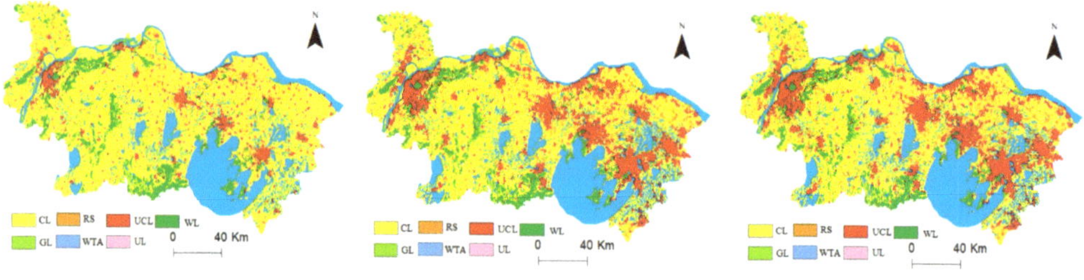

2000 year 2010 year 2020 year

Figure 4. Spatial distribution of land use in southern Jiangsu.

3.2. Simulating Land Demand in Southern Jiangsu Based on SD

3.2.1. Establishment of SD Model

The purpose of constructing the SD model is mainly to reflect the relationship between the set of socioeconomic factors and the demand for various land types. The most directly relevant factors must be selected as key variables. To avoid the over-sized and complex high-order models in the simulation system, this paper has conducted a more detailed classification of socioeconomic factors. Also, the key variables have been mainly constructed based on typical references. The prediction results of the SD model can provide data support for the next step of land use spatial simulation under different scenarios. The variables in the SD model mainly involve the table function with LOOKUP. The key role of the equilibrium constant is to fine-tune the variation trend and error. The adjustment coefficient is used to couple the relationship between the independent variable and the dependent variable, and the dimension is adjusted to satisfy the regression relationship between variables. The SD model mainly includes two parts of formulas for land area change and other variables, with the formulas involving key variables as follows:

Regional GDP = INTEG (GDP change, Initial value: 25,270.6) (unit: CNY 100 million)
Change in GDP = regional GDP * GDP growth rate (unit: CNY 100 million)
GDP growth rate = With LOOKUP (Time) (unit: %)
Total population = INTEG (population change, initial value: 2368.14) (unit: 10,000 people)
Population change = total population * population growth rate (unit: 10,000 people)
Population growth rate = With LOOKUP (Time) (unit: %)
Temperature = With LOOKUP (Time) (unit: °C)
Precipitation = With LOOKUP (Time) (unit: mL)
Agricultural mechanization power = With LOOKUP (Time) (unit: 10,000 kWh)
Per capita meat demand = With LOOKUP (Time) (unit: Kg/person)
Afforestation area = With LOOKUP (Time) (unit: Km^2)
Area of rural residential areas = INTEG (change in rural residential areas, initial value: 2486.43) (unit: Km^2)
Change amount of rural residential areas = rural residential area * change rate of rural residential areas—Rural residential area remediation intensity (unit: Km^2)
Area of arable land = INTEG (area of arable land change, initial value: 13,323.3) (unit: Km^2)
Area of arable land change = arable land area * arable land change rate—Intensity of returning farmland to forests (unit: Km^2)
Area of urban construction land = INTEG (area of urban construction land change, initial value: 4141.55) (unit: Km^2)
Area of urban construction land change = urban construction land area * urban construction land change rate (unit: Km^2)
The total area of regional land = the area of rural residential areas + the area of arable land + the area of urban construction land + the area of forest land + the area of grassland + the area of water area + the area of unused land

This paper sets rural residential areas, cultivated land, urban construction land, forest land, grassland, water areas, and unused land as stocks. There are minimal errors in the area of each land use type in the land use status map. In this study, the ratio of the simulated total area to the actual total area each year is used to correct the simulated area of each land use type. The relationship between the change rate of each land use type and the driving factor is adjusted using SPSS22.0. Then, the adjustment coefficient and constant are obtained and are further modified to meet the prediction accuracy. After setting all equations, a flow diagram of the SD model for predicting land demand in southern Jiangsu is obtained, as shown in Figure 5.

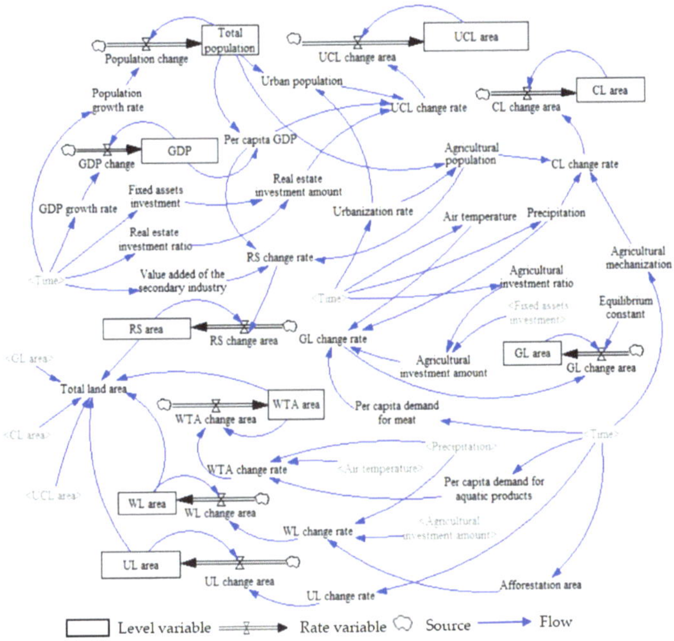

Figure 5. Flowchart of SD model for predicting land demand in southern Jiangsu.

3.2.2. Land Use Demand Forecast Under Different Scenarios

Based on the establishment of the SD model, this study compared the errors between the actual and simulated land quantities in the southern Jiangsu region in the past. The results are shown in Figure 6.

Due to the relatively small size of grassland and unused land and the large change rate of those land types in historical periods, the simulation error value of grassland is relatively large. Except for 2013, the error of unused land is within ±4%. The error values of the main land types in this paper are all within ±5%; thus, the simulation error is acceptable. The SD model of land demand in southern Jiangsu is affected by parameter settings. In order to explore the land use demand in southern Jiangsu under different development scenarios, we set up an inertial growth scenario: (A) a high-quality development scenario and (B) a rapid expansion scenario (C) based on socioeconomic factors. The specific parameter settings are shown in Table 2.

Table 2. Parameter settings under different development scenarios.

Types	Factor	Scenario (B)	Scenario (C)
Urbanization	Urbanization rate (%)	+5%	+8%
Economy	GDP growth rate (%)	−1%	+5%
Population	Population growth rate (%)	Unchanged	+2%
Policy	Rural settlement renovation (hm^2)	+20%	Unchanged
	Returning forests to farmland (hm^2)	+20%	Unchanged

The changes in the spatial reconstruction characteristics of land use in scenarios A–C are very obvious (Figure 7). In terms of the change in land demand for rural settlements, the demand in Scenario B is the lowest, while the demand area in Scenario C is the largest. The differences in the demand for different types of land in the three scenarios are relatively low at the beginning of the simulation, but the difference in land area demand gradually increases over time.

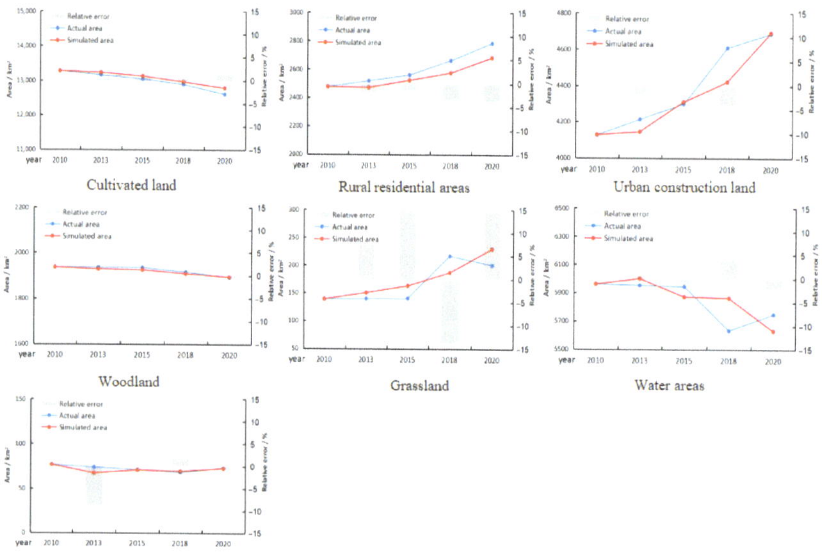

Figure 6. Relative error between actual area and simulated area.

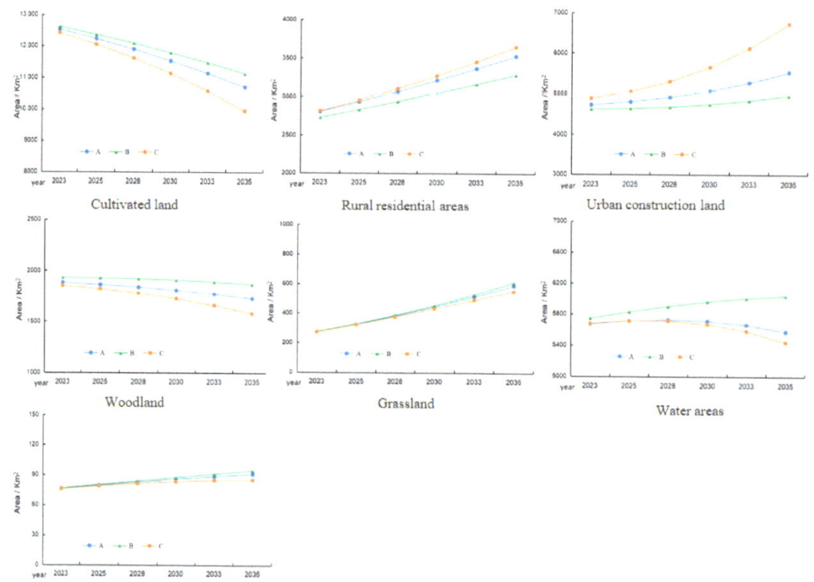

Figure 7. Simulated changes in the area of various land types in 2035.

In Scenario A, the quantity change of land use in southern Jiangsu presents the characteristics of continuous growth in urban construction land, rural residential areas, and unused grassland areas, while the areas of cultivated land, woodland, and water areas mainly show a series of decreasing trends. This study mainly analyzes the quantity and area of rural residential areas, cultivated land, woodland, and urban construction land. During the period from 2010 to 2020, the cultivated land in the southern Jiangsu region decreased by 4.91%, the woodland decreased by about 2.15%, and the water areas decreased by about 3.45%. The number and area of rural residential areas increased by 12.55%, and the area of urban construction land increased by 13.40%. During the simulation period, the annual average decrease in cultivated land was about 150 km^2, the annual average decrease in woodland was about 12 km^2, the number and area of rural residential areas increased by 50 km^2, and the area of urban construction land increased by 17.1%.

Scenario B is characterized by the construction of ecological civilization as the main goal. In this scenario, local governments formulate practical ecological protection policies based on territorial space planning. These measures have contributed to the formulation of urban–rural development policies that are in line with regional sustainable development. Under Scenario B, the change rate of various land areas in southern Jiangsu is relatively low. In addition, although rural residential areas will also increase in an inertial growth manner, the area's growth is the smallest among the three scenarios. Simultaneously, compared with the initial simulation period, the increase in urban construction land area is 7.3%, and the change in cultivated land area is −11.73%. The water areas and grasslands will increase slightly on the original basis, while the woodland area remains relatively stable, with a change rate of −3.32%.

In Scenario C, the main pursuit of social and economic development is to implement a relatively relaxed land use policy. Compared with Scenario A and B, the restrictions on land use in Scenario C are relatively few. The area of each land type is also significantly affected by regional population growth and economic development. This pattern is mainly manifested in a large decrease in the area of arable land and a large increase in the area of urban construction land in the later stages of the simulation. Compared to urban construction land, the simulated area of rural residential areas has the largest proportion of changes under the three scenarios. This may be due to the fact that, in the context of rapid socioeconomic expansion, contiguous rural settlements have gradually transited to urban construction land. The reduction in woodland is relatively large. In the early stage of the simulation, the proportions of cultivated land, rural residential areas, and urban construction land areas were 44.34%, 10.05%, and 17.46%, respectively. At the end of the simulation, the proportions of cultivated land, rural residential areas, and urban construction land areas were 35.54%, 13.05%, and 24.05%, respectively.

3.3. Multi-Scenario Simulation of Land Spatial Distribution Based on the FLUS Model

3.3.1. FLUS Model Parameter Settings

The land use conversion rules are represented in the form of a cost matrix in the self-adaptive inertia and competition mechanism CA module. The values are represented as 0 or 1. Specifically, 0 indicates that transformation cannot occur between two land types, and 1 indicates that transformation can occur between two land types. In Scenario A, this study sets a large-scale water surface as a restricted conversion zone. This study takes geological parks, forest parks, and nature reserves as point elements and sets a buffer zone with a radius of 1 km as a restricted conversion zone. The land conversion rules and restriction zone settings in Scenarios A and C are the same as those in this paper. Only the future spatial layout of land use under different socioeconomic development scenarios is compared. For Scenario B, the land use conversion rules are relatively strict, strictly controlling the conversion of various ecological land use to urban construction land and rural residential areas.

3.3.2. Land Use Spatial Layout Under Different Scenarios

The simulation results of the spatial reconstruction evolution characteristics of land use under Scenarios A–C are shown in Figure 8. Under Scenario A, the conversion of cultivated land and rural residential areas to urban construction land mainly occurs in the Jiangning district in the southern part of Nanjing and the border area between Jiangning district in Nanjing and Jurong in Zhenjiang. The land use transformation in Changzhou is concentrated in the eastern part of Liyang and Wujin district. Under Scenario C, urban construction land has expanded to the surrounding areas on the basis of the original scope. This is specifically manifested as urban construction land squeezing arable land and rural residential areas, while water areas land shows a certain degree of stability and no obvious change characteristics. The scope of urban construction land in Scenario C is the largest and most extensive. Also, urban construction land further squeezes ecological land, and this land transformation performance is highly active. In the following three scenarios, the change in cultivated land area is the least under Scenario B. In particular, the distribution of rural residential areas is similar to that under Scenario A. The woodland, grassland, and water areas remain stable (on the original basis in 2020), which is in line with the simulation goal of high-quality development.

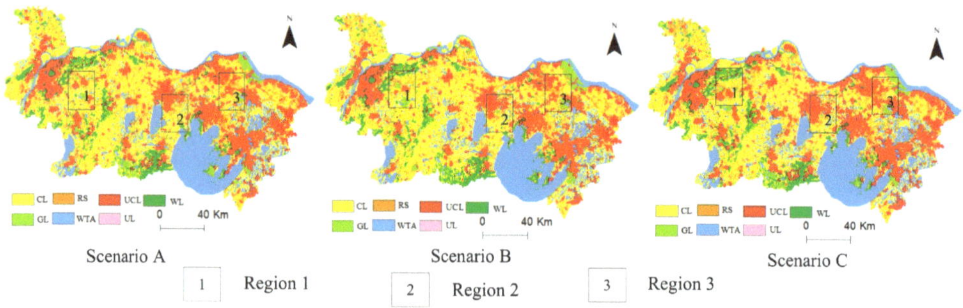

Figure 8. Land use spatial layout under different scenarios in southern Jiangsu in 2035.

4. Discussion

4.1. Simulation Analysis of the Spatial Evolution of Urban–Rural Land Use in Typical Regions

This study chooses the three typical regions above for further analysis and explores the laws and characteristics of urban–rural land use spatial reconstruction. To make the studied cultivated land and rural residential areas more prominent, land use is divided into agricultural living space, rural living space, urban living space, and ecological space (ECS). The three selected areas are marked in Figure 8, while the detailed land use spatial simulations are shown in Figure 9.

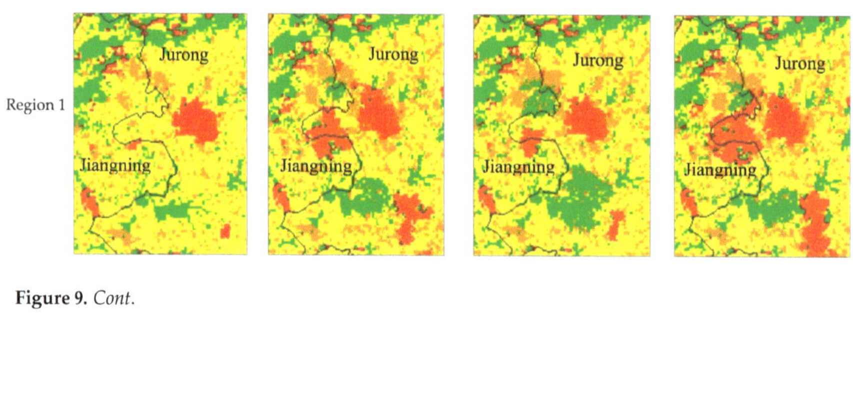

Figure 9. *Cont.*

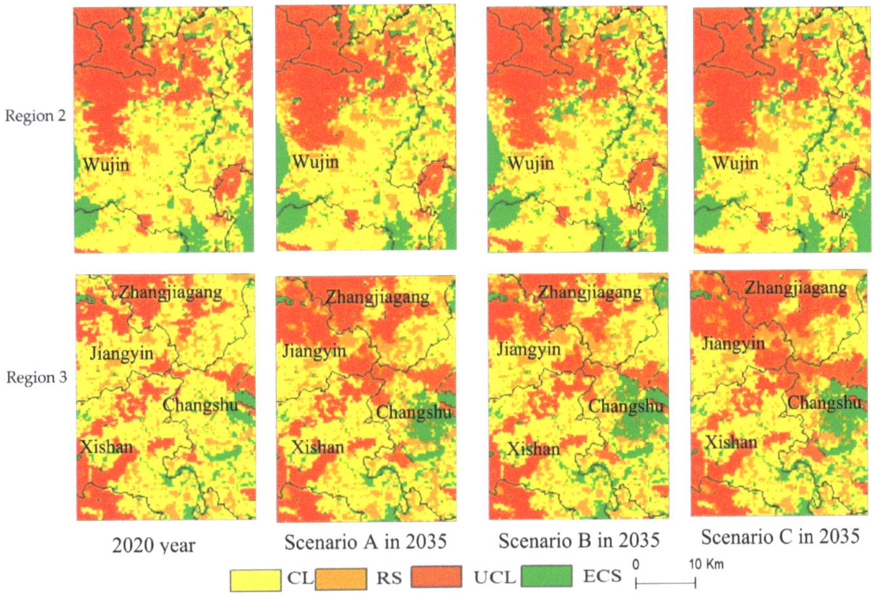

Figure 9. Simulation of rural land use spatial reconstruction in typical regions by 2035.

Region 1 is located at the junction of Jiangning district and Jurong City. According to the analysis of land use status in 2020, rural residential areas are concentrated in the northern border area of Region 1, distributed in a long strip shape. The rural residential areas in other regions are relatively scattered, while agricultural production spaces are mainly distributed in Jurong. The urban construction land is mainly concentrated in the urban area of Jurong, while the distribution proportion of urban construction land in Jiangning is low. In Scenario A, by 2035, the expansion of urban construction land will have severely squeezed rural residential areas. Based on the original urban area in 2020, northward expansion has squeezed agricultural production space. Scenario B is mainly characterized by the squeezing of agricultural living space by ecological space and rural residential areas. In Scenario C, the proportion of agricultural production space to urban living space in the southeast of Jurong is relatively large. In summary, for rural land, the rural residential areas in all three scenarios show a trend of squeezing agricultural production space, but the spatial evolution difference is relatively low in all three scenarios. Under the three development scenarios, significant differences are seen in the area and layout of agricultural production space. Scenario B strictly restricts the expansion of urban construction land, resulting in a relatively stable development trend of agricultural production space. However, in Scenario B, ecological space will squeeze agricultural production space.

Region 2 mainly includes most areas of Wujin in Changzhou. Wujin is adjacent to the urban area of Changzhou, with a large area of urban construction land in the northwest of Wujin. Agricultural production space is mainly distributed in the southeast of Wujin, while ecological space is mainly concentrated in the eastern and western regions of Region 2. From analyzing Scenario A, the transformation of agricultural production space into urban land is mainly located in the southern part of the actual distribution of urban construction land. Meanwhile, the proportion and spatial layout of land types in other regions are relatively stable. The squeezing of agricultural production space by rural residential areas is mainly concentrated in two areas. The first area is located at the junction of the northern area of Wujin and Tianning, adjacent to the urban area of Changzhou. The probability of rural residential areas squeezing agricultural production space in this area is relatively high.

The second area is located in most of the southeastern part of Wujin, especially in areas close to urban construction land. This is where the characteristics of rural residential areas transforming into urban construction land are more obvious. The number and structure of rural land use types in areas far from the urban area of Wujin are relatively stable. The ecological land in Region 2 exhibits stable characteristics under all three development scenarios with no significant changes.

Region 3 belongs to the Wuxi area, which is the border area between Zhangjiagang, Jiangyin, Xishan, and Changshu. In the 2035 land use simulation, the spatial reconstruction characteristics and laws of rural land use are extraordinarily complex. In Scenario A, the areas where urban construction land squeezes agricultural production space are mainly located in Jiangyin and Zhangjiagang and the border area between the four abovementioned cities. No significant change occurs in the agricultural production space between Xishan and Changshu. The areas where agricultural production space is transformed into rural residential areas are mainly concentrated in Jiangyin and Zhangjiagang. For Scenario C, urban–rural living spaces in Region 3 are severely squeezing agricultural production space. The construction land area in Jiangyin, Zhangjiagang, and Changshu is expanding and gradually converging at the administrative boundaries of the three regions. Notably, under Scenario C, the continuous expansion of rural residential areas gradually integrates with urban living spaces, forming contiguous construction land. In short, for rural land in Region 3, the urban–rural living spaces in Jiangyin and Zhangjiagang are severely squeezing agricultural production space in all three scenarios, with urban living space squeezing the most severely. The transformation of rural land in Xishan and Changshu is mainly concentrated at the junction of administrative boundaries.

4.2. Extracting the Model of Future Urban–Rural Land Use Reconstruction

There is relatively little research on simulating land use changes in the southern Jiangsu region. In land use simulation studies in other regions, few scholars have summarized typical characteristics. This study summarizes three development models, namely single-center-driving expansion, patchy expansion near the city center, and multi-center pole-driving expansion. The three urban–rural land development models are shown in Figure 10. Region 1 (single-center-driving expansion) represents the development of construction land in another area near the center of the original built-up area, driven by the central pole. The area of agricultural production space turning into urban living space is relatively concentrated. Rural living space shows expansion in its original spatial location, while other expansion areas are concentrated near urban construction land. Region 2 (patchy expansion near city center) represents the urban living space surrounding the built-up areas of large cities squeezing agricultural production space. Here, the urban living space expands in a patchy pattern. The areas where rural residential areas have undergone changes are concentrated near the built-up areas of large cities. Meanwhile, in areas far from existing built-up areas, the types of rural land remain stable. Region 3 (multi-center-driving expansion) can represent the complex characteristics of rural land use changes surrounding urban living spaces. Under the influence of multi-center poles, rural land gradually transforms into urban land, ultimately forming contiguous areas. In the future, the above three driving forms can be adopted to promote rural transformation. However, relevant departments should also take relevant measures to prevent the intensification of urban–rural land restructuring.

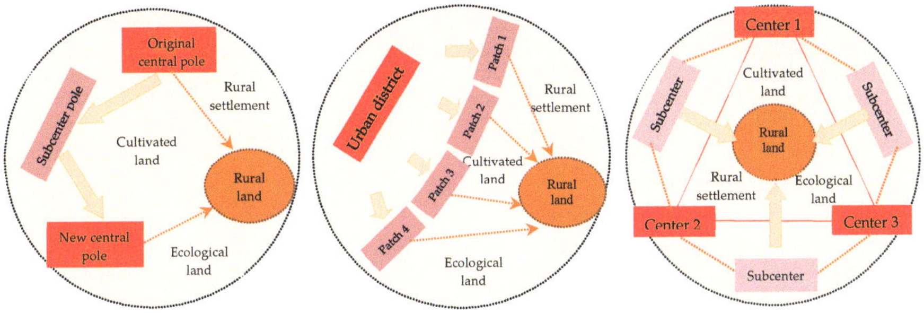

Figure 10. Urban–rural land use restructuring models in southern Jiangsu in 2035.

4.3. Policy Choices for Reducing Land Contradiction in Southern Jiangsu

This paper summarizes simulation results of urban–rural land evolution in various scenarios. The results mean to draw the following implications of policy choices that can be proposed regarding land development and ecological protection.

Firstly, efforts should be made to coordinate urban and village planning and to refine guidance strategies, from overall planning to detailed planning. County towns are the backbone of China's urbanization development and a key hub for the socioeconomic connection between urban and rural areas [34,35]. Government departments should establish a four-level spatial unit system of "county district—County functional area—Township—Administrative village", which would help to achieve systematic step-by-step guidance. In rural areas outside the boundaries of urban development, one or several administrative villages are currently used as units [36]. The township government organizes the preparation of a practical village plan that integrates multiple plans. For administrative villages that include a large number of natural villages for cluster improvement (or some administrative villages with high construction demand), village planning implementation could be accelerated. Administrative villages that include a large number of urban–rural integration of natural villages near the urban core living and cultural circle could also be included in the comprehensive planning of urban control and other detailed planning. The rural land use planning of administrative villages, which includes a large number of relocation, demolition, and merger villages, should adopt flexible management methods.

Secondly, the government should strictly control the scale of construction land and practically increase the quality and efficiency of cultivated land increments. For villages with urban–rural integration or agglomeration improvement, a "series", "cluster", and "satellite town" layout should be adopted. High-quality arable land should be strictly avoided, and the planned indicators for new construction land should be gradually reduced. On the premise of the strict protection of arable land, especially basic farmland, the government should fully leverage the regulatory role of the overall land use plan. According to the requirements of "industrial concentration in industrial parks, agricultural concentration in ecological areas, population concentration in urban areas, and residential concentration in communities", efforts should be made to coordinate the layout of land for various industries. Towns near core urban areas should strictly delineate and permanently protect basic farmland [37] and strengthen the balance between the quantity and quality of cultivated land occupation and compensation.

Finally, decision-makers should adhere to the bottom-line thinking of land planning and strictly implement the regional management of ecological space control. On the one hand, ecological space control areas in urban–rural areas should be both established and improved [38,39]. These ecological space control areas should focus on the protection of the rural ecological environment. Also, the delineation of ecological space should follow the characteristics of "concentration and contiguous, similar functions" in urban–rural ecological space control areas. This approach can make the ecological space control areas relatively complete. On the other hand, the red line of ecological protection [40,41] should ensure the systematization and integrity of ecological functions, guaranteeing that "ecological functions are not reduced, the area is not reduced". Additionally, there is an absolute need to fully leverage the regulatory role of land policies in industrial and economic restructuring and to strictly control high-emission projects.

5. Conclusions

As one of the more active regions in China's economic development, the evolution of land use patterns in the southern Jiangsu region is full of research significance. This paper uses a combination of the SD model and the FLUS model to simulate the spatial reconstruction of land use. Based on system theory, this study predicts land demand in 2035 by using a set of socioeconomic factors. Scenarios A–C are then set up according to different socioeconomic factors. Then, combined with the SD model to simulate land demand, the "point–axis" factors are used to simulate the spatial reconstruction characteristics of land use. Furthermore, typical regions are selected to conduct an in-depth analysis of the spatial reconstruction characteristics of rural land use, and policy choices are discussed. The conclusions drawn from the above research results are as follows:

From 2000 to 2010, there was a significant increase in urban construction land in the southern Jiangsu region, mainly focused on the expansion of urban areas in Nanjing, Wuxi, Suzhou, and Changzhou. From 2010 to 2020, the spatial structure of urban–rural land use was relatively stable. Under scenario A, the area of urban construction land and rural residential areas continues to grow up to 2035. The area of cultivated land, forest land, and water bodies will mainly show a decreasing trend. However, there is little difference in the changes in rural residential areas between Scenario A and Scenario C. In Scenario B, the change rate of land area in various regions is relatively low. The areas with significant changes in the spatial structure of urban–rural land use in 2035 mainly include the border area between Jiangning in Nanjing and Jurong in Zhenjiang, the border area between Wujin and Liyang in Changzhou and Wuxi, Taicang in Suzhou along the Yangtze River, and the border area between Taicang and Kunshan. Considering the simulation of land use, the future evolution of urban–rural land is summarized into three models: single-center-driving expansion, patchy expansion near the city center, and multi-center-driving expansion. In order to reduce the future conflicts and contradictions between urban and rural land use, this paper proposes policy recommendations around land use planning, basic farmland construction, and ecological space.

This research also has some limitations. The selection of driving factors is subjective, and although we have referred to a large number of studies in the literature, the choice of different driving factors can affect the research results slightly. In the SD simulation, due to the limitations of remote sensing technology, it is difficult to form accurate continuous time series data from existing land use data. In addition, the expansion of urban construction land is likely to result in rural residential areas being converted to urban land, which this study did not take into account. In the future, we will attempt to screen and construct a scalable driving factor system based on literature dose methods. Also, we will attempt to collaborate with government departments to obtain accurate land use data over a longer period of time. Exploring the interaction mechanism between rural residential areas and urban construction land at the urban–rural boundary through micro-case studies is also one of our future research plans.

Author Contributions: Conceptualization, C.J. and H.C.; methodology, C.J.; software, C.J.; validation, C.J. and H.C.; formal analysis, C.J.; investigation, C.J.; resources, C.J.; data curation, C.J.; writing—original draft preparation, C.J.; writing—review and editing, C.J. and H.C.; supervision, H.C.; project administration, H.C.; funding acquisition, H.C. All authors have read and agreed to the published version of the manuscript.

Funding: This research was funded by the Social Science Foundation of Jiangsu Province, grant number 23ZXZB022, and Supported by the Fundamental Research Funds for the Central Universities, grant number SKYZ2024014.

Data Availability Statement: Some or all data, models, or code that support the findings of this study are available from the corresponding author upon reasonable request.

Acknowledgments: The Corresponding author acknowledges survey support from the Agricultural and Rural Bureau, Natural Resources and Planning Bureau of Wujin District, Changzhou City.

Conflicts of Interest: The authors declare no conflicts of interest.

References

1. Jiang, C.J.; Li, J.T.; Liu, J.L. Does urbanization affect the gap between urban and rural areas? Evidence from China. *Socio-Econ. Plan. Sci.* **2022**, *82*, 101271. [CrossRef]
2. Wen, S.S.; Xiao, Q.; Li, J.J.; Li, J.P. The Impact of Agricultural Insurance on Urban-Rural Income Gap: Empirical Evidence from China. *Agriculture* **2023**, *13*, 1950. [CrossRef]
3. Jiang, C.J.; Li, Y.X. The Effect of Government Environmental Concern Strength and Differential Supply of Construction Land on Urban Environmental Protection Pressure. *J. Urban Plan. Dev.* **2023**, *149*, 05023033. [CrossRef]
4. Fan, Y.; Yu, G.M.; He, Z.Y.; Yu, H.L.; Bai, R.; Yang, L.R.; Wu, D. Entropies of the Chinese Land Use/Cover Change from 1990 to 2010 at a County Level. *Entropy* **2017**, *19*, 51. [CrossRef]
5. He, C.Y.; Zhang, J.X.; Liu, Z.F.; Huang, Q.X. Characteristics and progress of land use/cover change research during 1990-2018. *J. Geogr. Sci.* **2022**, *32*, 537–559. [CrossRef]
6. Chen, G.Z.; Zhuang, H.M.; Liu, X.P. Cell-level coupling of a mechanistic model to cellular automata for improving land simulation. *Giscience Remote Sens.* **2023**, *60*, 2166443. [CrossRef]
7. Varga, O.G.; Pontius, R.G.; Szabó, Z.; Szabó, S. Effects of Category Aggregation on Land Change Simulation Based on Corine Land Cover Data. *Remote Sens.* **2020**, *12*, 1314. [CrossRef]
8. Xu, Q.L.; Wang, Q.; Liu, J.; Liang, H. Simulation of Land-Use Changes Using the Partitioned ANN-CA Model and Considering the Influence of Land-Use Change Frequency. *Int. J. Geo-Inf.* **2021**, *10*, 346. [CrossRef]
9. Jiao, M.Y.; Hu, M.M.; Xia, B.C. Spatiotemporal dynamic simulation of land-use and landscape-pattern in the Pearl River Delta, China. *Sustain. Cities Soc.* **2019**, *49*, 101581. [CrossRef]
10. Liu, B.C.; Sun, X.B.; Liu, C.H.; Xiang, J. Supercapacitor module quality prewarning based on the improved whale optimization algorithm and GM (1,1) gray prediction model. *Control Eng. Appl. Inform.* **2023**, *25*, 52–59. [CrossRef]
11. Fu, F.; Jia, X.; Zhao, Q.J.; Tian, F.Z.; Wei, D.; Zhao, Y.; Zhang, Y.Z.; Zhang, J.; Hu, X.; Yang, L.C. Predicting land use change around railway stations: An enhanced CA-Markov model. *Sustain. Cities Soc.* **2024**, *101*, 105138. [CrossRef]
12. Han, Z.; Li, B.D.; Han, Z.P.; Wang, S.Y.; Peng, W.Q.; Liu, X.B.; Benson, D. Dynamic Simulation of Land Use and Habitat Quality Assessment in Baiyangdian Basin Using the SD-PLUS Coupled Model. *Water* **2024**, *16*, 678. [CrossRef]
13. Mei, Z.X.; Wu, H.; Li, S.Y. Simulating land-use changes by incorporating spatial autocorrelation and self-organization in CLUE-S modeling: A case study in Zengcheng District, Guangzhou, China. *Front. Earth Sci.* **2018**, *12*, 299–310. [CrossRef]
14. Zhong, Y.Q.; Zhang, X.X.; Yang, Y.F.; Xue, M.H. Optimization and Simulation of Mountain City Land Use Based on MOP-PLUS Model: A Case Study of Caijia Cluster, Chongqing. *Int. J. Geo-Inf.* **2023**, *12*, 451. [CrossRef]
15. He, Z.J.; Wang, X.B.; Liang, X.; Wu, L.; Yao, J. Integrating spatiotemporal co-evolution patterns of land types with cellular automata to enhance the reliability of land use projections. *Int. J. Geogr. Inf. Sci.* **2024**, *38*, 956–980. [CrossRef]
16. Ito, J. Impediments to efficient land reallocation in agriculture: Multi-agent simulation model of transaction costs and farm retirement. *Land Degrad. Dev.* **2024**, *35*, 1553–1566. [CrossRef]
17. Pawe, C.K.; Saikia, A. Simulating urban land use change trajectories in Guwahati city, India. *Int. J. Environ. Sci. Technol.* **2024**, *21*, 4385–4404. [CrossRef]
18. Liu, X.P.; Liang, X.; Li, X.; Xu, X.C.; Ou, J.P.; Chen, Y.M.; Li, S.Y.; Wang, S.J.; Pei, F.S. A future land use simulation model (FLUS) for simulating multiple land use scenarios by coupling human and natural effects. *Landsc. Urban Plan.* **2017**, *168*, 94–116. [CrossRef]
19. Liang, X.; Liu, X.P.; Li, X.; Chen, Y.M.; Tian, H.; Yao, Y. Delineating multi-scenario urban growth boundaries with a CA-based FLUS model and morphological method. *Landsc. Urban Plan.* **2018**, *177*, 47–63. [CrossRef]
20. Abdallah, A.-H.; Ayamga, M.; Awuni, J.A. Large-Scale Land Acquisition and Household Farm Investment in Northern Ghana. *Land* **2023**, *12*, 737. [CrossRef]
21. Wu, Y.; Nico, H.; Yu, L.H. Beyond the collateral channel: Real estate prices and manufacturing firm investment in China. *Int. Rev. Econ. Financ.* **2024**, *94*, 103352. [CrossRef]
22. Wei, Y.H.D.; Gu, C.L. Industrial development and spatial structure in Changzhou City, China: The restructuring of the Sunan model. *Urban Geogr.* **2010**, *31*, 321–347. [CrossRef]
23. Yuan, F.; Wei, Y.D.; Chen, W. Economic transition, industrial location and corporate networks: Remaking the Sunan Model in Wuxi City, China. *Habitat Int.* **2014**, *42*, 58–68. [CrossRef]
24. Gao, J.L.; Yang, J.; Chen, C.; Chen, W. From forsaken site to model village: Unraveling the multi-scalar process of rural revitalization in China. *Habitat Int.* **2023**, *133*, 102766. [CrossRef]
25. Ci, F.Y.; Wang, Z.H.; Hu, Q. Spatial pattern characteristics and optimization policies of low-carbon innovation levels in the urban agglomerations in the Yellow River Basin. *J. Clean. Prod.* **2024**, *439*, 140856. [CrossRef]

26. Katusiime, J.; Schütt, B.; Mutai, N. The relationship of land tenure, land use and land cover changes in Lake Victoria basin. *Land Use Policy* **2023**, *126*, 106542. [CrossRef]
27. Mao, W.; Jiao, L. Land-use intensification dominates China's land provisioning services: From the perspective of land system science. *J. Environ. Manag.* **2024**, *356*, 120541. [CrossRef] [PubMed]
28. Sharma, P.; Ajjarapu, V.; Vaidya, U. Data-driven identification of nonlinear power system dynamics using output-only measurements. *IEEE Trans. Power Syst.* **2022**, *37*, 3458–3468. [CrossRef]
29. Yu, B.; Zhang, C.; Kong, L.; Bao, H.L.; Wang, W.S.; Ke, S.P.; Ning, G.B. System dynamics modeling for the land transportation system in a port city. *Simul.-Trans. Soc. Model. Simul. Int.* **2014**, *90*, 706–716. [CrossRef]
30. Lin, Z.Q.; Peng, S.Y. Comparison of multimodel simulations of land use and land cover change considering integrated constraints-A case study of the Fuxian Lake basin. *Ecol. Indic.* **2022**, *142*, 109254. [CrossRef]
31. Wu, J.Y.; Luo, J.A.; Zhang, H.; Qin, S.; Yu, M.J. Projections of land use change and habitat quality assessment by coupling climate change and development patterns. *Sci. Total Environ.* **2022**, *847*, 157491. [CrossRef] [PubMed]
32. Wang, Y.; Shen, J.; Yan, W.; Chen, C. Backcasting approach with multi-scenario simulation for assessing effects of land use policy using GeoSOS-FLUS software. *MethodsX* **2019**, *6*, 1384–1397. [CrossRef]
33. Ma, R.R.; Zhou, W.; Ren, J.; Huang, Y.H.; Wang, H.Y. Multi-scenario simulation and optimization control of ecological security based on GeoSOS-FLUS model in ecological fragile area in northeast Qinghai-Tibet Plateau, China. *Ecol. Indic.* **2023**, *151*, 110324. [CrossRef]
34. Liu, X.R.; Ma, F.F.; Guo, T.Z.; Ding, Z.W. Spatial pattern of China's rural digital economy based on subjective–objective evaluation: Evidence from 2085 counties. *PLoS ONE* **2024**, *19*, e0292249. [CrossRef]
35. Zhang, H.; Chen, M.X.; Liang, C. Urbanization of county in China: Spatial patterns and influencing factors. *J. Geogr. Sci.* **2022**, *32*, 1241–1260. [CrossRef]
36. Zhang, L.Y.; Ma, X.C. Analysis on the path of digital villages affecting rural residents' consumption ppgrade: Based on the investigation and research of 164 administrative villages in the pilot area of digital villages in Zhejiang Province. *Comput. Intell. Neurosci.* **2022**, *2022*, 1–9. [CrossRef]
37. Wang, N.; Hao, J.M.; Zhang, L.; Duan, W.K.; Shi, Y.Y.; Zhang, J.Y.; Wusimanjiang, P. Basic farmland protection system in China: Changes, conflicts and prospects. *Agronomy* **2023**, *13*, 651. [CrossRef]
38. Li, S.N.; Zhao, X.Q.; Pu, J.W.; Miao, P.P.; Wang, Q.; Tan, K. Optimize and control territorial spatial functional areas to improve the ecological stability and total environment in karst areas of Southwest China. *Land Use Policy* **2021**, *100*, 104940. [CrossRef]
39. Chen, Y.D.; Xu, F.F. The optimization of ecological service function and planning control of territorial space planning for ecological protection and restoration. *Sustain. Comput.-Inform. Syst.* **2022**, *35*, 100748. [CrossRef]
40. Hu, P.; Zhou, Y.; Zhou, J.H.; Wang, G.X.; Zhu, G.W. Uncovering the willingness to pay for ecological red lines protection: Evidence from China. *Ecol. Indic.* **2022**, *134*, 108458. [CrossRef]
41. Zhang, Y.; Yang, B.G.; Liu, Y.; Wu, S.; Cui, Y.J.; Xu, T.H. High-precision Ecological Protection Red Line Boundary Optimization for Fangshan District, Beijing, China. *Sens. Mater.* **2023**, *35*, 835–854. [CrossRef]

land

MDPI

Review

A Comprehensive Bibliometric Analysis of Spatial Data Infrastructure in a Smart City Context

DMSLB Dissanayake *, Manjula Ranagalage, JMSB Jayasundara, PSK Rajapakshe, NSK Herath, Samali Ayoma Marasinghe, WMSB Wanninayake, HUK Dilanjani, ALWM Perera and Yukthi Herath

Department of Environmental Management, Faculty of Social Sciences and Humanities, Rajarata University of Sri Lanka, Mihintale 50300, Sri Lanka; manjularanagalage@ssh.rjt.ac.lk (M.R.); jmsb1610@yahoo.com (J.J.); pskr75@ssh.rjt.ac.lk (P.R.); samankherath4@ssh.rjt.ac.lk (N.H.); marasinghe@ssh.rjt.ac.lk (S.A.M.); sisirawanninayake@gmail.com (W.W.); dilanjani@ssh.rjt.ac.lk (H.D.); manoriperera95@gmail.com (A.P.); yukthi.h@gmail.com (Y.H.)

* Correspondence: dissanayakedmslb@ssh.rjt.ac.lk

Abstract: This study presents a bibliometric analysis of spatial data infrastructure (SDI) research and its application in city development. The fast urbanization and growing complexity of urban management recognize the importance of SDI in supporting sustainable urban planning and innovative city development. This study systematically reviews trends in the publications, key contributors, keywords, and thematic areas of SDI and urban settings. The study uses bibliometric tools such as VOSviewer and Biblioshiny, as well as data from 2003 to 2023. The results show that the number of publications has expanded, and the growth rate in publications has accelerated since 2013, increasing significantly due to geospatial technologies and broadening interest in the concept of smart cities. It identifies the key authors, countries, and collaborative networks that have recognized initiation in the research area. It puts forward the core contributions of Germany, Italy, and Croatia in this field. This research uses keyword co-occurrence and thematic mapping to illustrate dynamic areas of emphasis, including incorporating 3D city models with smart mapping and the application domains of Geographical Information Systems (GISs) and SDI in urban planning. This study further elaborates on other significant developing trends, such as implementing participatory sensing in environmental monitoring and securing SDI within smart city applications. It also highlights enhanced international collaborations toward strengthening the global knowledge base of the challenges in sustainable city development. Hence, this bibliometric analysis is supposed to be used for future research and policy decisions within SDI and city development. Overall, this study will support research by providing a direction for the literature on SDI and city studies and arranging bases for future studies that recommend developing urban resilience and sustainability using the effective practice of geospatial data.

Keywords: bibliometric analysis; environmental monitoring; geospatial technologies; spatial data infrastructure; urban planning

1. Introduction

Urban areas are highly complex and dynamic and thereby have a maximal socio-economic impact not only on urban communities but also on other communities (daily commuters) and peri-urban communities. Over the past few decades, urban areas have seen enormous developments in various aspects. Reshaping the built environment by transforming a rural area into an urban settlement, with a population shift from the rural to

https://doi.org/10.3390/land14030492

the urban area, is a highly complex socio-economic process [1]. An urban-to-rural boundary is demarcated based on the fraction ratio of Impervious Surfaces (ISs) from the city center and whether the composition of the IS fraction is less than 10 (<10%) when compared to the zone designated as a rural area [2]. The paradigm shift from rural to urban reflects a movement in the focus of society, typified by increasing industrialization, advances in technology, and moving from agrarian-based economies to service- and information-based systems. Some researchers [3–5] have consistently emphasized the current trend of urbanization worldwide as the gradual transformation of rural into urban settings. This includes the peripheral growth of cities and their suburbs, which further diffuse into the surrounding countryside and natural landscape. Urbanization has surged in recent decades, and the global urban population surpassed the rural population in 2007 [6]. Since then, cities have become the fastest-growing locality, more so than any other area, and if they continue to gain ground, 60% of the world's population will be living in cities by 2030, and this will progress up to 66% by 2050 [7]. Rapid urbanization causes some negative impacts on cities because urban areas can develop and sprawl without a proper plan [8]. The trends that have marked cities' development over the past few decades are rapid urbanization [9,10], technological development [11,12], and a "green" tendency [13,14]. The background of these developments is that the growing number of people worldwide drives the growth of populations in urban areas, which causes improvements in quality of life [15]. The rapid increase in the building of urban areas has made urban life easy. Still, some negative externalities have directly and indirectly challenged the quality of life of urban residents [10]. Good quality of life is usually associated with smart living [16]. The smart city has emerged as a modern concept to address perceptibly complicated urban development issues. The smart city can be defined as "an urban area which uses electronic data collection sensors, positioned throughout infrastructures, buildings, vehicles, institutions and any device so-called IoT (Internet of Things), to provide in real time information on the basic systems which characterize the operation of the main cities." It is a matter of energy, transport, water supply, sewage, waste, public order, and information and communication [17]. It is anticipated that the direction of smart city initiatives will shift from infrastructure-based strategies towards citizens' lifestyles and sustainable development [18]. Since the turn of the past century, cities have come a long way; they have undergone several phases of development, right from industrialization, which has sped up urban growth and city expansion. Smart cities encompass a plethora of multiple, interlinked, smart built environments, consisting of entities that learn from inter-day activity patterns and autonomously change concerning behavior alterations [19]. The smart city is a new concept and model that utilizes next-generation information technologies such as the Internet of Things (IoT), cloud computing, big data, mobile internet, and spatial geographic information integration to implement intelligent urban planning, construction, management, and service [20]. Indeed, smart city projects involve high investment in many respects: financial, technological, human, data management, and institutional ones [21]. A smart city should entail an urban area with accessible and secure ICT infrastructure, efficient and reliable physical systems that underpin maintenance and development, a productive, innovative economy, an equitable and inclusive society, a sustainable, resilient environment, and participatory transparent governance [22].

Spatial data infrastructure (SDI) is a core element of the smart city concept, as it enables the integration, management, and distribution of geospatial data that enhance the capacity of the city to compile, analyze, and visualize complex spatial relationships for improving urban planning, resource allocation, and the sustainability of smart and resilient cities [23]. Enhanced access, as well as the analysis and visualization of geospatial data, provides many potential benefits [24]. With rapid developments in spatial data collection and communication technologies, the role of SDI is pivoted in the way spatial data use is performed today within companies, governmental agencies, nations, regions, and even global contexts [23]. As a platform, SDI can help reduce duplication and preserve resources collected by various data in response to the common problem of different urban organizations not being aware of existing datasets [25]. Developed countries recognize the importance of this information and have already begun implementing initiatives to create SDI that facilitates the seamless sharing of such information [26]. SDI makes searching, gaining, and assessing information easier when finding necessities and infrastructure. Geographic information is critical in determining various valuable decisions regarding local, regional, and global affairs [27]. Knowing the trends and the current state of research in SDI is crucial in guiding further research on SDI and city development since it points out existing gaps, new emerging technologies, and developing challenges in integrating geospatial data within urban systems.

Studies based on the literature apply different methods of synthesizing and analyzing existing knowledge. Among them, PRISMA and bibliometric methods are the most outstanding. PRISMA is a structured approach designed to ensure transparency and reproducibility in systematic reviews by focusing on rigorous inclusion and exclusion criteria, often yielding qualitative or semi-quantitative insights. In contrast, bibliometric methods are quantitative analyses of large datasets of academic publications to uncover trends, citation patterns, and intellectual structures in a research field [28]. Further, it powers automated tools like VOSviewer and Biblioshiny with the R tool [29] for analyzing co-citation analysis, keyword clustering, and network visualization, streamlining the process and reducing manual effort [30]. Bibliometrics offers a singular advantage over PRISMA by allowing extensive bibliographic data examination and the identification of key contributors, collaborations, and thematic clusters and providing a macroscopic view of research dynamics [31]. This makes bibliometric analysis particularly valuable for understanding research landscapes and guiding future investigations, whereas PRISMA excels in synthesizing evidence for specific research questions. Both methods are complementary, but bibliometrics, with its data-driven approach, offers broader insights into the structure and evolution of knowledge, making it indispensable for fields requiring a holistic understanding of scholarly communication [28,31].

Bibliometric analysis is essential for tracing gaps in any particular research area and emerging concepts and developing future trends. Bibliometric studies provide systematic insights and collective wisdom on patterns in publications, keywords, and citation behaviors and identify the most influential studies, emerging trends, and research gaps, thereby guiding future investigations and ensuring that new research builds on a solid foundation of existing knowledge [28]. When conducting a systematic literature review, data extracting/data mining is crucial in producing science-oriented scholarly works [32–34]. Smart cities and sustainable urban planning have grown tremendously over the past decade, leading to directions for future research. One important analysis could study this discourse on SDI and local progress. At present, scholars have conducted quantitative analysis of the literature on SDI and city development-related research. Bibliometric analysis is essential to understand how research has evolved and what impact it has had in the field of SDI and city research. Most scholarly works available in the existing literature have generally studied city development, urban planning, and smart cities as concepts but have not focused on the role of SDI and its importance. This gap within the literature motivates this paper's research topic. This study aims to bridge the knowledge gap as well as the methodological gap regarding SDI in smart cities and urban development in a novel manner. Hence, our primary focus remains in this research area. Therefore, this analysis enables the discovery of knowledge gaps, the evaluation of the impact of themes, and guidance on future research trends.

The present study aims to explore the prevailing gaps and afford a new influence on SDI and city research. This study identifies the initial expansion phases in SDI and city development research. The study recommends bibliometric analysis to achieve this aim and offer valued comprehension. Scholars can gain a wide-ranging comprehension of the knowledgeable structure and influence of SDI and city research by analyzing publication records, collaboration networks, citation patterns, and other bibliographic measures. The study aims to discover trends and shifts in focus over time by this bibliometric analysis. When one is permitted to look at the trends, it allows the detection of emerging topics, growing or decelerating interest in any particular area, and the diffusion of concepts from one discipline to another. The capability to observe the path of a research field is significant in comprehending its future directions, what scholars can look forward to, and how their work may align with these upcoming developments. The proposed bibliometric analysis methodology will be described first, followed by an overview of the global research output on selected study areas covering publication trends, the collaboration patterns of authors and countries, and widespread keywords and themes [35]. Further, it can help identify the life cycle of the research themes from emerging to peaking and then declining, thus providing a roadmap of where to place reserves in research.

The "author and country analysis" is another primary tool for bibliometric analysis to understand the inputs of individual researchers and specific regions or countries in any given field. The author analysis permits research on the most active and influential scholars and, therefore, the nature of the collaboration network, identifying major experts driving innovation and knowledge creation. This is invaluable in grasping a research field's intellectual structure and identifying possible researchers, collaborators, or crucial people whose work is leading the discourse. In the meantime, co-citation analysis also gives a different aspect to this research. Science mapping is supposed to uncover the intellectual structure of a scientific field. Science mapping focuses on terms such as intellectual structure, scientific evolution, research front, invisible college, and domain analysis [36].

In bibliometric analysis, keywords and thematic analysis comprise core topics and develop SDI research field themes. Such research may permit the recognition of the most frequent terms and concepts at the heart of mapping the intellectual structure of the area under study and understanding the central topics driving the research. Keywords also allow for the potential to recognize emerging trends and shifts in focus within the data, whether in terms of how new topics grow in prominence or how established ones may change over time. This method enables the possibility of showing the patterns and trends by calculating the strength of associations between terms that represent relevant publications and patterns. These associations outline the conceptual structure of a specific scientific field [37]. This is supplemented by thematic analysis and grouping related keywords into broader themes, showing how research topics are intertwined and how they feed into bigger research agendas. Keyword analysis and thematic mapping techniques identify dominant topics and themes in the field, giving a more finely grained understanding of key research trends in the research area [35]. Knowledge of these themes can help strategic planning in research funding and policy-making by underscoring areas of current interest and potential future development.

2. Materials and Methods

In this study, bibliometric analysis was employed as a tool to provide an overview of the entire present scenario of scientific creation, assess the quality of prior studies, and gain extensive knowledge concerning particular subjects. Bibliometric analyses inform researchers about the history of a domain, highlight the present situation of the domain, and propose new research trends [38]. This bibliometric analysis study targeted publications dealing with spatial data infrastructure systems and city research, and the review only contains empirical and review publications. Studies produced in languages other than English were excluded from the data-gathering process in the Scopus database. Scopus is currently the largest existing multidisciplinary database [39]. Most bibliometric analyses involve an average of five stages: research design, data collection, analysis, visualization, and interpretation [40]. In this study, we focus on the following three main stages, which align with this research, as depicted in Figure 1.

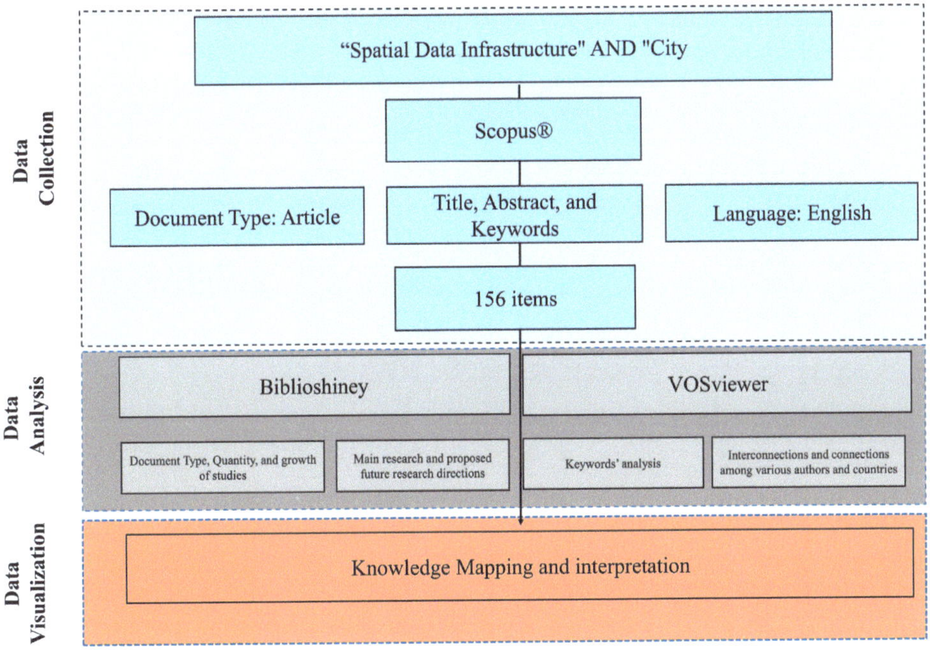

Figure 1. Schematic illustration of conceptual research flow.

2.1. Research Questions

The spatial dynamics of SDI and city research are solely mapped out from a worldwide perception. This sets the basic knowledge framework within SDI and city research, examining developmental trends in these fields and revealing the current SDI and city research landscape. This offers an objective view not only of the existing research in the area but also affords a scientific base for the future directions of studies on SDI and urban studies. More importantly, it has substantial reference value for information science studies in other areas. This study addresses the following questions:

1. What is the development of publications associated with SDI and its application in urban settings?
2. Who are the most cited authors in the SDI study field?
3. What is the total number of articles by country, and how is international collaboration linked to this?
4. What are the related author keywords associated with SDI and city research?

2.2. Data Sources

Data were collected from the Scopus database (https://www.scopus.com) under the keyword string ("City Planning" OR "Urban Development" OR "Smart Cities" OR "Urban Management") AND ("Spatial Data Infrastructure" OR "SDI" OR "Geospatial Infrastructure" OR "Geospatial Data Management") AND ("GIS" OR "Geographic Information Systems" OR "Geospatial Analysis" OR "Spatial Analysis") on 28 June 2024. Two steps were employed in the process, as follows:

First, inclusion and exclusion criteria were applied to select a reliable dataset. In this process, non-English-language articles and non-journal publications were removed to obtain a better review. Finally, we analyzed 156 English-language journal publications and scholarly works, carefully choosing the timespan between 2003 and 2023.

Second, we generated a comma-separated value (CSV) file by selecting appropriate meta-information and chose 156 scholarly works as usable documents for analysis.

2.3. Research Methods

Data records were analyzed using bibliometric tools, focusing on various aspects. We addressed publication trend analysis, countries, authors, and author keywords to gain an in-depth understanding of the collected data. Bibliometrics has become an essential technique for assessing research [41]. The statistical results of the large sample of the academic literature were visualized using bibliometric tools, including the Bibliometrix R package (www.Bibliometrix.org—Version 4.3.2) and VOSviewer (www.vosviewer.com—Version 1.6.20). The extracted data were efficiently organized in a bibliometric way and analyzed using the Bibliometrix software, thus supporting the thorough synthesis of the findings [35]. The Bibliometrix R package provides a set of tools for quantitative research [29].

This study used VOSviewer and Bibliometrix software for various reasons. VOSviewer is one visualization tool that focuses only on bibliometric maps. It has a functionality that is very appropriate for showing large bibliometric maps clearly and easily for interpretation [30]. According to Moral-Munoz (2020) [42], Bibliometrix is a robust tool for scientific mapping and data visualization. The use of Bibliometrix, an R package for bibliometric analysis, and VOSviewer, for network visualization, was based on the structured workflow of our literature review process. First, bibliographic data from Scopus were collected and subsequently analyzed with Bibliometrix to identify publication trends, country collaboration, keyword co-occurrence, and thematic evolution. This gave a comprehensive quantitative outlook of the research landscape. The steps involved mapping bibliometric networks of co-authorship structures, citation linkages, keyword clusters, and collaboration networks between countries by VOSviewer, which helped identify key research themes, influential authors, and collaboration networks among countries. This ensured a statistically solid and graphically represented methodology.

2.4. Data Visualization and Interpretation

Data visualization was accomplished using trend charts, top authors, country maps, and thematic maps [43]. Facts were envisaged via tables and figures, with trend graphs produced for top authors. Next, the tables and statistics were evaluated to draw relevant findings.

3. Results

This study is structured as follows to solve the research questions: First, the publication trends of SDI-related research and its applications in urban contexts are analyzed to understand influential works and their evolution. Next, it involves the analysis of authors and countries to explore the interrelationships among authors, countries, and affiliations, showing how SDI research in urban contexts has historically developed. Then, the study focuses on high-frequency author keywords and their evolution over time to highlight significant research trends and the impact of various contributors. Next, thematic evolution analysis visualizes the knowledge structure, research hotspots, limits, and emerging trends in SDI and urban studies to identify critical themes and future research directions. Finally, the highly cited articles and major contributions in the study area are qualitatively analyzed, and the structural gaps to be explored in the future are highlighted.

3.1. Trend of Publications on SDI and Cities from 2003–2023

Figure 2 presents an annual direction assessment for the number of publications (NoP) between 2005 and 2023. The NoP is divided into three stages; each development stage is provided to trace the trend. Only a few publications were found from 2005 to 2010, reflecting the lack of concern and interest in these concepts. The period from 2011 to 2016 provided results of approximately 15 publications for one year and fewer publications for other years. The maximum number of publications, i.e., 19, was reported for 2019. This denotes that researchers' concern on this topic was more significant this year than the others. From 2020, the aforementioned equal stage of development was observed, present in every year up to 2023, showing a considerable growth in publications. This growth may have arisen from discovering new scientific interests that have become popular with the broader public. This sharp increase in the NoP and the number of their citations is evidence that the "Spatial Data Infrastructure system of a City" can be called a hot topic in the research arena at present. There will be a continuous tendency for this line of investigation; on average, publications present 8.72% of growth yearly. Overall, 4229 references amounted to 12.79 citations in one document, on average.

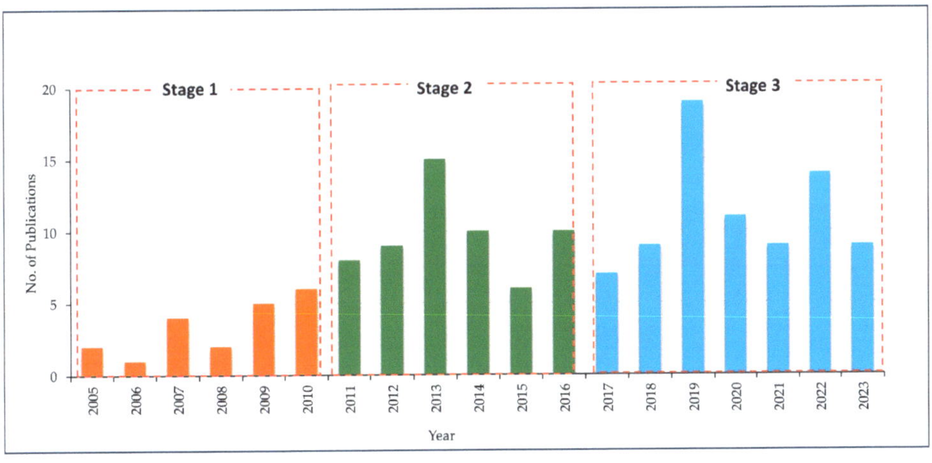

Figure 2. Publication trend in research on "SDI and Cities".

3.2. Authors

Four hundred and sixteen authors contributed to SDI- and city-related publications from 2003 to 2023. The five top-cited authors were Gorger, Gerhard (documents—three; citations—848); Plumer, Lutz (documents—three; citations—848); Kolbe, Thomas H, (documents—three; citations—404); Zipf, Alexander (documents—three; citations—50); and Dawidowicz, Agnieszka (documents—three; citations—13). This uncovers the relationships between authors using co-citation analysis and bibliometric coupling. Under author co-citation analysis, 6808 authors had a minimum of 19 citations for co-citation with other authors. The four clusters in the co-citation network are observable in Figure 3. The most critical cluster is the red one. In addition to having the highest number of authors, n = 06, it has some highly cited authors, such as Gorger G, who ranked first, and Plumer Lutz, who ranked first.

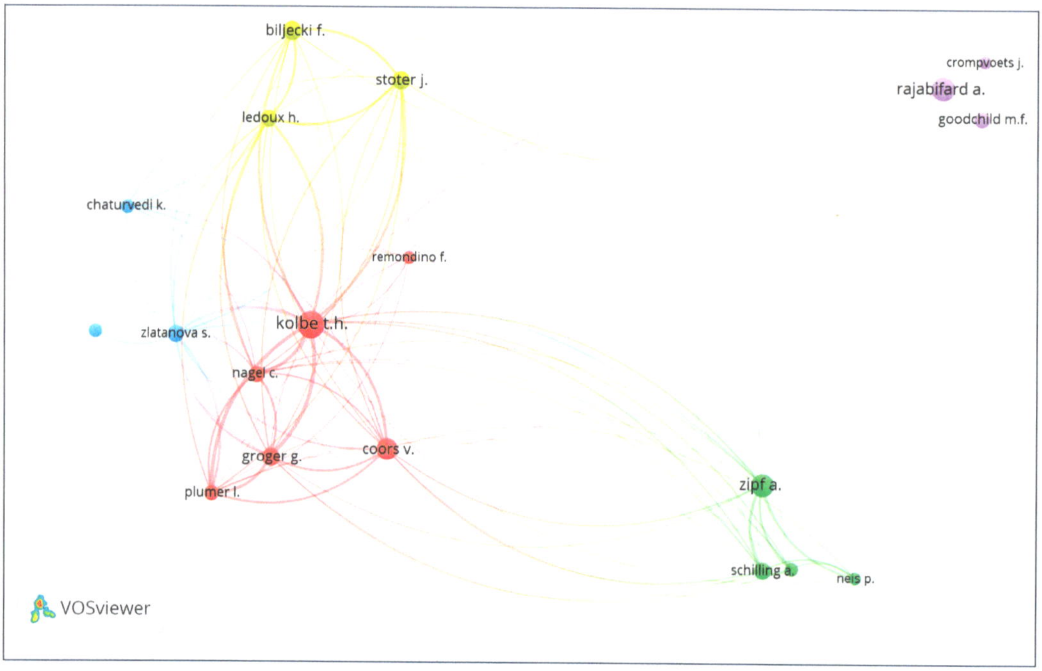

Figure 3. This network map displays clusters of authors (clusters of co-citations).

This cluster is central and connects to all other authors, implying that it enjoys high collaboration and complements research themes. The blue cluster acts more or less like an extension of the red cluster. The authors of this cluster are not entirely isolated from the central authors; therefore, their work relatively builds upon or complements the research conducted by the core cluster. The green cluster is the second-largest cluster, ranked second, and led by Luthe T. This cluster centers on urban analysis and spatial data. These stronger ties between the authors might indicate their interest in urban spatial data infrastructure and its uses, which likely results in closer contributions to GIS and urban planning methods. The yellow cluster is the smallest among the clusters and is also more detached and separate from the other researchers. It is more oriented toward data integration and 3D modeling.

By setting the minimum threshold for an author to seven for the study of author bibliographic coupling, seven authors met this threshold and were grouped into four clusters, as shown in Figure 4. The red cluster at the center of the map includes some of the most-cited authors. This centrality cluster indicates its core role in the field, contributing much to SDI research and urban studies. The bibliographic coupling analysis also represents the central figures and key clusters within SDI and urban research.

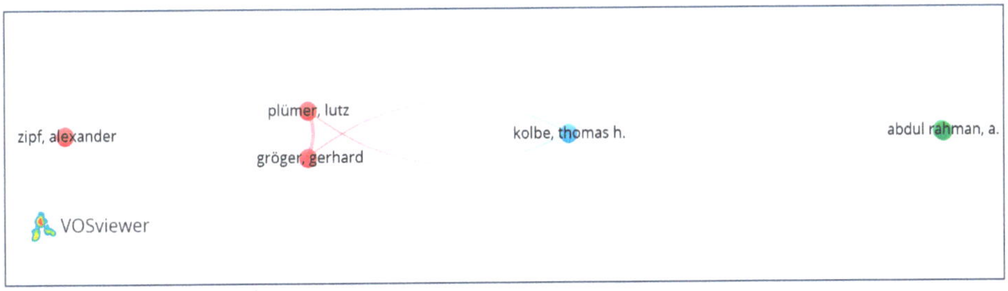

Figure 4. Network map presenting clusters of authors in spatial data infrastructure (SDI) and city research (2003–2023) (clusters of bibliographic coupling).

3.3. Countries

Based on the origin of the authors, forty-eight countries contributed to the publication of SDI and city research within the investigated period of 2003–2023. The country-level citation network was calculated by selecting countries with at least five publications reported in the investigated period (Table 1). In the process, 12 countries were selected and clustered into three sections, as shown in Figure 5.

Table 1. Most productive countries in SDI and city research (2003–2023).

S/N	Country	No. of Publications	No. of Citations
1	Germany	21	1184
2	Malaysia	12	16
3	India	11	31
4	Italy	10	158
5	Croatia	8	21
6	The Netherlands	8	54
7	China	7	90
8	Spain	6	15
9	Hong Kong	5	126
10	Brazil	5	7
11	Iran	5	34
12	Poland	5	13

Note: only countries with at least five published documents are included.

The red cluster represents five countries, led by Italy and Croatia. The visualization shows a global spread of collaborations or citations, representing countries from Asia (China, Hong Kong, Malaysia), Europe (Germany, Spain, Croatia, Italy), North America (United States), and South America (Brazil). The thickness of the connecting lines reflects the strength of these relationships, emphasizing the robust global academic network among these countries.

Figure 6 illustrates a collaboration map presenting the level of partnerships between countries in the SDI and city research domain. Within this context, notable collaboration is exhibited only twice in Croatia, Belgium, and Slovakia. Notable collaboration refers to significant connections between authors, institutions, or countries based on their co-authorship or shared research topics based on the SDI and city research context. Figure 6 depicts no existing collaborative networks among the countries analyzed. Figure 6 indicates the unique geographic distribution of global scientific production, with dark-blue regions, such as Germany, Spain, Croatia, and Italy in Europe and China, Hong Kong, and Malaysia in Asia, representing the highest amount of research being produced. In contrast, less research activity is represented by light-blue regions, and gray represents few or no recorded scientific papers in SDI and city research. This presents an international dimension, as this collaboration is permitted across countries or regions, enhancing information exchange and partnerships.

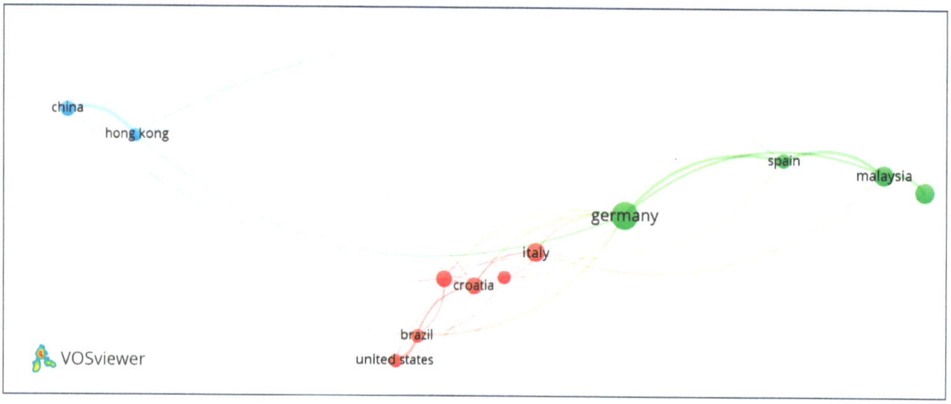

Figure 5. Network map showing clusters of countries in spatial data infrastructure (SDI) and city research (2003–2023): clusters of citations.

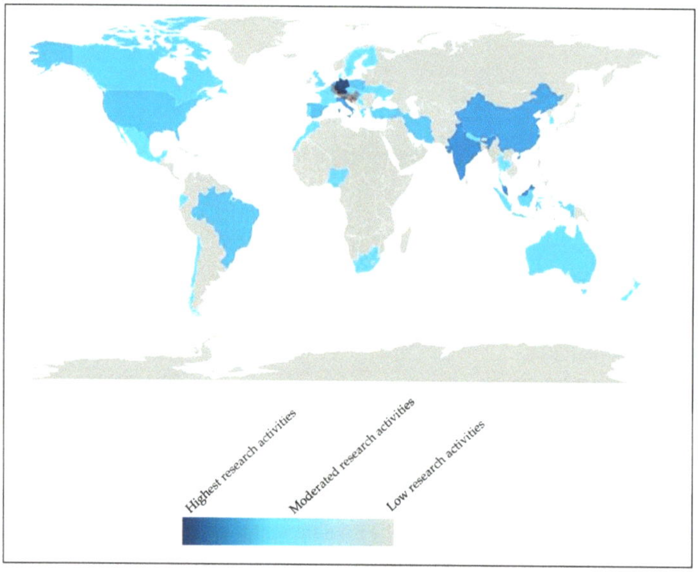

Figure 6. A country collaboration map for the SDI and city research across the world.

3.4. Keyword Analysis

Keyword analysis facilitates the identification of fundamental themes, enhances search tactics, and filters pertinent studies. Researchers may guarantee the inclusion of all relevant material through keyword analysis, enhancing the review process's comprehensiveness. It also facilitates uniformity and reproducibility, enabling others to follow the same process.

Keyword co-occurrence network analysis can identify research hotspots over time. This study used 1194 keywords in the literature on SDI and cities from 2003 to 2023.

By setting the keyword co-occurrence analysis criterion to a minimum frequency of five, 53 keywords satisfied this threshold. The 53 keywords were grouped into three clusters, indicating three different SDI and city domains. Figure 7a, concerning the keywords that belong to the red cluster, "Spatial data and urban," has 21. The list of them is spatial data infrastructure, spatial data, geographic information system, GIS, information management, SDI, decision making, local government, land use, sustainable development, inspire, urban area, urban growth, urban development, European Union, crowdsourcing, open data, volunteered geographic information, and Croatia. In the same way, the most significant cluster is the spatial data infrastructure and city cluster because it has the largest size and contains the keyword most important to this concept, that is, "spatial data infrastructure," which anchors the network center, while the green cluster represents the computer graphics system and contains a total of 19 keywords. Meanwhile, the blue cluster presents "data integration and new technology" with ten keywords.

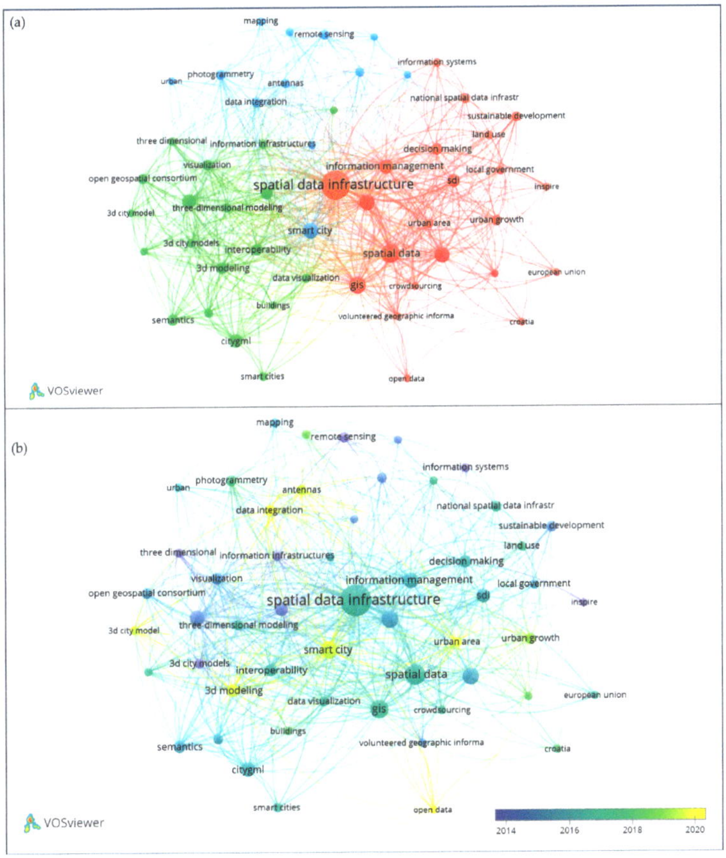

Figure 7. Keyword co-occurrence network map: (**a**) keyword clusters; (**b**) occurrence timeline.

The color gradient (from blue to yellow) shows a trend in research themes from 2014 to 2020. A standard period of keyword occurrence can be discovered for patterns of SDI and city research (Figure 7b). The earlier studies (darker blue) focused on areas like "information infrastructures", "interoperability", and "data integration" between

2014 and 2016. These themes changed, and research moved toward urban applications such as "smart cities", "open data", and "crowdsourcing". This resulted in an increased focus on practical implementations or community involvement. Further, "spatial data infrastructure", "smart city", and "GIS", among others, were the key elements resulting from the network of this study, which seemed to mark their centrality in both fields.

Tree mapping helps visualize the most important key areas based on their size and frequency. As shown in Figure 8, the most prominent term is "Spatial Data Infrastructure (12%)", which denotes a significant focus or interest in spatial data infrastructure and city. The ones that show a substantial concentration in this area are "Spatial Data (6%)", Urban Planning (4%)", "Smart City (4%)", "Geographic Information Systems (3%)", and "Three-Dimensional Computer Graphics". Other smaller but interesting clusters are "Information Management", "Web Services", "3D Modeling", "Urban Growth", and "Remote Sensing". This image is a tree map created to graphically represent the dispersion and the frequency of a set of terms or categories. The size of each rectangle is proportional to the count of occurrences or the significance of the term represented; colors help to discriminate between different categories.

3.5. Main Future Research Directions in the Field of Spatial Data Infrastructure (SDI)

Figure 9 shows the trending keywords used in SDI and city research from 2003 to 2023. These keywords were of prime importance in their use during the formative time of SDI and city research for enhancing sustainable city development. Further, Figure 9 illustrates trending keywords for "SDI and City", from the Biblioshiny application of the Bibliometric package in the R environment, which refers to the most frequent terms occurring in the field. These keywords represent the main topics that recently gained the most attention and influence in these SDI and urban studies. Based on Figure 9, keywords about cities' sustainability enhancement were more prominent during the later stages of growth from 2016 to 2022. Among them are spatial data, spatial data infrastructure, geographic information systems, smart cities, 3D modeling, urban growth, and urban areas, which are apparent in SDI and city research.

A keyword analysis related to the scientific contributions presented in this literature review was conducted to indicate further research areas. Figure 7 depicts the 1194 most meaningful keywords representing the results. The thematic analysis provides a pictorial presentation of the trending themes in this research area [43]. Figure 10, a thematic map, categorizes the developing topics in SDI and urban studies, showing the emergence of and increase in some regions of research. It maps out the relationships between themes and allows the visualization of how these focus topics have shifted through time. This map may reveal some clustering of related keywords or topics, showing the relationship between such concepts as smart cities, the interoperability of data, urban planning, GIS, and sustainability. As illustrated in Figure 10, the quadrant in the bottom right identifies the foundational themes representing the established research issues in the area. According to this, the main themes are "geographic information systems", "spatial data infrastructure", "spatial data", "3D modeling", "photogrammetry", and "urban planning." SDI, GIS, and urban planning are fundamental ideas in the research context. Through thematic analysis, the roles of "smart city initiatives" and "3D computer graphics" are expounded, which are essential in fields that deal with SDI improvement. These themes are relevant in the actual application and visualization of geographic data. They set the grounds for constructing sophisticated applications and technology in SDI. The upper-right corner of the map highlights themes that have gained significance in recent years [43]. The fact that "location-based services" and "mobile navigation systems" are only some of the more prominent specializations proves that they are niche areas of study by themselves. These may grow in the future as technologies and applications evolve.

Figure 8. Word treemap of high-frequency keywords in research field of SDI and cities.

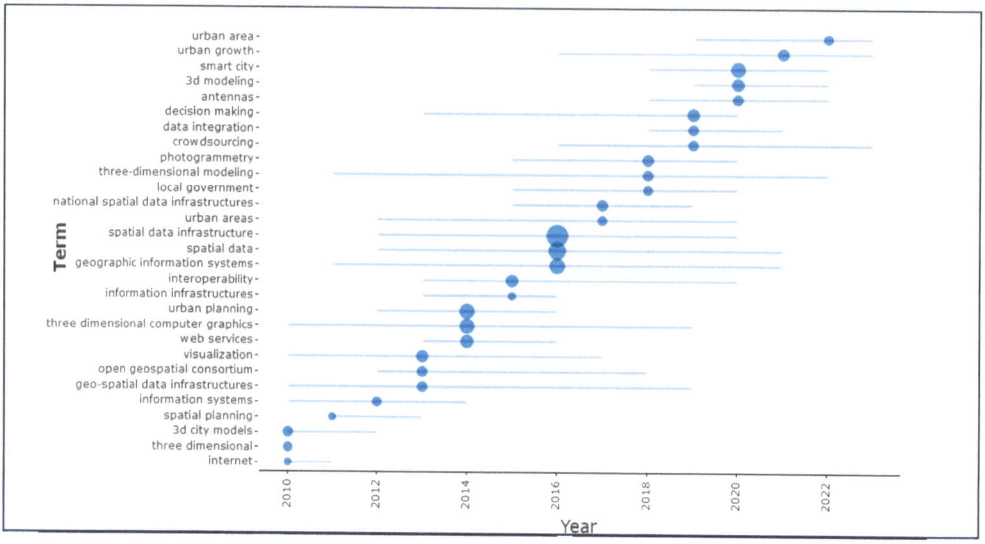

Figure 9. Trending keywords in SDI and city research derived from the Biblioshiny application of the Bibliometric package in the R software environment.

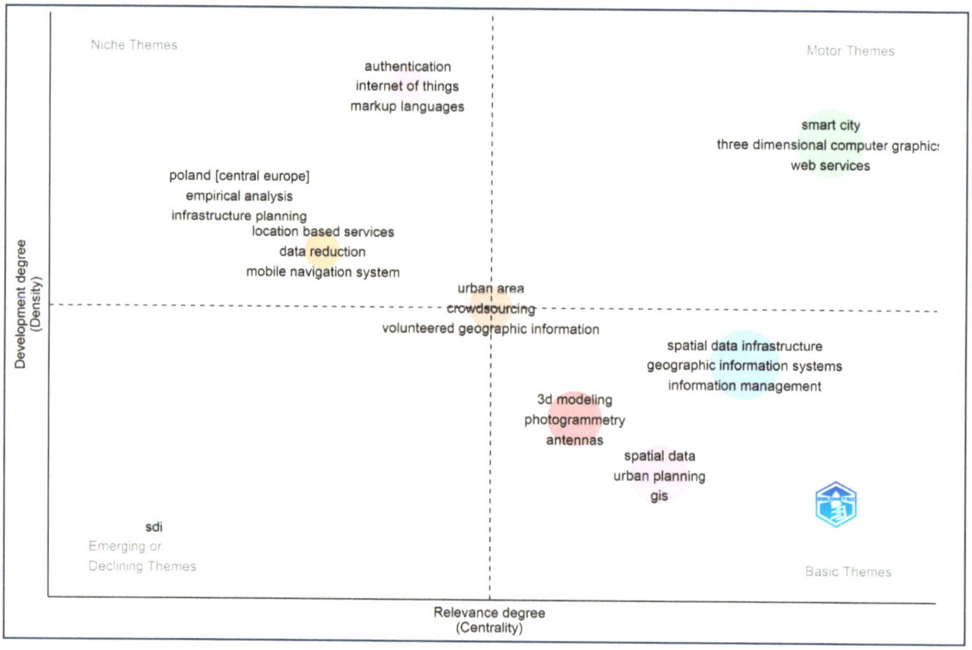

Figure 10. A thematic map of spatial data infrastructure and city research between 2003 and 2023, derived from the Biblioshiny applciation of the Bibliometric package in the R software environment.

The integration of results from Bibliometrix and VOSviewer fairly strengthened this study. Coupling the strong points of each, we executed a more far-reaching bibliometric analysis that could guarantee greater depth and accuracy regarding identifying publication trends, key research themes, influential authors, country-level collaborations, and thematic structures. In performing so, findings could be cross-validated using the combined approach to reduce possible biases arising from reliance on a single software tool. Such a dual-software methodology enhanced the robustness of the analysis and offers the reader a more rounded, fine-grained, detailed presentation of the SDI and city research landscape.

3.6. Qualitative Analysis of Highly Cited Scholarly Works

Highly cited papers are essential in advancing spatial data infrastructure and city research knowledge. These works are also influential contributions that shape research directions, provide insights into foundational ideas, and drive innovative applications. This section overviews the ten most-cited research articles on spatial data infrastructure (SDI) in urban studies from 2003 to 2023. Appendix A summarizes the highly cited articles in this field of study. As is the usual practice used by most past researchers, the title, objectives, methodology, and main findings are the common elements of qualitative analysis [35,43]. While considering these common elements, we add the *relevance of these studies* as a new element, providing the reader with additional insights beyond the traditional methodology. This method ensures a more comprehensive understanding of the subject matter by incorporating emerging research and expanding the scope of information presented.

Geography Mark-up Language (GML), which provides a standardized geometry model, is vital in creating and sharing 3D city models globally with Open Geospatial Consortium (OGC) standards. Gerhard Gröger and Lutz Plümer [44] studied "CityGML—Interoperable semantic 3D city models" and described how City GML could define the following:

> *"geometry", "topology", "semantics", and appearance of urban objects at multiple levels of detail; hence, it is semantically much more prosperous than simple geometric models. It is expected to be available for use in several sophisticated applications like urban planning, navigation, and environmental simulation or facility management. To what extent does geography mark-up language enable extensive application in supporting interoperable data exchange, stimulate broad adoption, and considerably impact scientific studies, such as noise simulation, energy assessment, or disaster management? The vast extent of usage of CityGML worldwide in most urban modeling projects and data exchange projects explains its key role in developing 3D city modeling technologies globally.*

Further, Kolbe et al. (2005) [45] researched "CityGML interoperable access to 3D city models", aiming to present CityGML as a standardized data model and exchange format for representing 3D city models. In this respect, the authors, Thomas H. Kolbe, Gerhard Gröger, and Lutz Plümer, motivated the need for interoperable access to up-to-date 3D city models, in particular for disaster management. Geography Mark-up Language enhances syntax and semantic interoperability in various systems. CityGML considers the geometric and semantic modeling of urban features in five levels of detail (LoDs). The application of CityGML is further demonstrated in disaster scenarios, such as flood simulations or evacuation planning, through its use in enabling precise spatial analysis and integrating multiple data sources. The authors finally concluded the assessment process for CityGML, which remains unfinished because it is part of a pilot project in several German cities; thus, it has higher applicability in urban planning and disaster response.

Over et al. [46] conducted a study based on CityGML's different aspects, titled "Generating web-based 3D city models from the open street map (OSM): The current situation in Germany". They used OSM datasets in the study for 3D visualizations in general, pointed out some of their potential, and summarized the pros and cons of using volunteered geographic information. Some major AI use cases in partial urban planning for street grid design and navigation are traffic management and mission-critical first-responder communications combined with SDI. The research reviewed some approaches to integrating OSM data with the digital elevation model (DEM). It showed the added value from synergy when combined with higher-resolution DEMs for improved visualization, which is essential for preparing SDI systems in any urban context. The authors demonstrated that sophisticated OSM attributes bring about accuracy in 3D city models.

A scholarly work titled "An open-science Crowdsourcing Approach for Producing Community Noise Maps Using Smartphones" [47] was developed in an open science framework using a smartphone app for noise capture to gather physical and perceptual data about noise in the environment (noise levels, GPS data) and how people perceive it (people's pleasantness rating of locations/times). The investigation revealed that the participatory strategy could handle usually constrained simulation boundaries and highly simplified emission models of traditional noise-mapping methods. The authors described (1) a stepwise methodological process with transferable protocols and (2) technological settings for a local multi-level data capturing system with a geographic information system (GIS) as a means of SDI for geo-data management. Preliminary analyses of the data collected to date generally supported the feasibility and utility of this approach despite concerns regarding data quality and user engagement. It provided a case for when community-driven data collection has potential added value in the scale, realism, and precision of environmental noise mapping for broader scientific or urban planning applications.

In another research work titled "Integration of aerial oblique imagery and terrestrial imagery for optimized 3D modeling in urban areas", Wu et al. [48] presented a novel method of incorporating two image types through automatic feature matching and joint bundle adjustment among them and used the combined results to optimize the geometry and texture of the 3D models generated from aerial oblique imagery. They used high-resolution aerial imagery and terrestrial imaging to improve 3D modeling in challenges within an urban setup. Three-dimensional modeling is a critical approach in SDI research that brings virtual reality to urban areas. Investigational analyses were conducted on two datasets of aerial and terrestrial images collected in Dortmund, Germany, and Hong Kong. The outcomes indicated that the proposed approach efficiently integrated images from the two platforms, thereby improving 3D modeling in urban zones [48]. The research handled occlusion problems between the data obtained from the street facades' sides and the geometry's complex structures under consideration. It addressed, therefore, the merging of aerial data with terrestrial data. The proposed method of automatic feature matching and combined bundle adjustment could improve the accuracy and details of 3D models. Real experimental studies based on Dortmund, Germany, and Hong Kong datasets demonstrated the improvement of the fidelity of the geometric shape and the textural quality. These urban data types hold massive potential in 3D city modeling to benefit domains like urban planning, management, and environmental studies by creating highly accurate photorealistic models.

Guillaume et al. [49] carried out a study titled "Noise Mapping based on Participative Measurements". It aimed to address the frustrations of conventional noise mapping techniques that are simulation-based, have poor temporal and spatial resolution, and are only statically mapped. The system filters raw measurements in real time and excludes outliers, synchronizing noise samples to geographical regions or statistical areas. The system is based on open-source software and on implementing the openness for spatial data interoperability standard recently released by the Open Geospatial Consortium to calculate advanced indicators related to environmental noise. They depict the ability of such a system to aid in improved accuracy and realism in noise maps while highlighting that the expectations for crowdsourcing use demand trustworthy approaches to information collection and input techniques by communities. The research indicates that participatory methods in data collection databases support traditional methods and further employ feedback to assist users in urban planning activities for environmental management programs. The proposed architecture rests on outsourced processes that filter outlier sensors and untrustworthy data, cross-reference delocalized noise measurements with geographical and statistical data to provide higher-level indicators, and map the collected and processed data based on web services (ref. [48]), which is similar to the functionality of the SDI system.

Volkan Cagdas [50] researched "Application domain extension to CityGML for immovable property taxation in Turkey". The study recognizes the need for an efficient inventory system that links the legal, administrative, and spatial information required for property taxation. The application domain extension (ADE) defines legal and economic aspects according to Turkish law in CityGML for representing the walls of cadastral parcels as they are in reality, therefore fitting real-world boundaries accurately for parcel building condominium units. It enables geo-information integration between municipal systems for urban planning and facilities management or another administrative services. The research highlighted that City GML is a strong foothold for other municipal, practical applications similar to property tax records. The study presents the design of an ADE for immovable property taxation, which extends the data model of CityGML to include concepts of immovable property coming from Turkish law. The study further shows that such an ADE could play the role of an SDI model for municipal information systems in supporting efficiency in the processes of taxation, urban planning, facility management, and other municipal services [50].

Chaturvedi et al. (2019) completed an important study titled "Securing Spatial Data Infrastructures for Distributed Smart City Applications and Services" [51]. The authors deliberately avoided generalizing the SDI and combined different systems, sensors, and applications under a unique framework. They considered the interoperability and safety of connections regarding different data sources. The research presented the concept of the security of distributed applications and services in such data infrastructure of smart cities. This proposed concept ensures privacy, security, and controlled access to all stakeholders and components through appropriate authorization and authentication mechanisms. The approach facilitates Single-Sign-On (SSO) authentication by a novel combination using state-of-the-art security concepts [51]. It is implemented as a framework in London's Queen Elizabeth Olympic Park, proving that it is a means to allow secure, standardized access to geographically distributed resources while attending to privacy regulations. Thus, this focus furthers the need for strong security mechanisms to protect sensitive information in smart city applications, which are also a hot topic in SDI research in the urban context. This demonstrates that their approach is more effective for managing urban data through the SDI system.

The "3D web visualization of huge City GML models" by Prandi et al. [52] also brings up some important facts about 3D cities and SDI. The authors presented the evolution of a 3D web-based visualization system oriented to large City GML models, describing the main challenges related to efficient web visualization and discussing interoperability issues. In this context, they proposed a service-oriented architecture (SOA) framework compliant with Open Geospatial Consortium (OGC) standards to manage and visualize large-scale 3D city models. The system keeps geometric and semantic data apart, and it could be an additional middleware layer at each level between the storage and the application, which would serve as a bridge to the data access. In the first stage, a framework was designed to meet the OGC's proposed 3D Portrayal Service specification. The second main focus was on developing an effective web client for the efficient rendering of 3D city models. The work constitutes an example of the potential to be tapped from integrating modern web technologies and open standards for improving visualization and access to 3D city models via web interfaces as an application of SDI.

Notably, the systematic review conducted by Loo and Tang [53] on "Mapping Smart Cities" articulates how digital maps and spatial data infrastructure evolved in the process of supporting smart cities, showing how digital maps and GIS moved from mainframe computers into web GIS and mobile applications, highlighting their key role in integrating, visualizing, and analyzing spatiotemporal data produced by IoT devices. They focus on four extensive research directions in innovative mapping: common spatial data standards and platforms, generating collective intelligence, empowering citizens in urban planning, and monitoring city performance. Such research is vital for providing the first comprehensive snapshot of the potential role of maps within smart cities, where valuable real-time socio-temporal data are continuously produced yet often remain underutilized or entirely unused by critical stakeholders in the pursuit of smart city goals [53].

These results were produced with the light limitation of extracting the data only from the Scopus database. However, the results or discussion will not have a significant effect because many scholarly works have followed the same methodology worldwide, and references can be found elsewhere [54–57].

4. Discussion

Our research explores the literature on spatial data infrastructure and urban research from 2003 to 2023, utilizing multi-level bibliometric approaches. The research identified the existing knowledge and methodological gaps in SDI and city research that may be of consequence to the academic discourse in this area. The unique output of this study provides a foundation for the following discussion.

4.1. Research Overview of Spatial Data Infrastructure and City Research

This review of the trends of scholarly works on "spatial data infrastructure systems and cities" from 2003 to 2023 indicates an upward tendency toward growing research activity and interest in the domain. From 2000 to 2012, this topic received less attention, as evidenced by the few articles published, which indicates that those were the early stages of its formation. This was further divided into three stages (Figure 3). Of these phases, Stage I in 2010 had the maximum publications. Stage II's peak of publications in the successive stages was in 2013, while Stage III's was in 2019. This ranged from 2013 to 2019, with steady expansion to an average of about ten articles per year and then a radical jump to 19 in 2019. The trend indicates a renewal of interest in the subject, probably propelled by both technological progress and growing requirements in terms of complexity for the management of towns and cities. Most studies indicate that the focus on developing sustainable and green cities has increased over the past five years [58].

The high and stable increase year by year since 2020 already denotes the development of this area and increasing awareness about its necessity. The excellent annual growth rate and number of references (4229), with an average number of 12.79 citations per document, indicate the growing academic relevance of spatial data infrastructure systems in an urban environment. This is a developing trend; hence, the future looks promising with continuous discoveries and improvements in the subject, led by scientific curiosity and critical urban feasibilities. Uncontrolled growth affects healthy urban development. It is of great practical importance to understand the trends in urban development and forecast future urban spatial patterns [59].

4.2. The Conceptual Foundation of Spatial Data Infrastructure and City

The analysis of author contributions and citation trends in the literature available on spatial data infrastructure and city studies from 2003 to 2023 depicts a highly collaborative and impactful research environment. Among the four hundred and sixteen writers, the five most cited have a considerable academic impact. Gorger and Plumer are in the first place, with the same number of citations, totaling 848. Co-citation analysis, including 6808 authors with at least 19 citations, shows the presence of four key groupings. Based on Figure 3, the red cluster is the largest group with the most frequent references, intensely collaborating in a network with focal influences on research subjects.

In contrast, the blue cluster extends the influence of the red one, meaning that both have complementary study focuses. The green cluster, led by Luthe T., is oriented toward urban spatial data and GIS applications, outlining the latter's role in urban planning approaches. The smaller, independent yellow cluster symbolizes specialized data integration and 3D modeling effort. This bibliometric coupling thus points out the key contributions made by the red cluster to the field, substantiating its critical contributions to SDI and urban studies. This interconnection, together with the prominence of key authors (Figure 3) and clusters (Figure 4), underlines this field's dynamic and collaborative character, which supports the development of urban spatial data infrastructure. With globalization growing, this trend of collaborative studies is anticipated to increase geographical diversity for collaborators, be it individuals, laboratories, or even institutes [60].

This research has substantiated that a strong academic community is also highly networked globally, with 12 countries represented and at least five publications organized into three distinct groups. The red cluster (Figure 5), dominated by Italy and Croatia, shows a significant contribution and collaborative strength. The plot visualizes one extensive network covering all continents, with countries from Asia (China, Hong Kong, Malaysia), Europe (Germany, Spain, Croatia, Italy), North America (the United States), and South America. Among these lines are solid academic ties representing remarkable cooperation, such as between Croatia and Belgium and between Croatia and Slovakia, both present twice. Although such connections exist, the low figure of wide collaborative networks among the countries under study indicates a need for stronger international cooperation.

Generally, international collaboration results in a considerably more significant and broader readership for the authors due to an increased extent of networking, a characteristic of the global nature of research. In the present case, this is a nationwide and united undertaking in the context of a leading, internationally oriented field of science with a high level of scientific quality [61]. This global distribution, paired with current but limited links, underscores the need for strengthened cross-regional cooperation to improve SDI and city research for better knowledge exchange and innovation. Based on Figure 7b, a timeline from 2014 to 2020 shows early topics of research, such as "information infrastructures" and "data integration", from 2014 to 2016, giving way to more recent interests in smart cities, open data, and crowdsourcing, marking an increase in practical implementations and emphasis on community involvement. The high ranking of the terms "spatial data infrastructure", "smart city", and "GIS" in the network underlines their crucial role and continuous relevance for SDI and city research, thus representing a vibrant, changing research environment driven by technological developments and urban applications. Spatial data are fundamental in the creation, management, analysis, and visualization of space with the use of geospatial technology. The management of smart cities and their functional applications all need a combination of geospatial data and technology [62]. Smart cities rely on smoothly integrating physical and digital infrastructure layers [63].

This points out the relevance of smart city initiatives and 3D computer graphics in the practical application and visualization of geographic information. Development towards smart cities has quickly emerged as the most current global trend, and it will likely continue shortly [18]. The improvement of the performance and usability of SDI systems is grounded in these two critical areas. SDI, GIS, and urban planning are principles that provide a basis necessary for developing intricate SDI technologies and applications.

Urban planners employ GIS in monitoring, developing, and implementing their projects for decision-making [64]. The spatial data become relevant to support decisions, planning, and managing cities. In this sense, safe, uninterrupted, and accurate spatial data must be available for city managers and planners to carry out sustainable management and planning [65]. Moreover, the dominance of specialized industries, including location-based services and mobile navigation systems, underlines their distinct but essential role. These industries are likely to grow along with technology; hence, these specialized applications will find increased implementation in SDI systems.

Further research is necessary to identify how best to develop and organize SDI. The establishment of SDI has significantly grown in several countries, particularly at the national level, though in most cases, this has been in regional or local government jurisdictional areas. However, an emerging trend exists in engaging various user communities, including public and non-governmental organizations, to realize the full benefits of spatial data development investments [66]. The current best solution to such spatial community concerns continues to be SDI. Finally, those involved in the SDI development must take optimum advantage of alternative information sources, including remotely sensed data and substitutes for traditional survey methods. Such an approach can help achieve considerable progress without experiencing long delays characteristic of the conventional database development process [67].

4.3. Research's In-Depth Analysis

The studies highlight that CityGML and SDI are essential in enabling forward-looking urban planning, management, and research. The works underline that integrating and sharing such complex spatial data through solid, interoperable frameworks like CityGML will allow a whole range of different domains and thus realize a broad scope of urban applications from traffic management to disaster response. The research also highlights participatory approaches where community-driven data facilitate the increased precision and relevance of urban assessments, particularly in noise mapping and environmental management. The findings also demonstrate the importance of implementing stringent security mechanisms within the SDI to protect sensitive data in urban areas, especially smart cities. These studies guide developing cost-effective, scalable, and more inclusive SDIs oriented around modern web technology interoperability, security, and integration. Hence, the findings will significantly contribute to the further evolution of the SDI, which is a practical tool for optimizing the use of resources, ensuring their sustainability, and improving the quality of life in urban environments by empowering more informed and effective decision-making processes.

This set of studies highlights the growing role of SDI, 3D modeling, and participatory approaches for urban spatial planning, environmental monitoring, and smart city applications. The articles jointly emphasize the flexibility and adaptability of frameworks such as CityGML to support various use cases that range from property taxation [50] to semantic 3D modeling [44]. These frameworks allow for the combination of geometry, topology, and semantic information, which are becoming crucial in urban management and disaster response. Integrating aerial and terrestrial imagery [48] and using crowdsourced OpenStreetMap data for 3D modeling [46] illustrate technological developments to improve the accuracy, accessibility, and cost-efficiency of urban data infrastructures. Studies on participatory noise mapping and community engagement, such as [47,49], provide novel ways of tackling environmental challenges through citizen contributions to enhance the temporal and spatial depth of data capture. Equally important are the security concerns, as noted in [51], which propose robust authentication frameworks for protecting sensitive data in smart city SDI.

Despite these valuable contributions, many considerable challenges and limitations have emerged for these studies. For instance, the application of CityGML requires great computational resources and expertise and can hardly be adopted in resource-limited situations [44,45]. Equally, the variable quality and coverage of crowdsourced data [46] and citizen-contributed noise data [49] raise concerns about their reliability and standardization. Integrating aerial and terrestrial imagery for 3D modeling [48] is promising but computationally demanding, potentially excluding smaller municipalities. Additionally, while participatory approaches foster inclusivity, they face challenges in maintaining engagement and ensuring data accuracy [47]. Security frameworks, although robust, may be complex and costly for smaller municipalities to adopt [51]. Lastly, the conceptual nature of studies such as [53] also has limited practical applicability in diverse global contexts. All these, put together, bring out the need for scalable, inclusive, and context-sensitive solutions in advancing SDIs and urban data management frameworks.

After mentioning the above high-impact scholarly literature, most scholars mention and focus their research on subjects regarding 3D modeling and its related technologies, such as CityGML. Based on an overview of their analysis, one can deduct an apparent lack of emphasis concerning integrating spatial data infrastructure systems within an urban planning context. Moreover, the concepts of SDI systems are emerging as a point of interest among scholars, as reflected in their research themes. This underlines the increasing recognition of the importance of SDI in addressing challenges in urban planning and smart city management.

Bibliometric analysis guides future SDI and city development research by showing trends, influential studies, and new areas of interest. The analysis extracts the most impactful publications, productive authors, and collaborative clusters that drive development in these fields by studying citation patterns, co-authorship networks, and keyword analysis. For example, understanding how the research focus evolved from the foundational concepts of SDI to the integration of 3D city models and smart urban planning may aid in setting future research agendas. In this respect, CityGML is investigated regarding the practical challenges of large city models and their interoperability with potential future extensions of GML. Further, bibliometric analysis might indicate gaps in the literature that set researchers on unexplored topics or innovative approaches with the promise of advancing state-of-the-art research. It will engender interdisciplinary collaboration and ensure the focus of research efforts on the most critical challenges in urban development and SDI.

Most research in the city development and urban management realms, including areas like spatial data sharing, is focused on the issues related to spatial data. Thus, spatial data management systems constitute a significant backbone in decision-making processes in several fields crucial for sustainable city development. SDI is one of the most widely deployed data management systems that efficiently manage geographical or spatial data. Therefore, our discussion of smart cities focuses on SDI.

5. Conclusions

This research reflects on a bibliographic analysis of spatial data infrastructure and city research, giving a fundamental view of trends and influential elements. It is vital to identify the most cited articles, top authors, and major academic journals because they reveal primary knowledge centers and main personalities that drive SDI and city research. Finally, the trends in rising themes in research related to spatial data infrastructure and city themes such as "spatial data infrastructure", "spatial data", and "geographic information system" have been understood to be of great importance. These characterize the active domain of research and innovation where progress is envisaged.

This can also demonstrate collaboration trends between scholars and institutions in SDI and urban studies. Planners, policymakers, and professionals in the city management dimension might relate to such data when searching for collaborations with academic institutions, technology companies, and other industry players to exchange information, access cutting-edge technologies, and create new solutions to urban challenges.

From the practical implications viewpoint, within the realm of spatial data infrastructures and urban management, city planners and policymakers will be better placed to translate the findings of their urban development strategies to improve the quality of life for citizens. This will give them a competitive edge in managing smart cities and sustainable urban environments. In summary, the acquired knowledge will allow city planners and policymakers to harness SDI and other related technologies in urban management for effective, responsive, and resilient cities. These identified trends, gaps, and recommendations are the rich sources that decision-makers use to design and implement strategies, combining technological advancements with data-driven approaches in urban planning and policy-making. This way ensures that policies comply with the dynamic needs of the urban environment for sustainable city development. Since urban development is a changing area, continuous attention is essential to remain up to date with emerging trends, new technological developments, and regulation changes. Such insights will thus enable stakeholders to make informed decisions, thereby improving collaboration in research and applications related to urban planning and smart city management.

Author Contributions: The corresponding and first author, D.D., proposed the topic, spearheaded the data processing and analysis, and wrote the manuscript with the collaboration of M.R. and H.D. Other authors, including J.J., P.R., N.H., S.A.M. and W.W., helped design this research, and A.P. and Y.H. supported research activities. All authors have read and agreed to the published version of the manuscript.

Funding: The authors acknowledge the financial support provided by the Research and Publication Committee, Faculty of Social Sciences and Humanities, Rajarata University of Sri Lanka, for the Environmental Management Department's research project (RJT/R&PC/2024/R/FSSH/03).

Acknowledgments: The authors express their gratefulness to the anonymous reviewers for their valuable comments and suggestions.

Conflicts of Interest: The authors declare no conflicts of interest.

Appendix A

Table A1. Main scientific research in the field of SDI and cities, categorized by authors, titles, objectives, methodology, main findings, and relevance to this study.

Authors	Title	Objectives	Methodology	Main Findings	Relevance to This Study
Gerhard Gröger, Lutz Plümer [44]	City GML—Interoperable Semantic 3D City Models	Present City GML as the OGC standard for 3D city models. Emphasize semantic aspects that should, more broadly, be on semantic issues of 3D city models and their structures, taxonomies, and aggregations.	Review of City GML with comparison to other models compared to purely geometrical/graphical models such as KML, VRML, and X3D.	City GML allows interoperable data exchange and has been applied in many applications. Advanced analysis and visualization tasks of the different levels of detail of application domains in which CityGML is being used at present date are urban planning, indoor and outdoor pedestrian navigation, environmental simulations, and cultural heritage preservation.	This paper points out the actual significance of City GML in the SDI development process and city research through the firm and interoperable framework, which enables a great range of urban applications and integration of complex spatial data across various domains.
Thomas H. Kolbe, Gerhard Gröger, and Lutz Plümer [32]	CityGML—Interoperable Access to 3D City Models	Introduce CityGML as a standardized data model and exchange format for 3D city models, especially in disaster management.	It uses Geography Mark-up Language (GML3) for syntactic and semantic interoperability. It provides five levels of detail about geometric and semantic modeling.	CityGML enables correct spatial analysis with data integration and is considered for evaluation in a pilot project involving a couple of German cities for urban planning.	It provides a standardized framework to integrate distributed data sources within spatial data infrastructures, which is crucial in urban planning, disaster management, and research to develop interoperability in cities.
M. Over, A. Schilling, S. Neubauer, A. Zipf [46]	Generating Web-based 3D City Models from Open Street Map: The Current Situation in Germany	Test how to generate 3D city models from OSM and public height data. Check the OpenStreetMap data for the appropriateness and quality for 3D visualizations of traffic infrastructure, buildings, and POIs.	Reviewing OSM data and the implementation of web services. The use of standards and draft specifications from the Open Geospatial Consortium for all applied web services of the 3D spatial data infrastructure.	OSM provides opportunities for 3D visualization; specialized web services developed for data processing are available. This research demonstrates how the differences in coverage and accuracy vary regionally within the OSM data. In some cases, the quantity of roads far outstrips that of any commercial or government dataset.	This contributes significantly to SDI development in a cost-effective manner and contributes to more participation in creating interoperable, multi-level city models, thus allowing a high coveragence of urban applications, from traffic management to disaster response.
Judicaël Picauta,, Nicolas Fortina, Erwan Bocherb, Gwendall Petitc, Pierre Aumonda, and Gwenaël Guillaumed [47]	An open-science crowdsourcing approach for producing community noise maps using smartphones	Harnessing crowdsourced data on citizens' noise recordings made using a mobile app. Alternative solutions for simulations for achieving the high temporal and spatial granularities recommended by European regulation in environmental noise mapping.	The development of cell-phone applications and open-source tools and the application of spatial data infrastructure. Adopting the spatial data infrastructure concept to modularize the business modules for sound-level data acquisition, analysis, and mapping.	Architecture filters out unreliable data and cross-referencing measurements and promotes citizens' participatory sensing. This allows the mapping of collected and processed data by web services, promoting participatory sensing by citizens with urban environmental observatories.	It contributes significantly to SDI through a transferable protocol and GIS-based data management framework, which can integrate local multi-level data into an assessment. Through this approach, urban noise assessments are be more accurate and participatory, offering a new avenue of citizen engagement in urban planning processes related to city research and bringing environmental noise management closer to reality and the public's perceptions.

Table A1. *Cont.*

Authors	Title	Objectives	Methodology	Main Findings	Relevance to This Study
Bo Wu, Linfu Xie, Han Hu, Qing Zhu, and Eric Yau [48]	Integration of Aerial Oblique Imagery and Terrestrial Imagery for Optimized 3D Modeling in Urban Areas	Integrate aerial oblique and terrestrial imagery to optimize 3D models. Crop the geometry and texture in 3D models generated from aerial oblique images in an urban area with densely distributed high-rise buildings.	Automatic feature matching, combined bundle adjustment. Experimental analysis on datasets of aerial and terrestrial images in Dortmund, Germany.	Integration improves the accuracy of 3D models and visual quality, enabling urban applications. The proposed technique can develop the 3D urban models into more accurate and visually good models. Hence, it would be more appropriate for urban planning, management, monitoring, and environmental studies.	The research supports SDI development by providing a fixed framework through integrative multiple sources of city thematic data, thus achieving improved interoperability and the usability of 3D models in many applications.
Gwenaël Guillaume, Arnaud Can, Gwendall Petit, Nicolas Fortin, Sylvain Palominos, Benoit Gauvreau, Erwan Bocher, and Judicaël Picaut [49]	Noise mapping based on participative measurements	To study the potential of participative measurement through smartphones for environmental noise mapping and develop a system to enhance the accuracy and relevance of these maps by using crowdsourced data.	The OnoM@p system is described in the context that the citizens are responsible for noise data collection, hence requiring them to use smartphones as an acquisition tool. By the standards of interoperability, the filtering and treatment exhibited by the open-source tools account for noise maps, hence sharpening the accuracy and realism of the noise maps. This has already been demonstrated with the added-on advanced data processing features by participatory sensing technology, upon which this method for deriving more precise, high-resolution noise maps is developed, compared to those based on traditional noise-mapping simulation techniques, from which several limitations follow. It presents such opportunities with crowdsourced data in environmental monitoring and urban planning.	It proposed the OnoM@p system, which is developed based on the idea of engaging citizens in collecting noise data with smartphones as an acquisition tool. OFCM uses open-source tools based on interoperability standards for filtering or processing collected data to develop noise maps that refine the precision and realism of the noise map. The research shows how participatory sensing, combined with sophisticated data processing, can generate more accurate high-resolution noise maps, hence avoiding the limitations of characterizing traditional noise mapping simulation-based methods. It shows such possibilities with crowdsourced data in environmental monitoring and urban planning.	The study depicts how integrating community-driven data into traditional methods can achieve more robust and detailed urban noise maps in SDI development. Such noise maps are highly instrumental in giving insights into city research for urban planning and environmental management programs by enhancing the possibility of making effective decisions with the full involvement of communities.

Table A1. *Cont.*

Authors	Title	Objectives	Methodology	Main Findings	Relevance to This Study
Volkan Ça ğdaş [50]	An Application Domain Extension to City GML for Immovable Property Taxation: A Turkish Case Study	Implement an ADE for property taxation and extend City GML by legal and administrative concepts.	The taxation system of Turkey and development of ADE in Turkey. The use of City GML in the context of the semantic and 3D geometrical representation of physical objects in urban areas.	ADE is a 3D national data model enabling efficient taxation and municipal services. It provides for the creation of comprehensive registers and inventories to make full and fair records concerning taxed property, including its legal and economic attributes.	It has also increased the potential of City GML—ADE to support SDI development through a robust framework that enables managing and sharing complex urban data for further research on cities and effective municipal governance.
Kanishk Chaturvedi, Andreas Matheus, Son H. Nguyen, and Thomas H. Kolbe [51]	Securing Spatial Data Infrastructures for Distributed Smart City Applications and Services	Secure distributed applications and services in smart city SDI. Provide privacy, security, and controlled access to all stakeholders and respective components through proper authorization and authentication mechanisms.	SDI concept development and the implementation of SSO authentication. The implementation of Single-Sign-On authentication based on a new combination of state-of-the-art security protocols, including SAML.	Ensures privacy, security, and controlled access with the assurance of security and enhancement of usability of smart city applications. The approach allows users and applications to conveniently interoperate with all systems using SSO authentication, which improves security and usability for distributed smart city applications and services.	It provides basic mechanisms of security in managing urban data and forms a significant basis for the advancement of SDIs with respect to smart cities. It plans a blueprint for enhancing data security and interoperability in city research and development.
F. Prandi, F. Devigili, M. Soave, U. Di Staso, and R. De Amicis [52]	3D Web Visualization of Huge CityGML Models	Create and develop a 3D web-based city model visualization system. Develop web-based efficient visualization systems and interoperable frameworks for access to city models in a standardized way.	Design an architectural framework. The development of web clients. Develop a suitable web client to represent the 3D city model in an appropriate form meaningful to the user.	Very efficient visualization, thus making the 3D models available. The view developed by the system clearly visualizes 3D city models and opens them to a broader audience than only a few experts.	This work establishes the backbone of city research. It will enhance the interoperability and usability of 3D city models, supporting more effective urban planning, management, and decision-making processes through advanced, web-based visualization techniques.
Becky P.Y. Loo and Winnie S.M. Tang [53]	Mapping smart cities	Review the role of digital maps and spatial data infrastructure in smart cities. Identify major research directions of smart mapping to support smart city initiatives further.	Conceptual framework: case studies. This concern examines the roles of digital maps and spatial data infrastructure in smart cities.	Digital maps and SDI are crucial for smart cities; four major research directions identified. This then involves smart mapping to optimize resources and maintain sustainability and quality of life within urban environments.	The work offers valuable insights to city planners and researchers on how SDI is potentially positioned to optimize resources, sustainability, and quality of life in urban setups.

References

1. UN Department of Economic and Social Affairs. *World Urbanization Prospects*; UN Department of Economic and Social Affairs, World Urbanization Prospects: New York, NY, USA, 2018; Volume 12, ISBN 9789211483192.
2. Estoque, R.; Murayama, Y. Monitoring Surface Urban Heat Island Formation in a Tropical Mountain City Using Landsat Data (1987–2015). *ISPRS J. Photogramm. Remote Sens.* **2017**, *133*, 18–29. [CrossRef]
3. Arundel, R.; Ronald, R. The Role of Urban Form in Sustainability of Community: The Case of Amsterdam. *Environ. Plan. B Urban Anal. City Sci.* **2017**, *44*, 33–53. [CrossRef]
4. De Vos, J.; Van Acker, V.; Witlox, F. Urban Sprawl: Neighbourhood Dissatisfaction and Urban Preferences. Some Evidence from Flanders. *Urban Geogr.* **2016**, *37*, 839–862. [CrossRef]
5. Ewing, R.; Hamidi, S.; Grace, J.B.; Wei, Y.D. Does Urban Sprawl Hold down Upward Mobility? *Landsc. Urban Plan.* **2016**, *148*, 80–88. [CrossRef]
6. Dissanayake, D. Land Use Change and Its Impacts on Land Surface Temperature in Galle City, Sri Lanka. *Climate* **2020**, *8*, 65. [CrossRef]
7. United Nations Development Programme (UNDP). *Sustainable Urbanization Strategy—UNDP's Support to Sustainable, Inclusive and Resilient Cities in the Developing World*; United Nations Development Programme: New York, NY, USA, 2016; p. 60.
8. Palevičius, V.; Burinskiene, M.; Antuchevičiene, J.; Šaparauskas, J. Comparative Study of Urban Area Growth: Determining the Key Criteria of Inner Urban Development. *Symmetry* **2019**, *11*, 406. [CrossRef]
9. Ranagalage, M.; Murayama, Y.; Simwanda, M. The Impacts of Landscape Changes on Annual Mean Land Surface Temperature in the Tropical Mountain City of Sri Lanka: A Case Study of Nuwara Eliya. *Sustainability* **2017**, *11*, 5517. [CrossRef]
10. Dissanayake, D.M.S.L.B.; Morimoto, T.; Murayama, Y.; Ranagalage, M.; Perera, E.N.C. Analysis of Life Quality in a Tropical Mountain City Using a Multi-Criteria Geospatial Technique: A Case Study of Kandy City, Sri Lanka. *Sustainability* **2020**, *12*, 2918. [CrossRef]
11. Surabhi Shanker, S.B. The Effect of Technological Advancement on Building a Sustainable City. *Stud. Indian Place Names (UGC Care J.)* **2020**, *40*, 935–940.
12. Mondschein, J.; Clark-Ginsberg, A.; Kuehn, A. Smart Cities as Large Technological Systems: Overcoming Organizational Challenges in Smart Cities through Collective Action. *Sustain. Cities Soc.* **2021**, *67*, 102730. [CrossRef]
13. Tirla, M.-L.; Manea, G.; Vıjulie, I.; Matei, E.; Cocoş, O. Green Cities—Urban Planning Models of the Future. In *Cities in the Globalizing World and Turkey: A Theoretical and Empirical Perspective*; University Press: Sofia, Bulgaria, 2014; pp. 462–479.
14. Pace, R.; Churkina, G.; Rivera, M. *How Green Is a "Green City"? A Review of Existing Indicators and Approaches*; IASS: Potsdam, Germany, 2016. [CrossRef]
15. Mouratidis, K. Urban Planning and Quality of Life: A Review of Pathways Linking the Built Environment to Subjective Well-Being. *Cities* **2021**, *115*, 103229. [CrossRef]
16. Ozkaya, G.; Erdin, C. Evaluation of Smart and Sustainable Cities through a Hybrid MCDM Approach Based on ANP and TOPSIS Technique. *Heliyon* **2020**, *6*, e05052. [CrossRef] [PubMed]
17. Moura, F.; de Abreu e Silva, J. Smart Cities: Definitions, Evolution of the Concept, and Examples of Initiatives. In *Industry, Innovation and Infrastructure*; Springer International Publishing: Cham, Switzerland, 2021; pp. 989–997. [CrossRef]
18. Myeong, S.; Park, J.; Lee, M. Research Models and Methodologies on the Smart City: A Systematic Literature Review. *Sustainability* **2022**, *14*, 1687. [CrossRef]
19. Kandt, J.; Batty, M. Smart Cities, Big Data and Urban Policy: Towards Urban Analytics for the Long Run. *Cities* **2021**, *109*, 102992. [CrossRef]
20. Wu, Y.C.; Sun, R.; Wu, Y.J. Smart City Development in Taiwan: From the Perspective of the Information Security Policy. *Sustainability* **2020**, *12*, 2916. [CrossRef]
21. Lim, Y.; Edelenbos, J.; Gianoli, A. What Is the Impact of Smart City Development? Empirical Evidence from a Smart City Impact Index. *Urban Gov.* **2024**, *4*, 47–55. [CrossRef]
22. Zhu, H.; Shen, L.; Ren, Y. How Can Smart City Shape a Happier Life? The Mechanism for Developing a Happiness Driven Smart City. *Sustain. Cities Soc.* **2022**, *80*, 103791. [CrossRef]
23. Rajabifard, A.; Williamson, I.P. Spatial Data Infrastructures: Concept, SDI Hierarchy and Future Directions. In Proceedings of the GEOMATICS'80 Conference, Tehran, Iran, 29 April–2 May 2001; Volume 10.
24. Ray, N.; Lacroix, P.; Giuliani, G.; Upla, P.; Rajabifard, A.; Jensen, D. Open Spatial Data Infrastructures for the Sustainable Development of the Extractives Sector: Promises and Challenges. In *Spatial Enablement in a Smart World*; GSDI Association Press: Reston, VA, USA, 2016.
25. Olfat, H.; Rajabifard, A.; Qureshi, F.M. Facilitating Urban Management through Local SDI Case Study: The Municipality of Tehran Facilitating Urban Management Through Local SDI Case Study: The Municipality of Tehran. In Proceedings of the GSDI 11 World Conference, Rotterdam, The Netherlands, 15–19 June 2009.

26. Qureshi, F.M.; Rajabifard, A.; Olfat, H. Facilitating Urban Planning and Management at Local Level through the Development of SDI (Case Study of Lahore-Pakistan). In Proceedings of the GSDI 11 World Conference, Rotterdam, The Netherlands, 15–19 June 2009.
27. Longley, P.A.; Goodchild, M.F.; Maguire, D.J.; Rhind, D.W. New Developments in Geographical Information Systems. In *Geographical Information Systems: Principles, Techniques, Management and Applications*, 2nd ed.; John Wiley & Sons Inc.: Hoboken, NJ, USA, 2001; p. 404.
28. Donthu, N.; Kumar, S.; Mukherjee, D.; Pandey, N.; Lim, W.M. How to Conduct a Bibliometric Analysis: An Overview and Guidelines. *J. Bus. Res.* **2021**, *133*, 285–296. [CrossRef]
29. Aria, M.; Cuccurullo, C. Bibliometrix: An R-Tool for Comprehensive Science Mapping Analysis. *J. Informetr.* **2017**, *11*, 959–975. [CrossRef]
30. van Eck, N.J.; Waltman, L. Software Survey: VOSviewer, a Computer Program for Bibliometric Mapping. *Scientometrics* **2010**, *84*, 523–538. [CrossRef]
31. Moher, D.; Liberati, A.; Tetzlaff, J.; Altman, D.G. Preferred Reporting Items for Systematic Reviews and Meta-Analyses: The PRISMA Statement. *J. Clin. Epidemiol.* **2009**, *62*, 1006–1012. [CrossRef] [PubMed]
32. Ferreira, D.R.; Valdati, J. Geoparks and Sustainable Development: Systematic Review. *Geoheritage* **2023**, *15*, 6. [CrossRef]
33. Németh, B.; Németh, K.; Procter, J.N. Informed Geoheritage Conservation: Determinant Analysis Based on Bibliometric and Sustainability Indicators Using Ordination Techniques. *Land* **2021**, *10*, 539. [CrossRef]
34. Zakharovskyi, V.; Németh, K. Systematic Literature Review of the Natural Environment of the Coromandel Peninsula, New Zealand, from a Conservation Perspective. *Conservation* **2021**, *1*, 270–284. [CrossRef]
35. Dissanayake, H.; Popescu, C.; Iddagoda, A. A Bibliometric Analysis of Financial Technology: Unveiling the Research Landscape. *FinTech* **2023**, *2*, 527–542. [CrossRef]
36. Chen, C. Science Mapping: A Systematic Review of the Literature. *J. Data Inf. Sci.* **2017**, *2*, 1–40. [CrossRef]
37. Delecroix, B. Co-Word Analysis for the Non-Scientific Information Example of Reuters Business Briefings. *Data Sci. J.* **2004**, *3*, 80–87. [CrossRef]
38. Durieux, V.; Gevenois, P. Bibliometric Indicators: Quality Measurements of Scientific Publication. *Radiology* **2010**, *255*, 342–351. [CrossRef]
39. Carrera-Rivera, A.; Ochoa, W.; Larrinaga, F.; Lasa, G. How-to Conduct a Systematic Literature Review: A Quick Guide for Computer Science Research. *MethodsX* **2022**, *9*, 101895. [CrossRef]
40. Župič, I.; Cater, T. Bibliometric Methods in Management and Organization. *Organ. Res. Methods* **2015**, *18*, 429–472. [CrossRef]
41. Xie, H.; Zhang, Y.; Wu, Z.; Lv, T. A Bibliometric Analysis on Land Degradation: Current Status, Development, and Future Directions. *Land* **2020**, *9*, 28. [CrossRef]
42. Moral-Munoz, J.; Herrera-Viedma, E.; Espejo, A.; Cobo, M. Software Tools for Conducting Bibliometric Analysis in Science: An up-to-Date Review. *Prof. Inf.* **2020**, *29*, 3. [CrossRef]
43. Dissanayake, H.; Iddagoda, A.; Popescu, C. Entrepreneurial Education at Universities: A Bibliometric Analysis. *Adm. Sci.* **2022**, *12*, 185. [CrossRef]
44. Gröger, G.; Plümer, L. CityGML—Interoperable Semantic 3D City Models. *ISPRS J. Photogramm. Remote Sens.* **2012**, *71*, 12–33. [CrossRef]
45. Kolbe, T.H.; Gröger, G.; Plümer, L. CityGML: Interoperable Access to 3D City Models. In *Geo-Information for Disaster Management*; Springer: Berlin/Heidelberg, Germany, 2005; pp. 883–899. [CrossRef]
46. Over, M.; Schilling, A.; Neubauer, S.; Zipf, A. Generating Web-Based 3D City Models from OpenStreetMap: The Current Situation in Germany. *Comput. Environ. Urban Syst.* **2010**, *34*, 496–507. [CrossRef]
47. Picaut, J.; Fortin, N.; Bocher, E.; Petit, G.; Aumond, P.; Guillaume, G. An Open-Science Crowdsourcing Approach for Producing Community Noise Maps Using Smartphones. *Build. Environ.* **2019**, *148*, 20–33. [CrossRef]
48. Wu, B.; Xie, L.; Hu, H.; Zhu, Q.; Yau, E. Integration of Aerial Oblique Imagery and Terrestrial Imagery for Optimized 3D Modeling in Urban Areas. *ISPRS J. Photogramm. Remote Sens.* **2018**, *139*, 119–132. [CrossRef]
49. Guillaume, G.; Can, A.; Petit, G.; Fortin, N.; Palominos, S.; Gauvreau, B.; Bocher, E.; Picaut, J. Noise Mapping Based on Participative Measurements. *Noise Mapp.* **2016**, *3*, 140–156. [CrossRef]
50. Çağdaş, V. An Application Domain Extension to Citygml for Immovable Property Taxation: A Turkish Case Study. *Int. J. Appl. Earth Obs. Geoinf.* **2013**, *21*, 545–555. [CrossRef]
51. Chaturvedi, K.; Matheus, A.; Nguyen, S.H.; Kolbe, T.H. Securing Spatial Data Infrastructures for Distributed Smart City Applications and Services. *Futur. Gener. Comput. Syst.* **2019**, *101*, 723–736. [CrossRef]
52. Prandi, F.; Devigili, F.; Soave, M.; Di Staso, U.; De Amicis, R. 3D Web Visualization of Huge CityGML Models. *Int. Arch. Photogramm. Remote Sens. Spat. Inf. Sci. ISPRS Arch.* **2015**, *40*, 601–605. [CrossRef]
53. Loo, B.P.Y.; Tang, W.S.M. "Mapping" Smart Cities. *J. Urban Technol.* **2019**, *26*, 129–146. [CrossRef]
54. Ng, S.L. Bibliometric Analysis of Literature on Mountain Tourism in Scopus. *J. Outdoor Recreat. Tour.* **2022**, *40*, 100587. [CrossRef]

55. Feitosa, B.F.; Barros, J.H.T.; Feitoza, J.V.F. Pseudocereals—A Bibliometric Analysis and Literature Review on the Potential for Manufacturing Flours, Bakery Products and Milk Analogues. *NFS J.* **2024**, *34*, 100163. [CrossRef]
56. Palumbo, R.; Manesh, M.F.; Pellegrini, M.M.; Caputo, A.; Flamini, G. Organizing a Sustainable Smart Urban Ecosystem: Perspectives and Insights from a Bibliometric Analysis and Literature Review. *J. Clean. Prod.* **2021**, *297*, 126622. [CrossRef]
57. Dissanayake, H.; Dissabandara, H.; Ajward, R.; Perera, W. SPIRITUALITY AND SUSTAINABILITY: A BIBLIOMETRIC REVIEW. *Agora Int. J. Jurid. Sci.* **2024**, *18*, 165–184. [CrossRef]
58. Javidroozi, V.; Carter, C.; Grace, M. Smart, Sustainable, Green Cities: A State-of-the-Art Review. *Sustainability* **2023**, *15*, 5353. [CrossRef]
59. Id, Y.Z.; Zhao, L.; Zhao, H.; Gao, X. Urban Development Trend Analysis and Spatial Simulation Based on Time Series Remote Sensing Data: A Case Study of Jinan, China. *PLoS ONE* **2021**, *16*, e0257776. [CrossRef]
60. Beaver, D. deB The Many Faces of Collaboration and Teamwork in Scientific Research: Updated Reflections on Scientific Collaboration. *Collnet J. Sci. Inf. Manag.* **2013**, *7*, 45–54. [CrossRef]
61. Van Raan, A.F.J. The Influence of International Collaboration on the Impact of Research Results: Some Simple Mathematical Considerations Concerning the Role of Self-Citations. *Scientometrics* **1998**, *42*, 423–428. [CrossRef]
62. Sharma, P.; Singh, R.; Srivastava, A. Analyzing the Role of Geospatial Technology in Smart City Development. In *Geospatial Technology and Smart Cities: ICT, Geoscience Modeling, GIS and Remote Sensing*; Springer: Berlin/Heidelberg, Germany, 2021; pp. 1–20. ISBN 978-3-030-71944-9.
63. Mortaheb, R.; Jankowski, P. Smart City Re-Imagined: City Planning and GeoAI in the Age of Big Data. *J. Urban Manag.* **2023**, *12*, 4–15. [CrossRef]
64. Xhafa, S.; Kosovrasti, A. Geographic Information Systems (GIS) in Urban Planning. *Eur. J. Interdiscip. Stud.* **2015**, *1*, 85. [CrossRef]
65. Rajabifard, A.; Sabri, S.; Chen, Y.; Agunbiade, M.; Kalantari, M. Urban Analytics Data Infrastructure: Critical SDI for Measuring and Monitoring The National and Local Progress of SDGs. In *Sustainable Development Goals Connectivity Dilemma*; CRC Press: Boca Raton, FL, USA, 2019; pp. 243–255. [CrossRef]
66. Kelly, P. Role of Spatial Data Infrastructures in Managing Our Cities. In Proceedings of the FIG Commission 3 2007 Annual Meeting and Workshop "Spatial Information Management Toward Legalizing Informal Urban Development", Athens, Greece, 28–31 March 2007; pp. 1–11.
67. Ilic, M.; Kilibarda, M.; Sekulic, D. The Role of Spatial Data Infrastructures (SDI) in Smart City Development. *Sustainability* **2018**, *10*, 3120. [CrossRef]

land

MDPI

Article

A Methodology for Using Dynamic Visualizations to Enhance Citizens Engagement in Mobility Planning in Thessaloniki

Maria Konstantinidou * and Josep Maria Salanova Grau

Centre for Research and Technology Hellas (CERTH), Hellenic Institute of Transport (HIT), 6th km Charilaou, Thermi Rd., 57001 Thermi-Thessaloniki, Greece; jose@certh.gr

* Correspondence: mariakon@certh.gr; Tel.: +30-23-1125-7629

Abstract: Many cities are developing semantic 3D models as digital representation of their environment to assess the outcomes of the implementation of new mobility services before their real-life piloting. So far, digital twins have reinforced evidence-based approaches for policy-making. The purpose of this paper is to present how digital twins can be used as a tool for vulnerable citizen engagement, improving efficiency of co-creation in mobility planning. An innovative methodological approach of integrating digital twins in engagement processes, by transforming them from real-time data-driven replicas, targeted to the facilitation of the decision-making process for the public authorities to a tool for enhancing citizen engagement in the mobility context has been developed and tested in Thessaloniki Smart Mobility Living Lab. Through a 3-steps workshops' approach, the mobility needs set by vulnerable users are collected, prioritized and matched with policy plans. By combining citizens feedback with real-time data incorporated in digital twins, citizens receive back visualized scenarios of the impact of specific mobility measures on their neighborhoods. By lifting the citizen engagement in mobility planning to the next participatory level, the results of the study create a bridge between participatory processes and digital twins.

Keywords: digital twin; citizens' engagement; co-creation; mobility planning; vulnerable users

1. Introduction

Citizens' engagement is fundamental in planning processes of forming an urban mobility strategy. End-users' feedback is an important feature for policy-makers to understand mobility services' acceptance and improve the provided transport system. Traditional ways of receiving feedback include actual public consultation in different forms. These methods of identifying the impact of a service are time-consuming in terms of data collection and interpretation and most of the time are not citizen-friendly or even accessible to all citizen groups. This leads to the exclusion of certain groups, such as elderly individuals, people with disabilities, and low-income communities [1], from the mobility planning process and, consequently, the non-responsiveness of the provided mobility services to their specific needs. Moreover, conventional methods are quite generic in terms of geographical focus as the feedback received by the citizens is interpreted as a call for implementing a specific service in the whole city, and rarely is a spatial correlation to target neighborhood-specific requirements performed [2]. Since the advances in technology have resulted in a new efficient representation of the transportation systems in the real world, new digital tools such as digital twins are available, creating a digital environment in which citizens and policy-makers can collaborate to provide better solutions [3].

https://doi.org/10.3390/land14040817

Digital twins first appeared in the product and manufacturing design industries and gradually have been integrated in industries such as aerospace [4], automation [5], shipbuilding [6], healthcare [7] and energy [8,9]. A broader application is now found at the city level. Cities are dynamic complex ecosystems with different services, operating simultaneously. Urban mobility is one of the main activities in urban environments. A vital component to ensure coordinated actions between services in this complicated environment is the representation of these activities in a holistic ecosystem to coordinate further actions to improve its performance [10]. Advances in technologies have resulted in a new representation of the systems in the real world. Big data and Internet of Things (IoT) [3], cloud computing [11] and artificial intelligence (AI) [12] collecting and processing data from different sources create a digital environment in which different groups of stakeholders, policy-makers and citizens can collaborate to select the right mix of new mobility solutions to be implemented.

Digital twins align directly with the objectives of smart cities. A smart city is defined as "a place where traditional networks and services are made more efficient with the use of digital solutions for the benefit of its inhabitants and business" [13]. It collects data through both information and communication technologies (ICT), as well as physical devices connected to the IoT network, enabling the collection, processing, storage, and transmission of data to facilitate communication and decision-making. All the collected data are used to optimize the efficiency of the city operations and services. A digital twin is a virtual model that replicates a real-world urban area/city, allowing the creator to determine the performance of its different systems [10]. Therefore, digital twins act as virtual representations of a physical environment that support smart city operations [14].

Digital twins' applications have demonstrated significant societal and economic value, benefiting municipalities, businesses and communities [8]. By accelerating communication between decision-makers and citizens, digital twins enable more inclusive planning processes. Their interactive features facilitate citizen engagement and empower the collaborative relationships with decision-makers to co-create a resilient and eco-friendly society.

A digital smart-city twin can be used as a tool to represent and engage all stakeholders in a project. It can also provide tools to incorporate feedback through comment sections and polling. Using digital twins, people do not have to be physically present and can access the city's services at their own convenience. Furthermore, a digital twin replica can be updated at any given moment, so the latest changes can always be displayed. Communication between the interested parties can be enhanced since the flow of information is facilitated through new channels [15].

This paper aims at presenting extended opportunities of the use of digital twins as tools in engagement processes. Through the implementation of a three-step methodology, a twofold goal will be achieved: (1) a higher level of engagement of citizens that are typically under-represented in participatory processes by the use of digital twin models, and (2) the validation of citizens' input through visualizations in digital twins, contributing to a more effective and accurate incorporation of citizens' feedback into policy plans. After introducing the benefits of digital twins in general and specifically in citizen engagement in Sections 1.1 and 1.2, respectively, Section 2 presents the methodological approach followed, including the target group and the area of study. Last, the results and their interpretation are described in Sections 3 and 4, respectively.

1.1. Benefits of Digital Twins

Large cities are usually characterized by a dynamism that conceals highly complex social structures and services. Thus, changes to some services affect the operational elements of others, increasing the level of uncertainty for proposed actions during planning processes. Digital twins allow for the simulation of plans before implementing them, exposing problems before they become a reality [16]. From land-use optimization to urban planning, policy-makers will have the power to govern an ecosystem in an effective way [17]. It is now possible to transfer the complicated system of a city to the digital world, enabling town planners to test innovative mobility schemes without taking too many risks, such as low operational performance. This capability of conducting physical experiments based on services while optimizing their efficiency under real-world complex conditions provide policy-makers with a valuable benefit: prior implementation risk management, reducing the uncertainty of the investments [14]. The element that distinguishes a digital twin from any other digital model is its connection to the process. Based on data from the physical system, a digital twin unlocks value by supporting improved decision-making, which creates the opportunity for positive feedback into the physical twin. Digital twins can assist decision-makers in evaluation for a variety of purposes, such as existing situations and future scenarios [18].

Digital twins produce better outcomes for the end users of the services, including increased satisfaction and improved mobility experience, through higher-performing services. Data computational models perform analytics and processing for the city's digital model, providing comprehensive results to policy-makers. Furthermore, digital twins contribute to financial sustainability by reducing infrastructure maintenance costs through predictive analytics, optimizing public transport operations for cost efficiency, and enabling new revenue streams such as monetized mobility data [19]. Digital twins aim to fundamentally transform the decision-making process; the innovative evaluation of a holistic system through a digital replica will be an accelerator key for decision-makers to adopt targeted solutions for a resilient urban environment [10].

Reduced implementation costs could be provided in a digital twin model as the system can be monitored and evaluated using real-time data. Based on information from the cities, industry can evaluate and place the infrastructure needed in specific locations, fueling financial revenue for business. New market corridors and investments will be established as new services are implemented in cities, creating job opportunities for operating the digital twin model and supporting all aspects of the services in the urban environment [10].

Built environments of future cities will need to cope with climate change and align with more intricate energy needs. Testing in a digital environment creates desire features and benefits externally. The reduction of physical experimentation concerning services in the real world will lead to minimized disruptions in the everyday life of citizens in the real world, as well as create the pillars of a more resilient urban environment [19]. Air pollution, congestion, delays and disruptions in the road network are features that decrease the level of resilience of a city. BING and SEA models primarily focus on spatial analysis and strategic environmental assessments, but they do not offer real-time replication, continuous feedback loops, or predictive analysis at the same scale as a digital twin [20]. Also, GIS-based transport planning tools are excellent for visualization and analysis, but they lack the real-time interaction that a digital twin provides. So, in contrast to the GIS-based planning tools, BING and SEA models, in digital twins, the real-time replication and evaluation of sustainable mobility solutions can be achieved, mitigating the levels of disruption occurring in cities. The operation of the proposed mobility interventions can be evaluated alongside other transportation modes in the city using real-time performance data. Parameters and features of the services can be examined, without additional implementation costs [15]. Historical data stored in the smart cities can be used to evaluate the current situation in an urban environment, such as congestion and quality of air, and evaluate future scenarios from the decision-making perspective.

Therefore, digital twins revolutionize traffic management and transport planning by integrating real-time data from IoT sensors, AI-driven decision-making and predictive modelling, unlike traditional tools that rely on historical data and static models, have mainly a reactive nature. By combining multiple data sources, digital twins enable dynamic simulations, scenario-based planning, and automated incident detection. This enhances traffic efficiency, safety, and environmental sustainability by optimizing signal timing, reducing congestion, and minimizing fuel consumption. Moreover, digital twins seamlessly integrate with smart city infrastructure, supporting connected vehicles and next-generation mobility solutions. Their scalability allows for applications in large cities' traffic control, autonomous vehicle coordination, and emergency response optimization, making them a cost-effective and future-ready solution for urban planning and transportation management.

The effectiveness of digital twins in improving traffic management are also recorded in a literature review: a digital twin technology based on Adaptive Traffic Signal Control (ATSC) showed significant reductions in control delays, ranging from 1% to 52% for low traffic demands, from 3% to 19% for moderate demands, and up to 45% for high demands, compared to traditional methods [21]. Research on digital twin-assisted cooperative driving systems at non-signalized intersections demonstrated a 20% reduction in travel time and a 23.7% decrease in energy consumption compared to traditional signalized intersections [22].

1.2. Digital Twins in Citizens' Engagement

The citizen engagement processes rely on three main pillars: connect, exchange, and empower. The connection between citizens and policy-makers is one of the major pillars of citizens' engagement processes. The main goal is to renew the relationship between policy-makers and citizens, to form an integrated transport network in a city. Raising public awareness about the new digital tools through campaigns allows for a more active form of participation from the citizens' perspective. The users of the services understand the importance of co-creation and engagement in the implementation process of mobility services. The policy-makers regain the trust of the citizens, as the new tools will allow the integration of the information received in the real world in a quicker way [2].

Since technological advances have resulted in a new efficient representation of the transportation systems in the real world that citizens have access to, they can collaborate with policy-makers to co-design better solutions [3]. Collaboration with the end users of a mobility service could reveal a handy visualization of a subjective point of view of the services. The extraction and creation of knowledge resulting from the fusion and processing of the information received from the end users can also feed dialogues and debates with citizens and create common understanding on mobility issues and the real needs of the society/neighborhood. Thus, the digital twins' function as an accelerator tool of the traditional time-consuming assessment activities of mobility services [3].

Currently, many cities are integrating digital twin technology into their daily operations for simulating potential scenarios and being able to address urban transportation issues proactively by following efficient decision-making procedures. New York City has adopted digital twin technology to optimize traffic flows and transportation networks [23]. In Helsinki, digital twins are used for public engagement, allowing residents to visualize and provide input into urban development projects before they begin. For example, the Kalasatama Digital Twin project in Helsinki in 2019 produced a high-quality digital twin city model of the Kalasatama area that was shared publicly, serving as a platform for designing, testing, applying and servicing the whole lifecycle of the built environment [24].

A public and open digital twin of the Docklands area in Dublin has been developed for the urban planning of skylines and green spaces. The digital twin serves as a virtual feedback loop where citizens can interact by reporting problems in specific areas and provide feedback on planned changes. It also serves as a co-creation space where citizens can also suggest their own ideas for green space planning. The goal is to engage citizens through digital twins and use the collected valuable feedback for key urban planning and policy decisions. However, the current digital twin does not include simulations related to mobility planning issues, so the evaluation of the impact that a potential urban mobility planning decision may have is not feasible at this stage [25]. Also, the prototype of an urban digital twin for Herrenberg in Germany was demonstrated to the public through a visualization platform for virtual reality and was positively assessed as a technology that could significantly contribute participatory processes [26].

Empowering citizens in decision-making addresses a high level of citizen engagement, where citizens advise and ideas are included in the decision-making and implemented to the maximum extent possible [2]. Digital twins can be directly linked to digital platforms dedicated to collect citizens' feedback, such as the vCity project in Barcelona, which integrates feedback from Decidim (a digital platform already used by over 200 cities for citizen participation) into the digital twin platform of the city, showcasing a structured and direct method for ensuring that urban planning is informed by the perspectives of the people it serves [27]. Citizens can share "objective" data-driven information through their first-hand experience with the implemented services.

Unlike static or simulation models, digital twins are continuously fed real-time data, enabling more informed and participatory decision-making processes, as stakeholders can see the direct impact of changes in real-time. Additionally, digital twins can predict future outcomes by running "what-if" scenarios. This is particularly valuable for participatory decision-making in mobility planning. Also, digital twins often incorporate 3D visualizations and AR/VR capabilities, presenting the city environment in an easily understandable way and thus making it easier for non-experts to contribute to discussions. The interactive environment of digital twins makes them more engaging than traditional reports or workshops.

However, while digital twins offer powerful tools for urban planning and efficiency, their success depends on their alignment with social values, ethical principles, and inclusive policies. A holistic approach that integrates technology with social, political, and cultural considerations will help the integration of social dimensions into urban digital twins. Ensuring equitable access to smart city benefits requires policies that integrate diverse community needs, including elderly individuals, people with disabilities, and low-income communities [1]. For example, real-time traffic monitoring should support accessible transportation options, ensuring that pedestrian pathways and public transport systems are inclusive. Moreover, digital literacy initiatives should accompany the implementation of digital twin systems so that all citizens, regardless of socioeconomic background, can engage with the technology.

The literature on digital twins that integrate social aspects identifies three focus areas, covering both their conceptual development and their practical application: (1) conceptualization and modelling, which aims at replicating human behavior in digital environment; (2) human participation and governance that promotes citizen involvement in the governance of digital twin for enhanced decision-making; and (3) social applications, where digital twins operate as tools to address social challenges [1]. For example, in the city of Bologna, the concept of a Civic Digital Twin has been introduced to support a citizen-centric approach to urban planning and the active engagement of city neighborhoods [28]. Additionally, it is noteworthy that advancements in urban digital twins and citizen engagement is a topic that gathers high interest in the internationally community and lot of workshops are organized bringing together researchers from various countries, including Greece (e.g., the international Digital Twin Cities Centre (DTCC) workshop on urban digital twins where researchers from the Aristotle University of Thessaloniki actively participated) [29].

Although in many cities digital twin users have the ability not only to be informed but also easily exchange information with decision-makers, there is no extended research of digital twins that are used as part of the ideation process and for jointly co-creating and testing suitable mobility solutions with the decision-makers. So, this paper aims to answer the following research questions: Can digital twins serve as a co-creation environment in mobility planning that enables policy-makers to implement improvements in transport systems? Can digital twins enhance citizen engagement and strengthen the communication between citizens and policymakers, leading to more responsive urban mobility strategies?

2. Methods

A Thessaloniki pilot project was implemented in the framework of the DVECE project in 2022 within Thessaloniki Smart Mobility Living Lab—a user-driven innovation environment where users and producers co-create innovation in a trusted, open ecosystem that enables business and societal innovation. The pilot was focused on a selected vulnerable population group in a specific neighborhood, thus representing a variety of mobility challenges and a geographical correlation with the citizens' needs. Through a series of ideation, co-creation and validation/evaluation workshops, the mobility needs set by the vulnerable users were collected, prioritized and matched with policy plans and real-time data incorporated in Thessaloniki's Digi Twin.

This approach allows for the possibility to feed the digital twin with "bottom up" data coming from the citizens and merge policy-driven mobility plans with actual perception and views of the neighborhood inhabitants, testing how the inclusiveness of the developed mobility solutions is improved. The ideation–co-creation–validation/evaluation workshops with the focus group are the core of the methodological approach.

2.1. Target Group

The Thessaloniki pilot focuses on pedestrians in the specific area of the Egnatia corridor. In the ideation and co-creation workshops, a group of people from the Pedestrian Association of Thessaloniki participated, discussing the specific characteristics of Egnatia road, the mobility problems in the area and specific suggested measures to improve mobility conditions for pedestrians moving in the examined area.

In the validation and evaluation workshop, the target group of pedestrians, policy-makers and other stakeholders participated to evaluate the results from the two previous workshops. The Thessaloniki Smart Mobility Living Lab lies at the center of an ecosystem of mobility-related stakeholders in the city of Thessaloniki, which consists of public administrations such as the Municipality of Thessaloniki, public and private transport operators, technology providers, research and academia. The long-term relationships of collaboration and trust between Thessaloniki Smart Mobility Living Lab and the mobility ecosystem enabled the engagement of a significant number of them for the workshops' purposes.

2.2. Selected Neighborhood

Thessaloniki's pilot focuses on Egnatia road—one of the main roads of the city center. Due to high levels of congestion, the lack of parking slots and insufficient infrastructure (e.g., inconsistencies in ramp corridors) it is one of the most challenging areas for pedestrians. The study area was Egnatia road's full length, but greater emphasis was given on the section of the road from Ethnikis Aminis to El. Venizelos Street (Figure 1), where most of the problems are appeared and the largest flows of the pedestrians are observed.

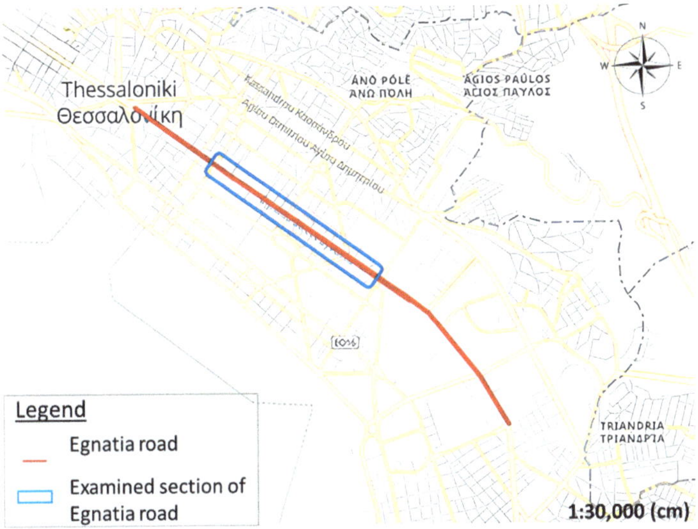

Figure 1. Thessaloniki's pilot area.

Also, Egnatia road is one of the focus areas of Thessaloniki's Sustainable Urban Mobility Plan (SUMP). Thessaloniki's SUMP suggests emblematic interventions in Egnatia road, aiming to reduce motorized traffic. Through interventions focused on the redistribution of the public space—such as the extension of the bicycle network and the available space for pedestrians, the reduction of traffic lanes, and the separation of the two traffic streams with a planted divided median strip—SUMP set as one of its main priorities an increase in the safety and accessibility of the pedestrians on Egnatia road [30].

2.3. Thessaloniki's Digital Twin

Thessaloniki's digital twin (Figure 2), developed by CERTH/HIT, is a cyber replica of the physical transportation system of the metropolitan area of Thessaloniki, allowing for the evaluation of the performance of various mobility solutions of the city's transport system. It consists of static data such as transportation infrastructure and dynamic data from heterogeneous sources [31], with the functionality to formulate predictions throughout the network. It leverages real-time data for predictive traffic management, achieved by the combination of multi-resolution simulations and artificial intelligence algorithms for solving specific transportation problems, such as traffic flow forecasting, managing fleets and providing routing services, ensuring that the transportation network of Thessaloniki operates smoothly under various conditions.

Through the virtual representation, a new technology or/and solution can be examined, and possible behaviors can be predicted to optimize the results and the performance and better respond to citizen's needs. Applied to a system or process, Thessaloniki's digital twin aims to eliminate the need for extended physical experimentation while optimizing performance under different conditions. The employment of Thessaloniki's digital twin allows for the exact simulation of the real transportation system and road network, as well as the implementation of cutting-edge tools and methodologies based on predictive analytics and artificial intelligence for making decisions on the development of the transport network, such as performing calculations to estimate its congestion [32].

The cyber replica of the road network of Thessaloniki Is utilized for combining real data obtained by heterogeneous sensors with simulated data generated through the digital twin for training artificial intelligence algorithms capable of capturing complex spatiotemporal patterns of transportation data [33]. Thessaloniki's digital twin is also harnessed for training agents on challenging tasks and testing their performance in real conditions. Thessaloniki's digital twin can precisely forecast the future status of the transport network, providing management strategies by combining real transport data feeds, artificial intelligence (AI) and machine learning (ML) techniques, as well as high-speed simulations with the emulation of congestion/incident-mitigation measures. It allows us to study fictional scenarios and predict how external actions affect the system while offering added value to policy-makers, since they gain basic insights into aggregated behavior to prevent losing control and understanding of the system. To that end, it can be utilized as a test bed for assessing cutting-edge tools and technologies and examining the effects of potential soft or hard interventions in the transportation system.

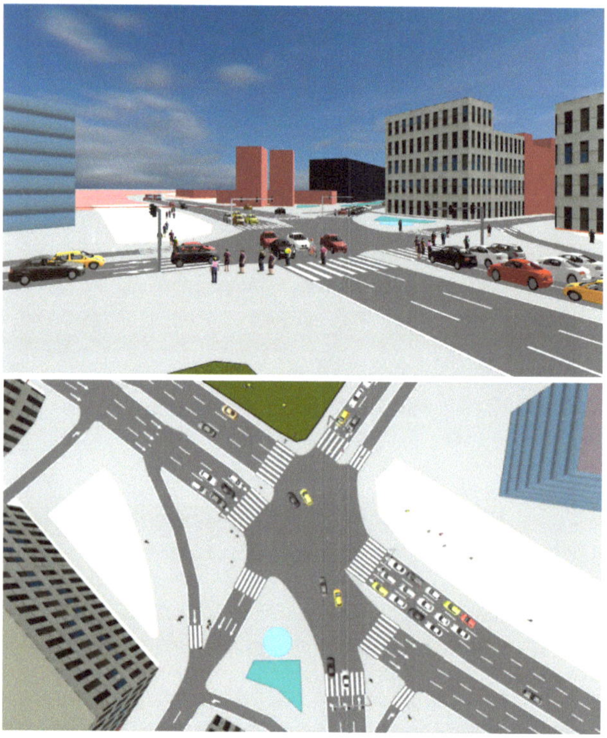

Figure 2. Thessaloniki's digital twin.

The capacity of Thessaloniki's digital twin to translate analytical and simulated outputs into distinctive performance metrics is essential for facilitating data-driven decision-making. Having access to delays, travel times, pollution emissions, speed and routes, performance metrics can be developed, enabling the development of evidence-based transportation management strategies. For performance indicators that cannot be directly measured, such as emissions, Thessaloniki's digital twin can formulate predictions throughout the network, even where detectors are not present [34].

Different types of data are integrated in Thessaloniki's digital twin: digitized road networks, floating car data, Bluetooth sensor data, conventional traffic counter data, public transport schedules and timetables, traffic control systems and shared mobility data. These datasets are used for developing different traffic simulation scenarios.

Thessaloniki's digital twin is mainly used to address issues related to traffic management and routing. Since it simulates traffic patterns in real time, it enables city planners and traffic management centers to implement adaptive signal controls and dynamic routing strategies. It is also used for managing public transportation systems, sustainable urban logistics and emergency response.

The use of Thessaloniki's digital twin as a citizen engagement tool required the integration of new types of data, such as crowdsourced citizen data collected during the ideation and co-creation workshops (feedback on the current mobility situation and suggested improvement measures) and data from field measurements and observations (including measurements of vehicle/bus/motorcycle/bicycle/scooter flows, pedestrian flows on crossings, pedestrian origins-destinations (Ods) and observations of traffic lights intervals). The field measurements of vehicle flows were additionally integrated into Thessaloniki's digital twin for further validating and calibrating the traffic simulation scenarios. The series of simulated observations of the digital twin were statistically compared with the real measurements of the actual system to determine whether reproducing the system behavior is of desirable accuracy and reliability.

Additionally, simulation outputs of the area of interest related to pedestrian flows, numbers of stops for pedestrians, pedestrian speed, pedestrian stop time, pedestrian travel time, and pedestrian walking time were extracted (Figure 3).

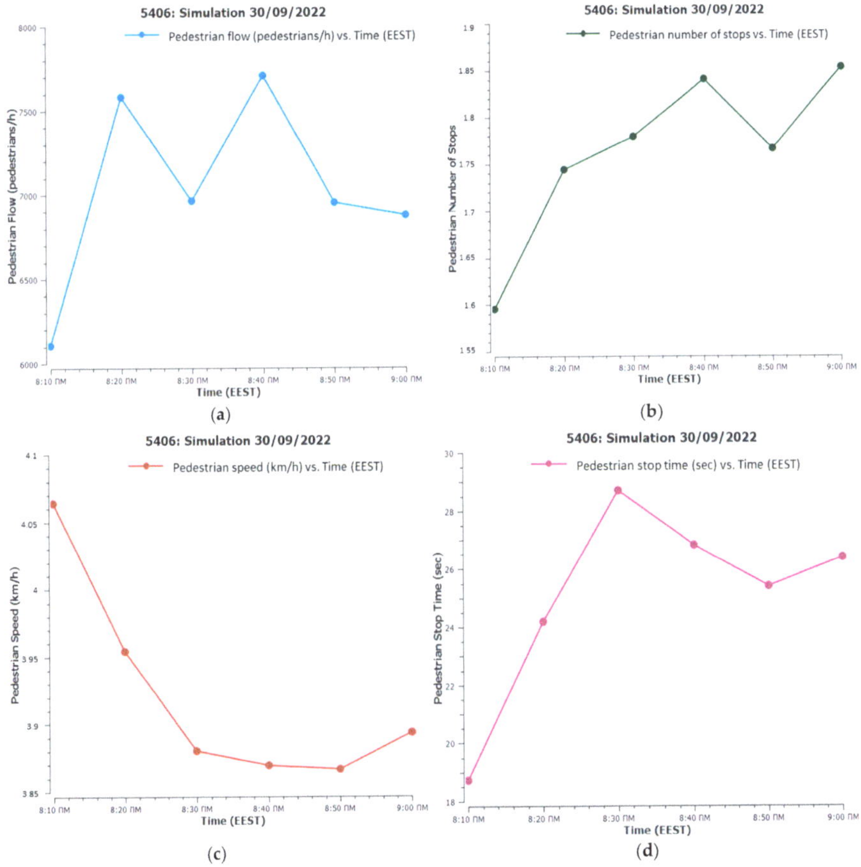

Figure 3. *Cont.*

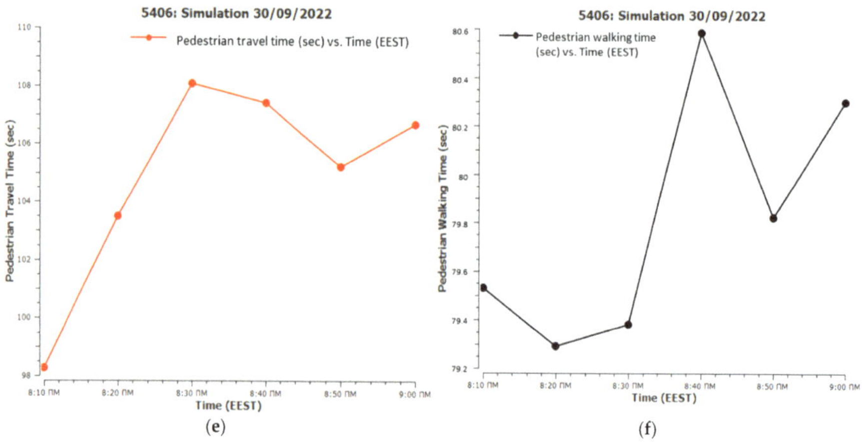

Figure 3. Simulation outputs of (**a**) pedestrian flows; (**b**) number of stops; (**c**) speed; (**d**) stop time; (**e**) travel time; (**f**) walking time during the morning peak time slot.

The integration of crowdsourced data into Thessaloniki's digital twin was performed following three phases (Figure 4).

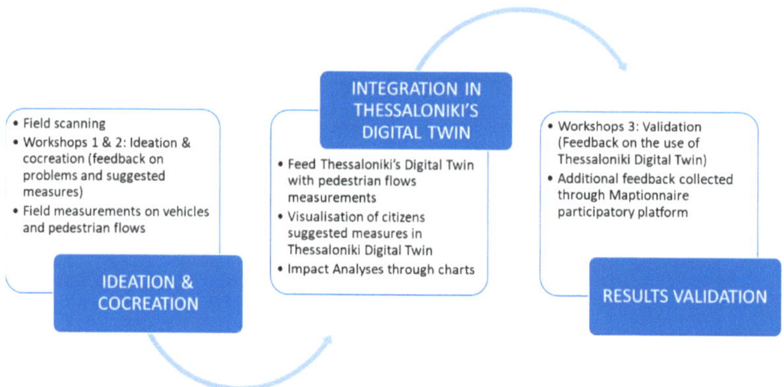

Figure 4. The process of crowdsource integration into Thessaloniki's Digital Twin. Note: The arrows show the sequence of the process in which the outputs of each phase fed the next one.

2.4. *Ideation, Co-Creation and Validation-Evaluation Workshops*

The ideation, co-creation and validation/evaluation workshops were performed in person and were at the core of the methodological approach.

The main objectives of the three workshops were as follows:

- To collect the mobility needs and requirements of the citizens within the ideation workshop;
- To match these needs and requirements with policy plans and real-time data incorporated into the digital twin;
- To co-create neighborhood scenarios within the co-creation workshop;
- To bring designs from the mobility plans and citizens' opinions/feedback into the digital twin;
- To discuss and validate with the policy-makers the results from the visualizations of the proposed measures in the digital twin.

The ideation workshop took place in the presence of individuals from the Pedestrian Association of Thessaloniki. The main objective of the ideation workshop was to engage the participants in a discussion about the specific characteristics of Egnatia road, the definition of their specific mobility needs and the mobility related problems that they face as pedestrians on this street. A detailed report including the main problems that were identified in the field "scanning" was prepared by CERTH/HIT team before the workshop. These problems included bad pavement conditions at several points in the examined area, which makes it difficult for pedestrians to walk safely on the pavements, and short intervals of green traffic lights at specific intersections. More specifically, the green interval of the traffic light for pedestrians at the Kamara crossing was 10 s, which is not sufficient considering the pedestrian load that passes through this specific crossing. The moderators presented these problems along with the related photos and used them as the basis for the discussion with the workshop participants. The participants were encouraged to write down additional problems on Post It notes and put them on the specific location on printed maps to facilitate the next phase of the co-creation workshop.

The co-creation workshop took place with the same group of participants from the Pedestrian Association. The objective of this workshop was to confer with the participants on a list of prioritized mobility-related measures and discuss the impact they could have on the mobility conditions of the examined area. Considering the suggested interventions of the Sustainable Urban Mobility Plan in Thessaloniki and building on the problems discussed during the ideation workshop, participants were asked to propose specific measures to address the main problems of the Egnatia corridor, to improve the safety and the accessibility of the pedestrian network.

The validation/evaluation workshop took place with a wider group of participants, including representatives from the Municipality of Thessaloniki, the Major Development Agency of Thessaloniki, the Pedestrian Association, transport operators (shared bike operators and taxi association) and Thessaloniki Transport Authority.

The main purpose of the third workshop was the validation and evaluation of the measures visualized by the participants. The measures co-created by the participants were visualized in the digital twin, and different impact scenarios were illustrated in charts and discussed with the participants. This gave participants the opportunity to develop a common understanding on the main problems of Egnatia road and the measures that could be implemented to address these problems. In parallel, they were able to gain quantitative insights into the impact (e.g., decrease in waiting time and near-miss incidents) that the suggested measures would have in terms of improving moving conditions for the pedestrians in the examined area. The alignment of the suggested measures that were visualized through the digital twin with the priorities and interventions of Thessaloniki's SUMP was validated by the policy-makers, who agreed to consider them in the revised version of Thessaloniki's SUMP.

3. Results

The key output of the ideation workshop was a list of problems in the examined area, which created a clear image of where the suggested measures of the second co-creation workshop should be focused. The main problems as summarized through the discussion were clustered in five groups:

1. Pedestrian crossings: Absence of a fourth pedestrian crossing at some intersections and a long distance between successive pedestrian crossings;
2. Pavement condition and geometric characteristics: Fluctuations in pavements' width and damaged surfaces;

3. Obstructions and illegal parking: Occupation of the pavements by goods and illegally parked vehicles, and existence of barriers such as signs;
4. Infrastructure for disabled people: Absence of tactile paving and acoustic traffic signal for visually impaired people, and the ramps for impaired people do not comply with the requirements;
5. Signing: Insufficient green intervals for pedestrian crossings and existence of useless signs at specific locations that cause confusion for pedestrians.

During the co-creation workshop, specific measures of high importance were suggested to improve the safety and accessibility of the pedestrian network of Egnatia road; they were clustered as follows:

1. Measures related to intersections: Addition of a fourth crossing (where missing) at some intersections and longer green intervals of traffic lights for pedestrians;
2. Policy measures: Reduction of the speed limit for vehicles from 50 km/h to 30 km/h and creation of low-speed/-emission zone;
3. Infrastructure-related measures: Separation of the two traffic streams with a planted divided median strip;
4. Multimodality measures: Restructuring of the public transport network (bus lanes) due to the operation of the metro, creation of a bicycle lanes in the examined area and connection with the existing bicycle network, aligning with the promotion of micro-mobility and active mobility modes.

Additionally, a list of more generic measures was developed, including interventions related to the regular maintenance of sidewalks and removal of the obstacles; the existence of tactile paving and acoustic traffic signal for visually impaired people along the entire length of the sidewalks; the placement of protective (trees, etc.) parallel to the sidewalks; yellow grid lines where needed in roads; the creation of parking spaces for cars and motorcycles; widening sidewalks; and reconsidering the location of traffic lights and their placement before pedestrian crossings.

During the validation/evaluation workshop, the priorities, needs and suggested measures for Egnatia (as they derived from the ideation and co-creation workshops) were discussed and validated with the policy-makers collecting feedback on the technical and economic feasibility of the measures' implementation.

Some of the suggested measures were visualized in Thessaloniki's digital twin, and their impacts were captured in related charts. Two of the main interventions simulated were (a) increasing the green interval of the traffic lights for pedestrians at the Kamara crossing (as shown in Figure 5) that has important impact on pedestrian waiting time and on near-misses (as shown in Figure 6) and (b) adding a fourth crossing at the Iasonidou and Egnatia intersections (as shown in Figure 7) that decreases the pedestrian waiting time and the near-miss incidents (as shown in Figure 8).

The measure of adding a fourth crossing was considered particularly important by the participants. However, it was mentioned that the fourth crossing would in some cases create a problem for vehicles turning left and right, as it would significantly limit the time available to drivers for these movements. The increase in green interval time for pedestrians, observing the charts presented, was considered a measure that must be implemented, considering the assumption that such a measure would cause an increase in vehicle traffic congestion. Regarding the illegally parked vehicles on the sidewalks, the responsibility of the Municipality in their removal (especially motorcycles parked illegally on the sidewalks) was emphasized.

Figure 5. Increasing the green interval of the traffic lights for pedestrians at the Kamara crossing.

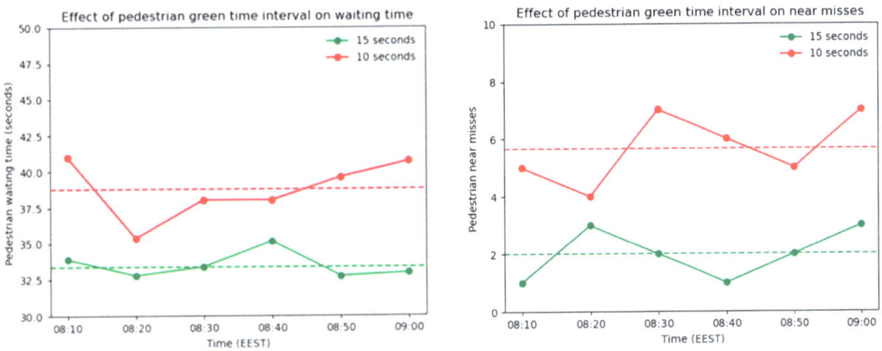

(**a**) (**b**)

Figure 6. Effect of increasing the green interval of the traffic lights for pedestrians at the Kamara crossing (**a**) on waiting time; (**b**) on near misses.

(**a**) (**b**)

Figure 7. Iasonidou and Egnatia intersection (**a**) before the addition of a fourth crossing; (**b**) after the addition of a fourth crossing.

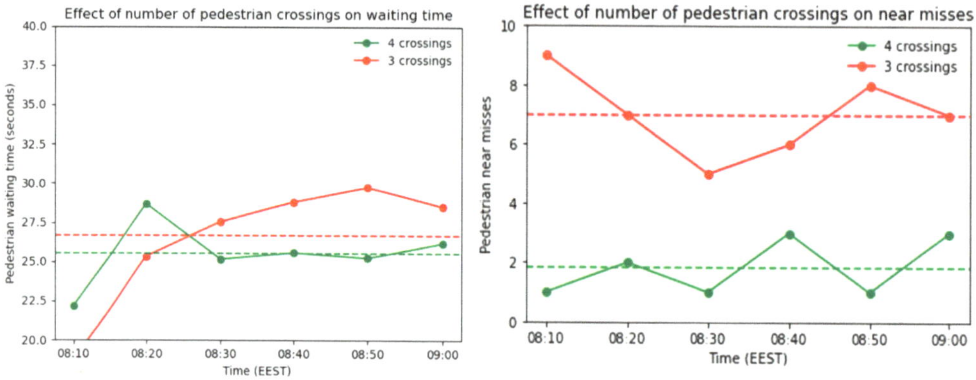

(a) (b)

Figure 8. Effect of addition of a fourth crossing at Iasonidou and Egnatia intersection (**a**) on waiting time; (**b**) on near misses.

4. Discussion

The current paper presents an innovative methodological approach of using digital twins by transforming them from real-time data-driven replicas targeted to the facilitation of the decision-making process for the public authorities to a tool for enhancing citizen engagement in a mobility context. Collaboration with the end users of a mobility service could reveal a handy visualization of a subjective point of view of the services. Thus, digital twins, when combined with structured participatory processes, can function as an accelerator tool for the traditional time-consuming assessment activities of mobility services.

By using Thessaloniki's digital twin and a participatory three-step workshop methodology (ideation, co-creation, and validation), citizens were able to articulate their needs, evaluate mobility solutions, and visualize the potential impact of proposed interventions. The use of digital twins in co-design processes was positively assessed by all the participants in the workshops, and the knowledge extracted from the methodological approach on how to update and bring designs from the mobility plans and citizens' opinions into the digital twins was considered added value.

The findings underscore the ability of digital twins to visualize the direct impact of proposed measures, thereby fostering transparency, trust, and more informed decision-making among all stakeholders. The suggested measures will be taken into account for the revision of Thessaloniki's SUMP.

An important conclusion extracted from the three workshops was the need of the different stakeholders to communicate regularly—not only within specific projects but also in broader conversations about the mobility conditions in the city of Thessaloniki and what is really needed for improving them, considering different groups of end users, including vulnerable users. The need for scheduling regular meetings between all stakeholders (citizens and policy-makers) was highlighted as a critical point to achieve ecosystem cooperation, moving towards the achievement of better transport conditions for all citizens, including vulnerable groups. Key findings of the applied methodology indicate that digital twins not only improve the efficiency of participatory planning but also foster stronger communication between citizens and policymakers, leading to more responsive urban mobility strategies.

The extraction and creation of knowledge resulted from the fusion and processing of the information received from the participants of the workshops within Thessaloniki Smart Mobility Living Lab fed dialogues and debates between the participants. The use of a digital twin created a common understanding about the real needs of the pedestrians on Egnatia road, the mobility issues and the measures that need to be implemented to facilitate the communication between the interested parties and keep them engaged during the whole process.

The findings of the present study corroborate and go beyond the existing literature that proposes digital twins as promising tools for participatory urban planning. In contrast with the Kalasatama digital twin project in Helsinki [24] and the Dublin Docklands project [25] that focused mainly on infrastructure-change visualization and feedback collection, the present study shows how digital twins can also support the co-creation of mobility measures and validate citizens' input through simulation. The integration of citizens' needs into the digital twin is aligned with civic digital twin research, such as that in Bologna [28], but goes further by applying the digital twin for the quantitative evaluation of specific pedestrian-oriented interventions (e.g., pedestrian green light timing).

Moreover, this study provides results on digital twins' ability to improve mobility conditions (e.g., travel time and safety), confirming previous findings from Wang et al. [22] and Dasgupta et al. [21]. The outcomes contribute to this part of the literature by demonstrating how even small-scale pedestrian-focused measures (e.g., green interval adjustments) can be evaluated through digital twins using co-produced data.

The application of the methodological approach in Thessaloniki Smart Mobility Living Lab has two main policy implications. This study demonstrates how digital twins can examine the impact of specific mobility-related interventions (e.g., improvement in pedestrian waiting times and decrease of near-miss risks) and directly inform and revise policy strategies and plans such as SUMPs. It also highlights the need for institutional coordination among municipalities, researchers, mobility operators and civil society. This calls for governance frameworks that support regular data-sharing and joint decision-making. Enabling real-time participatory feedback loops through the integration of participatory platforms such as Decidim in Barcelona [27] will ensure the replicability of the suggested methodology in other cities.

Based on these implications, five main policy recommendations were also extracted:

1. Incentivize citizens to participate in planning processes.

Citizen engagement is a critical point for policy-makers to be able to address mobility challenges at local scale. Participation in workshops does not mean active engagement in planning processes. A certain level of frustration was expressed many times by the citizens, as they are regularly invited to workshops organized by the cities, which requires investing part of their personal time, but their expectations of what should be done with the results of the workshops are not met at the end. Citizens should participate in all the steps of the planning processes—from the design to the implementation and evaluation. Giving them incentives to participate (e.g., by using gamification and/or rewarding schemes) and offering feedback on the procedure results after the end of the project/initiative will keep them engaged long term.

2. Cooperation of all mobility-related stakeholders.

The implementation of radical measures that improve mobility conditions in local areas for vulnerable users requires the cooperation of all related stakeholders and citizens. It is important for all the key stakeholders involved to have continuous contact and exchange views—not only within specific projects but also in broader conversations about mobility conditions in the city, to identify common goals and action points. The establishment of a regular dialogue and communication ensures a better understanding of what is really needed for improving mobility for different groups of end users, including vulnerable users, as well as the potential and limitations in the adoption of new engagement tools.

3. Adoption of the digital twin for citizen engagement with regards to accessibility and/or participation in the city planning processes.

It is recommended to have a proper introduction of what a digital twin is and the purposes it serves, especially for people who hardly or never work with a data-driven approach. Depending on the policy life cycle, digital twins can be used differently and have different functionality needs, e.g., in the planning phase, more focus might be necessary on numbers and facts, while the engagement phase might benefit from better visualization and a gaming component. Digital twins also enable the visualization of planning processes in 4D, with time being the fourth dimension alongside 3D. A digital twin is a support tool for a policy-making or engagement process and it is not the complete answer to a discussion. Also important is finding key stakeholders, securing resources, commitment to long-term adoption, maintenance, moderation and marketing the citizen engagement tool.

4. Inclusion of relevant stakeholders and citizen groups in the development of new engagement tools.

All citizens should be informed about the existence of digital twins or any available dynamic visualization tool and the opportunity to visualize the impact of their decisions through it, actively participating in decision-making. It is important to include various user groups in the design and development process of new tools in the early phase. With regards to accessibility, persons with first-hand experience of accessible routes and services should be included in the ideation process of tools for engagement. Workshops and webinars/seminars should be conducted with different user groups and stakeholders, to collect and exchange ideas and information and to raise citizens' awareness about existing/new digital tools. The commitment of local authorities, interest organizations and national authorities to secure adoption, development and moderation of new engagement tools is imperative.

5. Use of the digital twin as an engagement tool accessible to all.

The digital twin as an engagement tool does not offer fully equal and accessible possibilities for participation if considering all possible vulnerable user groups, such as visually impaired persons or persons with cognitive disabilities. It is a valuable addition to the current set of tools available, but in its current form, it is not able to replace more accessible tools, such as 2D maps. Further developments are required for the digital twin to become an engagement tool accessible to all.

By adjusting their recommendations regarding their needs and integrating them into their local strategies, cities can benefit by building on the potential of digital twins as an engagement tool in participatory processes in the mobility field.

However, the present study has some limitations mainly related to the local scope and the automated integration of the citizens' feedback into the Thessaloniki's digital twin. The study focused only on a single urban corridor in Thessaloniki. While this allowed for detailed analysis, it limits the generalization of the study's insights. Also, at this stage, the direct link of Thessaloniki's digital twin with a participatory platform that would enable the automated integration of citizens' feedback into the digital twin was not feasible.

Therefore, to further advance this research, investigating the scalability of this methodology in different urban contexts and urban typologies (e.g., in suburban areas) would provide valuable insights for its replicability across cities with different infrastructure and policy frameworks. Future research could also focus on leveraging AI and ML to automate the analysis of citizen feedback, enhancing the responsiveness of planning processes. Additionally, the long-term impact of digital twin-driven engagement on urban policy development should be assessed.

From a technological perspective, the integration of standardized digital twin platforms across municipalities could facilitate more consistent and data-driven policy decisions. Policy-makers should prioritize the development of user-friendly interfaces that allow seamless citizen participation, ensuring accessibility for diverse social groups, including individuals with disabilities.

Strengthening the alignment between technology and societal needs will ensure that transport planning in smart cities is both innovative and inclusive, ultimately improving life quality for citizens and all urban residents.

5. Conclusions

This study develops and validates a methodological framework that suggests a digital twin as a participatory planning tool, moving beyond its traditional role in infrastructure monitoring and simulation. This approach was applied in Thessaloniki's Smart Mobility Living Lab, focusing on enhancing citizen involvement through a structured three-step process involving ideation, co-creation, and validation workshops.

The application of this methodology shows that digital twins can serve as collaborative environments where citizens and decision-makers jointly identify urban mobility challenges, suggest solutions, and visualize their potential impacts. The integration of crowdsourced qualitative data with real-time data enabled the evaluation of specific pedestrian-oriented interventions that, when simulated within the digital twin, were shown to significantly reduce pedestrian waiting times and near-miss incidents, offering quantitative evidence-based insights to support their implementation.

A major contribution of this study lies in demonstrating the role of digital twins in bridging the gap between top-down policy planning and bottom-up citizen feedback, as the inclusion of citizen-generated insights into Thessaloniki's SUMP validates the practical utility of the suggested methodological approach. Although the methodology developed in the context of Thessaloniki, it can be applied also in other cities, particularly those with existing digital twins. Cities such as Helsinki, Dublin, and Barcelona that are already exploring digital tools for public engagement could benefit from adopting this participatory digital twin approach to enhance citizen involvement in mobility and urban planning processes. Future research should build on these findings by exploring long-term impacts of digital twin-driven engagement on urban policy development.

Author Contributions: Conceptualization, J.M.S.G.; methodology, M.K.; software, J.M.S.G.; validation, J.M.S.G. and M.K.; formal analysis, M.K.; investigation, M.K.; resources, J.M.S.G.; data curation, J.M.S.G.; writing—original draft preparation, M.K.; writing—review and editing, J.M.S.G.; visualization, J.M.S.G.; supervision, J.M.S.G.; project administration, J.M.S.G.; funding acquisition, J.M.S.G. All authors have read and agreed to the published version of the manuscript.

Funding: This research is performed within the framework of DVECE-"Dynamic Visualizations to Enhance Citizens Engagement" project (ID 22086), co-funded by EIT Urban Mobility (EIT Urban Mobility Addendum to the Internal Agreement 2021–2022), an initiative of the European Institute of Innovation and Technology (EIT), a body of the European Union and has been supported by the DigiTwin4Mobility2 project (Digital Twins for Smart Mobility—edition 2) (ID 240553), funded under the EIT Urban Mobility Academy—Competence Hub Open Call 2024 (EIT Urban Mobility Financial Support Agreement 2023–2025).

Data Availability Statement: The data presented in this study are available on request from the corresponding author.

Acknowledgments: The authors of the paper would like to thank all the participants of the ideation, co-creation and validation/evaluation workshops for the valuable input provided.

Conflicts of Interest: The authors declare no conflicts of interest.

Acronyms

The following abbreviations are used in this manuscript:

IoT	Internet of Things
AI	Artificial Intelligence
ICT	Information and Communication Technologies
ATSC	Adaptive Traffic Signal Control
DTCC	Digital Twin Cities Centre
SUMP	Sustainable Urban Mobility Plan
ML	Machine Learning
OD	Origin-Destination

References

1. Qanazi, S.; Leclerc, E.; Bosredon, P. Integrating Social Dimensions into Urban Digital Twins: A Review and Proposed Framework for Social Digital Twins. *Smart Cities* **2025**, *8*, 23. [CrossRef]
2. Fitzgerald, C.; McCarthy, S.; Carton, F.; O Connor, Y.; Lynch, L.; Adam, F. Citizen participation in decision-making: Can one make a difference? *J. Decis. Syst.* **2016**, *25*, 248–260. [CrossRef]
3. Kaur, M.J.; Mishra, V.P.; Maheshwari, P. The convergence of digital twin, IoT, and machine learning: Transforming data into action. In *Digital Twin Technologies and Smart Cities*, 1st ed.; Farsi, M., Daneshkhah, A., Hosseinian-Far, A., Hamid, J., Eds.; Springer Nature Switzerland AG: Cham, Switzerland, 2020; pp. 3–17.
4. Liu, S.; Bao, J.; Lu, Y.; Li, J.; Lu, S.; Sun, X. Digital twin modeling method based on biomimicry for machining aerospace components. *J. Manuf. Syst.* **2021**, *58*, 180–195. [CrossRef]
5. Talkhestani, B.A.; Jazdi, N.; Schlögl, W.; Weyrich, M. Consistency check to synchronize the digital twin of manufacturing automation based on anchor points. *Procedia CIRP* **2018**, *72*, 159–164. [CrossRef]
6. Mondoro, A.; Grisso, B. On the integration of SHM and digital twin for the fatigue assessment of naval surface ships. In Proceedings of the Twelfth International Workshop on Structural Health Monitoring, Stanford, CA, USA, 10–12 September 2019.
7. Jimenez, J.I.; Jahankhani, H.; Kendzierskyj, S. Health care in the cyberspace: Medical cyber-physical system and digital twin challenges. In *Digital Twin Technologies and Smart Cities*, 1st ed.; Farsi, M., Daneshkhah, A., Hosseinian-Far, A., Hamid, J., Eds.; Springer Nature Switzerland AG: Cham, Switzerland, 2020; pp. 79–92.
8. Enders, M.R.; Hoßbach, N. Dimensions of digital twin applications—A literature review. In Proceedings of the 25th Americas Conference on Information Systems (AMCIS), Cancun, Mexico, 15–17 August 2019.
9. O'Dwyer, E.; Pan, I.; Charlesworth, R.; Butler, S.; Shah, N. Integration of an energy management tool and digital twin for coordination and control of multivector smart energy systems. *Sustain. Cities Soc.* **2020**, *62*, 102412. [CrossRef]
10. Furst, J.; Cheng, B.; Hebgen, B. Realizing the Digital Twin Transition for Smart Cities. *Open J. Internet Things (OJIOT)* **2021**, *7*, 32–42.
11. Khatib, M.M.E.; Al-Nakeeb, A.; Ahmed, G. Integration of cloud computing with artificial intelligence and its impact on telecom sector—A case study. *iBusiness* **2019**, *11*, 1–10. [CrossRef]
12. Kharchenko, V.; Illiashenko, O.; Morozova, O.; Sokolov, S. Combination of digital twin and artificial intelligence in manufacturing using industrial IoT. In Proceedings of the 2020 IEEE 11th International Conference on Dependable Systems, Services and Technologies (DESSERT), Kyiv, Ukraine, 14–18 May 2020; IEEE: Piscataway, NJ, USA, 2020. [CrossRef]
13. European Commission, Smart Cities. Available online: https://commission.europa.eu/eu-regional-and-urban-development/topics/cities-and-urban-development/city-initiatives/smart-cities_en (accessed on 10 March 2025).
14. Wan, L.; Nochta, T.; Schooling, J.M. Developing a city-level digital twin–propositions and a case study. In Proceedings of the International Conference on Smart Infrastructure and Construction 2019 (ICSIC): Driving Data-Informed Decision-Making, Cambridge, UK, 8–10 July 2019; pp. 187–1193. [CrossRef]
15. Gao, Y.; Qian, S.; Li, Z.; Wang, P.; Wang, F.Q.; He, Q. Digital Twin and Its Application in Transportation Infrastructure. In Proceedings of the 2021 IEEE 1st International Conference on Digital Twins and Parallel Intelligence (DTPI), Beijing, China, 15 July–15 August 2021; pp. 298–301. [CrossRef]
16. Deren, L.; Wenbo, Y.; Zhenfeng, S. Smart city based on digital twins. *Comput. Urban Sci.* **2021**, *1*, 1–11. [CrossRef]
17. Jin, G.; Deng, X.; Chu, X.; Li, Z.; Wang, Y. Optimization of land-use management for ecosystem service improvement: A review. *Phys. Chem. Earth Parts A/B/C* **2017**, *101*, 70–77. [CrossRef]
18. Mohammadi, N.; Taylor, J.E. Smart city digital twins. In Proceedings of the 2017 IEEE Symposium Series on Computational Intelligence (SSCI), Honolulu, HI, USA, 27 November–1 December 2017; pp. 1–5. [CrossRef]
19. Gerlach, B.; Zarnitz, S.; Nitsche, B.; Straube, F. Digital Supply Chain Twins-Conceptual Clarification, Use Cases and Benefits. *Logistics* **2021**, *5*, 86. [CrossRef]

20. Hildén, M.; Furman, E.; Kaljonen, M. Views on planning and expectations of SEA: The case of transport planning. *Environ. Impact Assess. Rev.* **2004**, *24*, 519–536. [CrossRef]
21. Dasgupta, S.; Rahman, M.; Jon, S. Harnessing Digital Twin Technology for Adaptive Traffic Signal Control: Improving Signalized Intersection Performance and User Satisfaction. *IEEE Internet Things J.* **2024**, *11*, 36596–36618. [CrossRef]
22. Wang, Z.; Han, K.; Tiwari, P. Digital Twin-Assisted Cooperative Driving at Non-Signalized Intersections. *IEEE Trans. Intell. Veh.* **2022**, *7*, 198–209. [CrossRef]
23. Columbia University. The Digital Twin of New York City, Smart City Project Named Winner in IDC Smart Cities North America Awards. 2023. Available online: https://datascience.columbia.edu/news/2023/the-digital-twin-of-new-york-city-smart-city-project-named-winner-in-idc-smart-cities-north-america-awards/ (accessed on 18 March 2025).
24. KIRA-Digi Pilot Project. The Kalasatama Digital Twins Project. The Final Report of the KIRA-Digi Pilot Project. 2019. Available online: https://www.hel.fi/static/liitteet-2019/Kaupunginkanslia/Helsinki3D_Kalasatama_Digital_Twins.pdf (accessed on 18 March 2025).
25. White, G.; Zink, A.; Codecá, L.; Clarke, S. A digital twin smart city for citizen feedback. *Cities* **2021**, *110*, 103064. [CrossRef]
26. Dembski, F.; Wössner, U.; Letzgus, M.; Ruddat, M.; Yamu, C. Urban Digital Twins for Smart Cities and Citizens: The Case Study of Herrenberg, Germany. *Sustainability* **2020**, *12*, 2307. [CrossRef]
27. Observatory of Public Sector Innovation. vCity, a Human-Centric Platform for Urban Digital Twins. 2024. Available online: https://oecd-opsi.org/innovations/vcity-a-human-centric-platform-for-urban-digital-twins/ (accessed on 18 March 2025).
28. Luca, M.; Lepri, B.; Gallotti, R.; Paolazzi, S.; Bigi, M.; Pistore, M. Towards Civic Digital Twins: Co-Design the Citizen-Centric Future of Bologna. *arXiv Preprint* **2024**. [CrossRef]
29. Digital Twin Cities Centre (DTCC). International DTCC Workshop Brings Together Digital Twin Researchers. 2023. Available online: https://dtcc.chalmers.se/2023/06/ (accessed on 25 March 2025).
30. Thessaloniki's Sustainable Urban Mobility Plan. In *Thessaloniki's Sustainable Urban Mobility Plan*, 1st ed.; Centre for Research and Technology Hellas-Hellenic Institute of Transport: Thessaloniki, Greece, 2019. Available online: https://www.svakthess.imet.gr/Portals/0/Diavoulefseis/Diavoulefsi03/SVAK_Thessalonikis.pdf (accessed on 15 March 2025).
31. Salanova, J.M.; Toumbalidis, J.; Chaniotakis, E.; Karanikolaos, N.; Aifadopoulou, G. Correlation between digital and physical world, case study in Thessaloniki. *J. Locat. Based Serv.* **2017**, *11*, 118–132. [CrossRef]
32. Salanova Grau, J.M.; Mitsakis, E.; Tzenos, P.; Stamos, I.; Selmi, L.; Aifadopoulou, G. Multi-source data framework for network traffic state estimation and prediction. *J. Adv. Transp.* **2018**, *2018*, 9078547. [CrossRef]
33. Mitsakis, E.; Chrysohoou, E.; Salanova, J.M.; Iordanopoulos, P.; Aifadopoulou, G. The sensor location problem: Methodological approach and application. *Transport* **2017**, *31*, 1–7. [CrossRef]
34. Mitsakis, E.; Stamos, I.; Salanova Grau, J.M.; Chrysohoou, E.; Aifadopoulou, G. Urban Mobility Indicators for Thessaloniki. *J. Traffic Logist. Eng. (JTLE)* **2013**, *1*, 148–152. [CrossRef]

land

Article

Rural Migrant Workers in Urban China: Does Rural Land Still Matter?

Huiguang Chen [1], Wojciech J. Florkowski [2,*] and Zhongyuan Liu [2]

[1] College of Public Administration, Nanjing Agricultural University, Nanjing 210095, China; chenhuiguang@njau.edu.cn
[2] Department of Agricultural and Applied Economics, University of Georgia, Athens, GA 30602, USA; zhongyuan_liu@hotmail.com
* Correspondence: wojciech@uga.edu

Abstract: The purpose of this study is to test the response of rural–urban migration to land endowment while recognizing the heterogeneity of land tenure security perceptions. Based on the survey data of 751 migrant workers in Nanjing City, the latent class model identifies the secured group and unsecured group, with a particular focus on how heterogeneous concepts of land tenure security mediate the above relationships. Empirical modeling explores how land endowment affects household labor distribution and individual migration decisions, taking the heterogeneity of tenure expectations into account. The key findings reveal the following: (1) an inverse U-shaped relationship between land endowment and household labor distribution, but not between land endowment and individual migrant decisions; (2) farm households who perceive land tenure as unsecured tend to send fewer household members as job-seeking migrants, even as their land endowment increases; (3) however, individual migration decisions under conditions of a secured land tenure reduce the risk of losing land and induce migrant workers to stay in cities. The findings contribute to advancing the institutional analysis of the impact of land endowment on household labor allocation and how land tenure security affects immigration decisions, providing empirical evidence for China's rural revitalization policies, which would support reforms that secure land tenure, such as the second-round contracting renewal policy and the rural revitalization plan in China.

Keywords: land endowment; urban migration; household labor allocation; land tenure; U-shaped response

1. Introduction

As an emerging economy, China has been experiencing an unprecedented labor migration from rural to urban areas since the late 1980s. The outmigrant population surged from 30 million in 1989 [1] to 177 million in 2023 [2]. This has attracted attention, focus, and some related research. Centering on the topic of internal migration, the initial literature in China, while extensive, mainly focuses on constraints from the labor market, including the household registration system, i.e., the hukou system [3,4], and the labor market discrimination [5,6]. In fact, despite China's implementation of hukou reform and the removal of labor market discrimination, the migrant population increased less than one percent in the three years preceding the COVID-19 pandemic and was negatively affected during the pandemic. In other words, when the institutional and non-institutional constraints from hukou were gradually eliminated, the outmigration from rural areas to urban areas did not increase as expected. Migration may not develop if constraints

https://doi.org/10.3390/land14040901

from other factor markets remain. This is supported by the evidence that the extension of rural land rights to peasants has driven urbanization more effectively than the relaxation of migration restrictions [7]. One of its root causes lies in the fact that, at the inception of the rural household contract responsibility system, land allocation was linked to the number of labor or farmer members. In addition, the rural contracted law does not allow contracted land to be abandoned. Therefore, at the end of the last century, rural labor flow was constrained by rural land institutions. The experience of most developing countries also shows that farm households are systematically exposed to market imperfections and constraints other than labor markets [8]. With the explosive growth of migrant workers in China, off-farm labor market activities are the main catalyst of rising land rental activity [9]. In the face of population outflows, China's government has put forward a large number of policies to guide the transfer of farmland management rights [10]. In theory, if the land and labor markets work well, labor allocation decisions should be independent of land endowment [8,11]. Nevertheless, the manner in which rural land endowment affects the decisions of household labor allocation is an extremely complex process.

The literature has further provided insights into the relationship between household land endowment and rural–urban migration. Specifically, several studies found evidence of the U-shaped relationship between land endowment and the probability of migration [12–14], implying that the probability of migration decreases as the area of land operation increases but that eventually the probability increases as the area of land operation continues to increase. Conversely, other research reports an inverse U-shaped relationship [15–17]. The critiques of the U-shaped relationship may stem from the absence of a formal, rigorous procedure for testing the U-shaped relationship. Past research almost exclusively included a quadratic term in its standard regression models and relied on the sign and significance of the level term and the quadratic term parameters to establish the U-shaped relationship's existence [18]. However, this criterion can be misleading when the estimated extremum is located at the end of the data range [18]. Indeed, for a farm household, the operated land includes the contracted land (i.e., initial endowment) and the rent-in/rent-out land (i.e., increased endowment). The above studies, whether supporting a U-shaped or an inverse U-shaped relationship, have failed to distinguish between the impact of these two types of endowments on labor migration. Whether a U-shaped or an inverse U-shaped relationship exists, fundamentally, is a matter of a spatial mismatch between labor mobility and farmland fixation. The mismatch, in turn, leads to the coupling imbalance of rural–urban migration and rural land-use transition [19].

More importantly, existing studies of the U-shaped relationship in terms of land endowment oversimplify to provide an economic explanation of the identified relationships. On the one hand, the studies ignore the lack of well-defined land tenure in rural China, which can create considerable distortions in the relationship between land endowment and labor allocation. An unsecured land tenure increases the need for "guard labor", which was previously principally devoted to agricultural production, to solidify informal land tenure [20,21]. Even considering the marketization of land leasing, land renting-in tends to increase the labor involved in agricultural production [22]. However, land renting-out may reduce the labor allocation toward agricultural activities, and in cases of unsecured land tenure, it may also consume some of the time of outmigrants to tackle leasing affairs. On the other hand, studies confuse the perspectives of individual decisions and joint decision-making in households. While the early migration models emphasized the migration decisions made by an individual household member [23], more recent research has focused on household labor allocation [12,15,17]. Given the fundamental distinction between the household and individual agency, it is worthwhile to distinguish between those decisions. Against this backdrop, the purpose of this paper is to focus on the relationship between land endowment and rural labor migration in the case of unsured land tenure. By addressing the institutional complexities of land rights and differentiating between individual and household decision frameworks, this research seeks to provide a more comprehensive explanation of the observed U-shaped relationship.

This paper examines how rural land endowment shapes household and individual labor allocation decisions in an attempt to provide an explanation for the pervasive, incomplete rural-to-urban migration. To do so, this paper uses survey data collection through a specifically designed questionnaire administered to migrant rural workers with urban jobs. The latent class model identifies distinct groups of migrants with different perceptions of land tenure, specified as a "secured group" (those confident in their land rights) and an "unsecured group" (migrants unsure of the security of their land rights). The results show an inverse U-shaped relationship between land endowment and migrant workers' household labor distribution in the secured group. For households in the unsecured group, the larger the land endowment is, the less likely it is that a household sends members in search of urban jobs. Further, the results show land that endowment has a limited impact on a rural individuals' migration plan, suggesting that land endowment affects household labor distribution rather than individual migration plans. Additionally, land tenure insecurity hinders rural labor outmigration at the household and individual level. The findings are robust under alternative models and variable specifications. Complementary evidence from the survey sample's demographic profile strengthens the analysis: the average age was approximately 43 years, projected to rise to 57 by 2025. As a result, the competitiveness of this group above the average age in the urban labor market has declined, while fostering a general return-to-rural trend. Significantly, most of the participants went back to engage in farming, reflecting their relatively stable attachment to land. Their perceptions of land tenure security thus continue to shape rural labor allocation dynamics at both the household and individual level, underscoring the institutional and demographic complexities influencing migration patterns.

This study complements previous studies on labor allocation across sectors [24–27] through two distinct contributions. Firstly, it may extend the research scope of labor allocation by focusing on migrant workers and their cross-market constraints from land endowment. Unlike prior studies, this paper systematically differentiates the effects of land endowment between two groups of migrants with different perceptions of land tenure security, thereby uncovering nuanced institutional influences on migration behavior. Secondly, methodologically, the study helps to identify heterogeneous migrant subgroups by employing a latent class cluster model, and helps to test for U-shaped relationships by implementing a robust analytical approach. These methodologies enhance the rigor of our labor allocation dynamics analysis, addressing limitations in earlier research. Collectively, this work will help to deepen our understanding of how land endowment, moderated by tenure security perceptions, shapes rural–urban migration and cross-sectoral allocation.

2. Theoretical Framework

2.1. Labor Migration and Land Tenure

The early literature on development economics documented that the growth path of most advanced economies was accompanied by the process of structural transformation involving the reallocation of the labor force across sectors and/or regions [23,28]. While rural households in China are able to migrate, the household registration system (i.e., the hukou system in China) [3,4], labor market discrimination [5,6], and other institutional and non-institutional constraints result in the phenomena of temporary and circular migration [29–31]. Temporary migration leaves migrants largely excluded from public services such as education, health care, and other forms of social support in cities [4,32].

Access to land and land ownership is a major household asset in most developing countries, shaping the socioeconomic stratification within rural communities. Land area affects the labor (or working time) distribution of rural households, as well as that of household members who have already migrated in search of urban jobs. For example, when farming labor runs short, households operating with sizable land areas do not necessarily recall household members back from migration because they can lease out their land in order to lower labor requirements or hire a worker to substitute for their migrating household members. However, farm households in most developing countries are systematically exposed to various market imperfections and constraints [8]. Market imperfections arise from high transaction costs and limited labor, including the nonexistence of local labor markets [8]. In China, local labor markets and land leasing markets have been underdeveloped [33,34]. The exception to this is that the relationship between suburban and rural areas and land transfer is much stronger [35]. Therefore, the understanding of how migrant workers allocate labor in the context of market failures should consider the inseparability between land endowment and the labor market. The migration literature, while extensive, provides little insight into the effects of land endowment on migrant workers' allocation of labor between farms and urban jobs. Therefore, it is important to rebuild the rural household labor allocation model to consider land endowment.

2.2. Rural Household Model for Labor Allocation

Consider a rural household that has two available resources: labor and farmland. The household can allocate labor to farming and urban jobs. Let the household have a fixed total labor resource normalized to size 1 and allocate l of the labor to rural-to-urban migration and $1-l$ to farming activities. The household has a land endowment of A, which is exogenously determined. Farmland can generate income by raising crops, be left idle, or be underutilized. Land rental activities are ignored in this study because the prevalence of the land rental market in rural China remains low [36]. In other words, the land leasing market is not active. If farm households perceive their land rights to be secure, they will invest production factors, such as labor and fertilizer, into their farmland in expectation of additional returns. Otherwise, farmers underutilize their land to avoid the risk of a loss of investment under insecure tenure.

Let the development of land tenure be given by $\lambda \in [0,1]$. As land tenure is viewed as more secure, the value of λ increases. There are two extremes of the land tenure system: $\lambda = 0$ indicates the "public" and unprotected land tenure regime, whereas $\lambda = 1$ represents "private" and protected land tenure. Any situation in between implies imperfect land tenure. In the absence of the land market, λ may be considered as land adjustment or reclamation. The Rural Land Contracted Law in 2002 stipulated that farm households' contracted land should be withdrawn if they migrated and registered in urban areas (see Article 26). The law was amended in 2018, as well as the withdrawal article was abolished. If A^f is the amount of land available to the amount of household labor left to farm and A^u is the underutilized farmland distributed to household farm labor, we assume the following:

$$A^f = A^f(l,\lambda) \tag{1}$$

$$A^u = A^u(l,\lambda), \tag{2}$$

where $A^f_\lambda(l,\lambda) > 0$, $A^f(1,0) = 0$, $A^u_\lambda(l,\lambda) < 0$, $A^u(l,1) = 0$, and $A^f_\lambda(l,\lambda)$ and $A^u_\lambda(l,\lambda)$ are the derivatives of $A^f(l,\lambda)$ and $A^u(l,\lambda)$ with respect to λ, respectively. The positive sign indicates that, as land tenure becomes secure, more land will be farmed than left underutilized. We assume two simplified expressions for land distribution:

$$A^f(l,\lambda) = \frac{\lambda(1-l)}{l}A \tag{3}$$

$$A^u(l,\lambda) = \frac{(1-\lambda)l}{1-l}A \tag{4}$$

where $\frac{\lambda(1-l)}{l} \in [0,1]$, and $\frac{(1-\lambda)l}{1-l} \in [0,1]$, because this study assumes that land rental activities are ignored. If the land rental market works, the results of $\frac{\lambda(1-l)}{l}$ and $\frac{(1-\lambda)l}{1-l}$ will be greater than 1.

Household labor can generate income from off-farm activities and/or farming. Each activity is characterized by a twice-differentiable production function, with P_a denoting the farming output and P_{na} representing the migration output:

$$P_a = f\left[l, A^f(l,\lambda)\right] \text{ and} \tag{5}$$

$$P_{na} = g[l, A^u(l,\lambda)], \tag{6}$$

where $f'(l) < 0, f''(l) > 0, g'(l) > 0$, and $g''(l) < 0$. The derivatives show that the farming output is a continuous and convex function in terms of migrant labor, while the output function of migration is continuous and concave in terms of labor for off-farm activities. Meanwhile, as land area increases, households increase farming output; that is, $f'\left(A^f\right) > 0$. If households leave farmland idle or underutilized, the risk of loss of farmland exists when household members outmigrate; thus, $g'(A^u) < 0$.

2.3. Locally Stable Equilibrium and Conditions

Farm households allocate labor between urban and rural areas. Their family members engage in agricultural or non-agricultural activities in pursuit of more substantial income. Household members capable of working face the choice of migrating or farming the land. The equilibrium state arises when household working members lack the incentive to switch between migration and farming:

$$pP_a = wP_{na}, \tag{7}$$

where p and w are the exogenous prices for agricultural output and migrant labor, respectively. Two inequalities rule out the corner solution to labor distribution (or specialization in production): "$l = 1$" and "$l = 0$". Thus, $pf\left(0, A^f\right) < wg(1, A^u)$ implies that if all household working members remain farming, the income from farming will be lower than the income from an urban job. There is an incentive for household members to migrate. This inequality establishes the first condition for a locally stable equilibrium.

Condition 1: $pf\left(0, A^f\right) < wg(1, A^u)$.

Another requirement for a locally stable equilibrium is that any transfer of household working members to urban jobs should lead to a smaller marginal income from off-farm activities than from farming. Otherwise, the transfer will persist until all the rural household labor force migrates to urban areas. The condition is as follows:

Condition 2: $p\frac{\partial P_a}{\partial l} > w\frac{\partial P_{na}}{\partial l}$ or $pf_l + pf_{A^f}\frac{\lambda}{l^2}A > wg_l + wg_{A^u}\frac{1-\lambda}{(1-l)^2}A$.

However, both conditions do not necessarily map onto a unique solution to a locally stable equilibrium. According to Equation (7), both migration to urban jobs and farming are functions of land tenure security perception, λ. Figure 1 shows a locally stable equilibrium for two scenarios: homogeneity and heterogeneity in land tenure perceptions. Under the assumption of the homogeneity of land tenure perceptions, there exists a unique, locally stable equilibrium (Figure 1a).

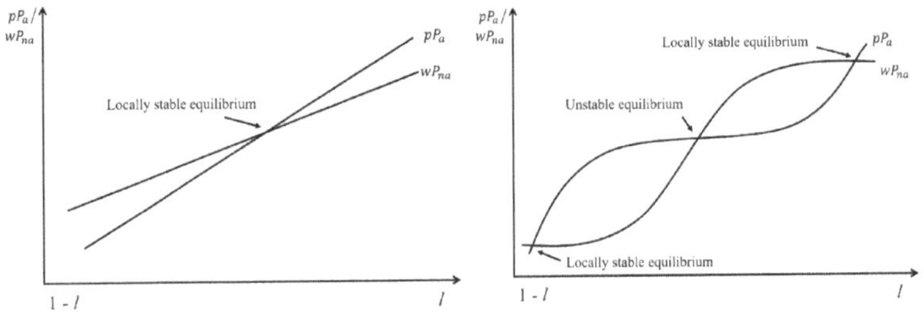

Figure 1. Locally stable equilibriums in two scenarios of land tenure perceptions. (**a**) Homogeneity in land tenure. (**b**) Heterogeneity in land tenure.

In contrast, relaxing the assumption allows for a heterogeneity of land tenure perceptions. There are two locally stable equilibria and one unstable equilibrium, shown in Figure 1b. In Figure 1b, the upper interaction point represents a locally stable equilibrium under the expectations of secured land tenure, while the lower interaction point is a locally stable equilibrium with the expectation of unsecured land tenure. The interaction shown in the middle of Figure 1b is an unstable equilibrium since the slope of migration income is greater than the agricultural income's slope. Thus, under these conditions, higher migration will result in additional income from migration, which violates the second condition of a stable equilibrium.

2.4. Land Endowment and Labor Supply: Heterogeneity in Land Tenure Security

When the equilibrium state represented by Equation (7) occurs for a farmer family, its land endowment will also correspond to the result of an equilibrium state $\overline{A}$. Obviously, it will also be determined by l and λ, i.e., $\overline{A} = \overline{A}(l, \lambda)$. In this way, $\overline{A}$ is an exogenous variable. Thus, the equilibrium Equation (7) can be written as an identity:

$$F = p * f\{l, \overline{A}(l, \lambda)\} - w * g\{l, \overline{A}(l, \lambda)\} \equiv 0 \quad (8)$$

The critical causal effect in the current study is the impact of land endowment on rural labor migration to urban jobs. First consider the overall impact, assuming that rural laborers are homogeneous in terms of their perceived land tenure security. That is, i.e., how land endowment affects rural household labor distribution while holding other factors constant. According to the implicit function theorem [37], using Equations (3)–(8) and two conditions, the locally stable equilibrium is as follows:

$$\frac{dl}{d\overline{A}} = -\frac{\partial F/\partial \overline{A}}{\partial F/\partial l} = \frac{wg_{\overline{A}}\frac{\lambda(1-l)}{l} - pf_{\overline{A}}\frac{(1-\lambda)l}{1-l}}{pf_l + pf_{\overline{A}}\frac{\lambda}{l^2}\overline{A} - wg_l - wg_{\overline{A}}\frac{1-\lambda}{(1-l)^2}\overline{A}} \quad (9)$$

therefore, the overall sign of the deviation of household migrant labor over land endowment is negative.

However, Equation (9) also indicates that the deviation of household migrant labor over land endowment is a function of land tenure security, λ. Thus, the research question of this article is expressed as how farm households balance the allocation of land endowment and labor under the constraints of land tenure security/insecurity; that is, how much land endowment to allocate in the agricultural sector and how much labor to allocate in the non-agricultural and agricultural sectors. Therefore, the extreme cases, $\lambda = 1$ and $\lambda = 0$, provide a simple and intuitive illustration.

<u>Scenario 1: λ = 1.</u>

This case involves secured land tenure with well-defined, individualized land allocation and legislative protection. Equations (3)–(9) lead to the following equation:

$$\frac{dl}{d\overline{A}}|_{\lambda=1} = \frac{wg_{\overline{A}}\frac{1-l}{l}}{pf_l + pf_{\overline{A}}\frac{1}{(1-l)^2}\overline{A} - wg_l} > 0 \quad (10)$$

The denominator of Equation (10) is the difference in marginal incomes between farming and employment in urban areas. According to China's National Bureau of Statistics, the average annual income per migrant worker was CNY 57,360, while the average annual income per farmer was CNY 21,691 in 2023 [2,38]. This significant difference implies a negative sign for the denominator. Because the sign of the numerator is negative, the overall sign of the deviation of migrant labor over land endowment under the expectations of secured land tenure is positive. That is, the increase in land endowment inducing migration allows us to state the first hypothesis in terms of the relationship between land endowment and migration decisions.

Hypothesis 1: *In a secured land tenure setting, a large land endowment induces migration from rural areas to urban areas.*

Scenario 2: $\lambda = 0$.

This case indicates a "public" and unprotected land tenure system. From Equations (3)–(9), we obtain the equation below:

$$\frac{dl}{d\overline{A}}|_{\lambda=0} = \frac{-pf_{\overline{A}}\frac{l}{1-l}}{pf_l - wg_l - wg_{\overline{A}}\frac{1}{(1-l)^2}\overline{A}} < 0 \quad (11)$$

where the sign of the numerator on the right side of Equation (11) is negative, as mentioned in Section 2.2. The sign of the denominator is not easily determined. It contains the following: (1) pf_l, the marginal income from farming; (2) $wg_{\overline{A}}\frac{1}{(1-l)^2}\overline{A}$, the opportunity cost from the loss of land tenure because of migration to urban areas; (3) wg_l, the marginal income from migration. The China Migrants Dynamic Survey indicates that only 11.8% of migrants are willing to convert a rural hukou to an urban hukou if they are required to return the land they are allowed to farm. It appears that migrants place more value on the sum of income obtained from farming, pf_l, and the opportunity cost of the loss of land tenure because of migration, $wg_{\overline{A}}\frac{1}{(1-l)^2}\overline{A}$, than the benefits package associated with the urban hukou and the income from migration (wg_l). Thus, one can infer that the overall sign of Equation (11) is negative, leading to the second hypothesis:

Hypothesis 2: *If land tenure is unsecured, the greater land endowment, the more likely it is to hinder migration from rural to urban areas.*

3. Data

3.1. Questionnaire Survey

The data used in the current paper were collected via a structured questionnaire survey in Nanjing, Jiangsu Province, in 2011. Nanjing is the capital city of Jiangsu Province and located in the Yangtze River Delta economic zone, which is one of the most developed regions in China. According to the Nanjing Statistics Bureau, Nanjing hosted more than 1.78 million migrant workers from different areas of the country by the end of 2013. Several locations frequented by migrants were identified in Nanjing, such as the Andemen job seeker market, Shenyang Village, dining halls in universities and restaurants where most of the servers were migrants, and dormitories at construction sites. To capture representative migrant perspectives, the survey team targeted these migrant locations. Survey data were collected during weekends and holidays from September to November 2011, employing a convenience sampling method. The enumerators randomly approached individuals at the listed locations and, following a customary greeting, asked screening questions. Upon agreeing, the respondent was presented with a consent form and asked to read it and sign it or withdraw from the survey. The interview took about 20 min, and the participants were rewarded with a token gift such as a towel, toothpaste, or a bar of soap upon the completion of the interview.

The five-part questionnaire included closed and partly open questions. The first part included questions about household labor distribution and respondents' future migration plans. The second set of questions asked the respondents to share information about their household and individual characteristics, such as income, gender, age, and education. The third part of the questionnaire involved questions pertaining to land, such as land endowment, the land rental market, land contracts, and attitudes towards land adjustment. Finally, the fifth part probed the respondents about residing in urban areas as migrants, covering subjects such as labor contracts, job training, and integration into city life. This study used fully completed questionnaires obtained from 751 individuals.

3.2. Variable Explanation and Descriptive Analysis

Table 1 shows that 55% of the working individuals in households had jobs as migrant workers in urban areas. Among the migrants with jobs, 47% of the respondents wanted to stay there for the foreseeable future. In contrast to other studies, where the migration decision was investigated regarding the intent to migrate from rural areas to urban areas, this study asked respondents whether they wanted to stay in the city or return to their village.

Table 1. Selected descriptive statistics of variables used in estimation.

Variable Name	Definition	Mean	Std. Dev.	Min	Max
	Migration variables				
Share	Share of migrants in household	0.55	0.24	0.17	1
Plan	1 = stay in cities; 0 = return to villages	0.47	0.50	0	1
	Land endowment				
Acre	Land endowment in *mu*	5.39	3.99	0.6	23
Acreind	Land endowment per household laborer	1.96	1.67	0.23	10
Parcel	Land parcel number	3.94	2.81	1	15
	Land tenure variables				
Landcont	1 = land contract with collective	0.65	0.48	0	1
Landadjustatt	Attitude towards land adjustment [a]	2.29	1.32	1	5
Landprivate	Land ownership assessment [b]	2.09	1.40	1	4
Landrent	1 = land rental market	0.18	0.38	0	1
	Control variables				
	Individual-level				
Male	1 = male	0.82	0.38	0	1
Educ	Respondent education level [c]	2.72	0.85	1	5
Age	Age in years	42.66	9.86	16	72
Incomediff	Annual income difference between city and farming job(s), in 1000 CNY	34.97	20.74	7.2	138
Integration	1 = integrated into cities	0.67	0.47	0	1
	Household-level				
Hhedu	Household-head education [c]	2.67	0.83	1	5
Hhsizenl	Number of non-working household members	3.04	0.95	1	6
Hhincomediff	Household annual income difference between city and farming job(s), in 1000 CNY	54.15	40.36	−15	237.6
	Instrumental variable				
Moneygift	Monetary gift-giving in 1000 CNY	4.15	3.82	0	36

Note: N = 736. a 1 = strongly agree; 2 = agree; 3 = neither agree nor disagree; 4 = disagree; 5 = strongly disagree. b 1 = state ownership; 2 = city-owned; 3 = village-owned; 4 = personally owned. c 1 = no formal education; 2 = elementary school; 3 = middle school; 4 = high school; 5 = college or postgraduate education.

The average area of land endowment was (Acre—see Table 1) 5.39 mu (0.88 acres) per surveyed household, and the land endowment per working household member (Acreind) was 1.96 mu (0.32 acres). To explain migration decisions, a variable that related a household's land area to the number of laborers in the household was more appropriate than one that related land area to the number of household members. Both figures correspond to the national agricultural census results. The survey results showed that agricultural land is very fragmented in the households where the migrants come from, with an average household forming four parcels (*Parcel*). Each parcel averaged about 1 mu.

The survey included land variables that distinguished secured and unsecured land tenures. Instead of using a single proxy measuring land tenure setting, such as land readjustment risk [39], this study measured farmers' understanding, perceptions, and expectations of land tenure security using four indicators. The first indicator was the land contractual relationship between farmers and their village collective (*Landcont*). Farmers in China are not the landowners of their contracted land. Instead, the village collective owns the land and signs land-use contracts with every household to authorize the cultivation, management, lease, and revenues from land use. While most farmers signed land-use contracts with their collectives, some of the signed contracts were lost or damaged over time. Table 1 shows that more than 65% of the respondents had signed land-use contracts.

Another indicator of land tenure security perception is the so-called adjustment. Land distribution requires periodic readjustment as the populations in villages change over time [40]. Generally, land adjustment has been initiated by villagers and implemented by the collective. Because land adjustment affected farmer expectations of land tenure security, the Rural Contracted Land Law in 2002 restricted land adjustment, although small-scale land adjustment was allowed to continue. The current study uses the land adjustment index (*Landadjustatt*) to measure the farmer's attitude towards land adjustment. Specifically, the respondents indicated their own attitudes along a five-step scale, where 1 stood for "strongly agreeing" with land adjustment, 5 referred to "strong disagreement", and 3 indicated a "neutral" view. Generally, the respondents agreed with land adjustment as the average value is 2.29.

The farmers' privatization assessments of land ownership was another indicator. *Landprivate* is a categorical variable that ranges from state ownership (coded as 1) to private ownership (coded as 4). The average value was 2.09 (Table 1). More than three out of five (61.28%) respondents indicated state ownership of the land, and 243 (33.02%) respondents believed that the land they cultivated was private. Village community land ownership differs from the ownership of urban land, which is deemed to be state-owned. Regarding the existence of a land rental market in their villages (*Landrent*), only 18% of respondents answered in the affirmative.

The control variables relevant to individual migration decisions included demographic characteristics, income, migrants' opinion about integration into urban life, and willingness to abandon rural land-use rights. Among those variables, 82% of respondents were male (*Male*), the average respondent was 42 years old (*Age*), and middle school was the average education attainment level (*Educ*) (Table 1). On average, the annual income earned as a migrant worker exceeded that obtained from farming by CNY 34,970. The annual income differences (*Incomediff*) were top-coded to eliminate outliers that would exert a disproportionate influence on the estimation results. The lower and upper bounds were set at the 1st percentile and 99th percentile of the original data to reduce the influence of outliers. About 67% of respondents declared that they had integrated into urban life (*Integration*).

The household-level control variables characterizing the household included the education level of the household head (*Hhedu*), the number of non-working household members (*Hhsizenl*), and the annual household income from urban wages and farming

(*Hhincomediff*). On average, household heads had a middle school education and there were three non-working members per household. The average difference in annual household income earned during migration exceeded the income from farming by CNY 54,150. The household annual income difference was top-coded, and the lower and upper bounds were set at the 1st percentile and the 99th percentile of the original data.

4. Methodology

4.1. Classification and Latent Class Model

Latent Class Analysis identifies the underlying data structure to assign each observation to a particular cluster [41,42] and has been widely used to characterize heterogeneity among groups, such as farmers and consumers [43–45]. Compared with traditional cluster models, such as those using factor analysis, the structural equation model, or the random effect model, the latent class cluster model is superior in dealing with categorical latent variables. A key advantage of the latent class cluster model lies in its flexibility to accommodate complex distributions and mixed data types (i.e., continuous, categorical, or a combination of the two), enabling the nuanced segmentation of heterogeneous groups. Generally, basic latent class cluster models are structured to estimate the probability of an observation belonging to a specific latent class, conditional on its observed indicators, thereby capturing underlying subgroup structures that may not be apparent through manifest variables alone. The models have the following form:

$$f(y_i|\theta) = \sum_{k=1}^{K} \pi_k f_k(y_i|\theta_k) \tag{12}$$

where y_i refers to the scores on a set of indicator variables and the θs are the model parameters. The term $f(y_i|\theta)$ represents the distribution of y_i, which is assumed to be a mixture of cluster-specific densities; $f_k(y_i|\theta_k)$ K is the number of clusters; and π_k denotes the prior probability of belonging to latent cluster k. A maximum-likelihood estimation is used to estimate the parameters.

The criteria applied to select the best-fitting model included the Bayesian Information Criterion (BIC), the number of parameters in the estimation, and a *p*-value associated with the Chi-squared statistics. The current study used Latent GOLD 4.5 software because it includes model selection criteria and probability-based classification.

This study classified migrant workers according to their perceptions of the security of their land tenure. Four land-tenure-related measures were coded, including assessments of land ownership (*Landprivate*), attitudes towards land adjustment (*Landadjustatt*), land contract presence (*Landcont*), and the market for land renting (*Landrent*). Hypothetically, as the land tenure system evolves, households perceive more individualized land tenure and refuse land adjustment to avoid the risk of the loss of the land allocated to them. Additionally, having a land contract with the collective is a measure of legislative protection, and the existence of a local land rental market implies well-established land tenure.

Table 2 shows four different cluster models in the latent class analysis, with cluster numbers ranging from 1 to 4. The index Npar indicates the number of parameters in the estimation. The BIC and Chi-squared statistics (L^2) were used to identify the best fitting model. The L^2 statistical significance is indicated by the *p*-values. The model selection was guided by *p*-values higher than 0.05, indicating an adequate fit, and the fewest number of parameters. Model 2, with two clusters (cluster 1 and cluster 2), outperformed the other models.

Table 2. Model summary from four cluster models.

Model	Number of Clusters	Npar	BIC (LL)	L^2	*p*-Value
Model 1	1 Cluster	9	4950.61	132.44	9.6×10^{-6}
Model 2	2 Clusters	14	4935.45	84.28	0.054
Model 3	3 Clusters	19	4948.17	63.99	0.34
Model 4	4 Clusters	24	4971.27	54.08	0.51

Notes: N = 736. Npar is the number of parameters in the estimation, BIC is the criteria from the Log-likelihood statistics, and L^2 and *p*-value are criteria from the Chi-squared statistics.

Table 3 shows the number of observations, means of land-tenure-related variables, and *t*-tests for each cluster. There were 416 observations (57%) in cluster 1 and 320 observations (43%) in cluster 2. The means of most land-tenure-related variables calculated for cluster 1 were larger than the means of cluster 2, and three of the six were statistically significant. Specifically, the respondents in cluster 1 tended to oppose land adjustment, indicated the existence of a land rental market in their village, and were more likely to have a land-assigning certificate and knowledge of land law than those in cluster 2. Thus, hereafter, cluster 1 is renamed as the secured group and cluster 2 is the unsecured group.

Table 3. Land-tenure-related variables by group, with secured and unsecured land tenure expectations and cluster classification.

Variables	Cluster 1—No of Obs.	Cluster 2—No of Obs.	Cluster 1	Cluster 2	*p*-Value
Landadjustatt	416	320	3.27 (0.90)	1.00 (0.00)	0.00
Landprivate	416	320	2.09 (1.40)	2.08 (1.41)	0.94
Landcont	416	320	0.67 (0.47)	0.62 (0.49)	0.21
Landrent	416	320	0.26 (0.44)	0.08 (0.26)	0.00
Landadjust	416	320	0.69 (0.46)	0.72 (0.45)	0.46
Landlaw	416	320	0.15 (0.36)	0.09 (0.29)	0.02

Notes: Numbers under cluster 1, col. 4, and cluster 2, col. 5, are indicator means. Standard deviations are in parentheses.

A potential identification issue could arise, because farmers who have secured tenure expectations may have stronger motivations to migrate than those who do not. Should this be true, the causality between the land endowment and household labor allocation will be biased. Table 4 shows the comparison and *t*-test between the two groups with different land tenure expectations regarding the dependent and explanatory variables. None of the covariates were statistically significant between the two groups, suggesting that land tenure security was an exogenous factor for both groups in household labor allocation. However, individual migration decisions statistically differed between the two groups. Therefore, the models were specified differently to address the endogeneity in land tenure in their estimations.

Table 4. Results of *t*-test on differences between means for migrant workers and their land tenure expectations.

Variables	Secured Group	Unsecured Group	*p*-Value
	Migration variables		
Share	0.55 (0.24)	0.54 (0.24)	0.65
Plan	0.51 (0.50)	0.41 (0.49)	0.01 **
	Control variables		
	Individual-level		
Acreind	2.04 (1.73)	1.84 (1.59)	0.11
Parcel	4.02 (2.91)	3.83 (2.69)	0.37
Male	0.83 (0.38)	0.82 (0.39)	0.69
Educ	2.77 (0.86)	2.66 (0.85)	0.09
Age	42.31 (10.09)	43.11 (9.53)	0.28
Incomediff	34.74 (21.19)	35.27 (20.18)	0.73
Integration	0.67 (0.47)	0.68 (0.47)	0.66
	Household-level		
Acre	5.53 (4.04)	5.21 (3.92)	0.28
Hhedu	2.68 (0.83)	2.66 (0.82)	0.83
Hhsizenl	3.01 (0.98)	3.08 (0.90)	0.37
Hhincomediff	53.25 (40.81)	55.32 (39.81)	0.49

Note: Standard deviations in parentheses. *** $p < 0.01$; ** $p < 0.05$; * $p < 0.1$.

4.2. Empirical Models and Identification

The empirical modeling explored how land endowment affected household labor distribution as well as individual migration decisions, taking the heterogeneity of tenure expectations into account. The household labor distribution model was specified by a fractional probit function:

$$E(y_{ik}|\boldsymbol{x}) = \Phi(\boldsymbol{x\varphi}), \tag{13}$$

where y_{ik} is the share of household labor that navigates to urban jobs divided by the total labor force in household i in group k and characterized by land tenure perception. Given that the sample consisted of migrant workers, the ratio of off-farm laborers in households had the range of $0 < y_{ik} \leq 1$. The vector of the explanatory variables, $\boldsymbol{x}$, includes land endowment and the number of land parcels in household i and (secured or unsecured group) group k, and characteristics such as household head education, the number of non-working household members, and annual household income from urban jobs and farming. $\boldsymbol{\varphi}$ is the parameter to be estimated. The reasons for choosing a fractional probit function over the alternatives—such as a two-limited Tobit model, a log-odds transformation, or a Beta distribution—were motivated by two key considerations: the absence of zero observations and our analytical focus on estimating the conditional mean effects. It is well suited to modeling continuous outcome variables bounded between 0 and 1 when zeros are absent and mean-based interpretations are prioritized.

Given that the mechanism of individuals' time allocation may be different from that of household labor distribution, a separate model examined how land endowment affects an individual's decision to migrate to work urban job. The model for individual decisions is as follows:

$$E\left(d_{jk}\middle|z\right) = \frac{exp(z\omega)}{[1 + exp(z\omega)]}, \quad (14)$$

where d_{jk} is the decision to migrate for individual j in group k. The explanatory variable (z) in Equation (14) includes a vector of personal characteristics, such as gender, age, education, integration in migrant cities, and household-level variables. The land tenure security perception indicator is an explanatory dummy variable, and ω is the parameter to be estimated.

A potential endogeneity arises in estimating the effect of the income difference between farming and urban jobs on the share of household members employed outside agriculture and individual decisions to migrate. Instead of the income difference changing the share of household members with urban jobs or individual migration decisions, it is possible that greater shares of laborers with urban jobs or positive migration decisions cause income differences to be higher. Ignoring this issue could result in inconsistent estimators of all parameters, φ and ω.

The application of instrumental variables (IVs) is widely recommended to assure consistent estimators. However, obtaining a valid IV is quite difficult and a weak IV will lead to considerable efficiency loss compared to OLS. This study selects monetary gift-giving as an IV, which is highly correlated with income difference but does not affect household labor allocation. China has a long tradition of money gift-giving for interpersonal connectedness, such as at weddings, funerals, and specialized packets for holidays and other big social events or occasions [46,47]. The sample data show that over 97% of respondents gave or received a monetary gift in the previous year. A valid IV must be relevant to the endogenous variable, income difference, at either the household or individual level. The data show that the gross correlation between *Moneygift* and *Hhincomediff* is 0.30 (p-value = 0.00) and the correlation between *Moneygift* and *Incomediff* is 0.19 (p-value = 0.00). The strict testing for endogeneity is based on the following equation:

$$y_1 = f(y_2, X, v), \quad (15)$$

where y_1 is the dependent variable, i.e., the share of off-farm laborers or individual migration decisions, y_2 is the income difference at either the household or individual level, X is a vector of the exogeneous covariates, and v is the fitted residual vector from the first-stage OLS regression of income differences on exogeneous covariates plus the IV, *Moneygift*. A test of exogeneity would be a test of whether the coefficient of v, defined as ρ, is zero. If the null hypothesis of exogeneity is rejected, then the income difference is endogenous.

Tables A1 and A2 in Appendix A show that the coefficients of *Moneygift* in both estimations are statistically significant, which validates the relevance of IV. However, the coefficient of the fitted residual, $\hat{v}$, is not statistically significant. Thus, in the case of the two used data subsamples, the income differences, *Hhincomediff* and *Incomediff*, are exogenous in estimating household labor allocation or individual migration decisions, respectively.

4.3. The Test for a U-Shaped Relationship

This study follows the framework for a test for a U-shaped relationship [16] between labor allocation and land endowment in a Chinese farm household. The most common model specification in testing a U-shaped (or an inverse U-shaped) relationship the inclusion of a quadratic term in the standard regression model:

$$y_i = \alpha + \beta w_i + \gamma w_i^2 + \delta' C_i + \varepsilon_i, \quad (16)$$

where w_i is the key explanatory variable, C_i is a vector of the control variables, and ε_i is the normally distributed error term. Rather than focusing on the sign and significance of β and γ to determine the U-shaped relationship, the current study applies Lind and Mehlum's method [18], which provides an enhanced test of the relationship. Specifically, if Equation (16) has only one extreme point (either a peak or bottom point), the requirement for a U shape is a negative slope at the starting point (w_l) and a positive slope at the endpoint (w_h). The start- and endpoints can equal the minimum and maximum values of the data range. A U shape is implied by the condition that $\beta + 2\gamma x_l < 0 < \beta + 2\gamma x_h$. To test the condition, the following composite hypothesis is tested:

$$H_0 : \beta + 2\gamma w_l \geq 0 \; and/or \; \beta + 2\gamma w_h \leq 0$$

$$H_a : \beta + 2\gamma w_l < 0 \; and \; \beta + 2\gamma w_h > 0.$$

The null hypothesis is rejected if either one or both of the conditions are rejected. However, given the combination of the null hypotheses, the standard test methods are no longer suitable. Sasabuchi [48] provided a framework to construct the rejection area as follows:

$$\mathrm{R}_\alpha = \left\{ (\beta, \gamma) : \frac{\hat{\beta} + 2w_l\hat{\gamma}}{\sqrt{s_{11} + 2w_l 2s_{12} + 2w_l s_{22}}} < -t_\alpha, \text{ and } \frac{\hat{\beta} + 2w_h\hat{\gamma}}{\sqrt{s_{11} + 2w_h 2s_{12} + 2w_h s_{22}}} > t_\alpha \right\}, \tag{17}$$

where $\hat{\beta}$ and $\hat{\gamma}$ are the estimated parameters. The estimated variance and covariance between the $\hat{\beta}$ and $\hat{\gamma}$ parameters are s_{11}, s_{22}, and s_{12}, and α is the confidence level. Meanwhile, the estimated extreme point can be calculated as $\hat{w}_{min} = -\frac{\hat{\beta}}{2\hat{\gamma}}$ and the confidence interval for the extreme point can be constructed using the Fieller method [49].

5. Results and Discussion

5.1. Household Labor Allocation

Table 5 shows the estimation results of two specifications for household labor distribution with and without control variables, i.e., land and household characteristics. Each specification is estimated using the whole sample and the two subsamples with different expectations of land tenure security. The results based on the whole sample represent homogeneity in terms of expectations of tenure security. The estimations without control variables indicate that the estimated models marginally fit the data, as indicated by the low values of the pseudo-R^2s. Including control variables increases the goodness-of-fit markedly. Thus, the following discussion focuses on the model with control variables.

In the model with homogeneous expectations of land tenure security, the relationship between land endowment and the share of migrant labor in a household follows an inverse U-shaped trajectory. Initially, the households allocate additional labor to migration in urban areas as their land endowment increases, driven by the high wages from urban employment. However, when the land endowment exceeds a certain threshold, any additional increase in land endowment decreases the amount of labor allocated to jobs in the city, reducing migration. This phenomenon is explained by the complementarity between land and labor input in farming, especially within the family farm system, where larger land endowments require more labor input.

However, land endowment does not significantly affect the share of migrating household members for households with secure expectations of land tenure. In contrast, land endowment is significantly negatively associated with the share of off-farm labor in households with unsecure expectations of land tenure. The results reveal that land endowment hinders rural labor outmigration if land tenure is poorly protected. These findings are consistent with the reports that unsecured land tenure increases the need for "guard labor" to claim land access [20,21].

Table 5. Estimation results of land endowment's influence on migration for two model specifications using overall sample and subsamples for secured and unsecured tenure perceptions.

Variable	Full Sample	Secure Group	Unsecure Group	Full Sample	Secure Group	Unsecure Group
Acre	0.014	0.026	−0.005	0.033 **	0.030	0.038 *
	(0.015)	(0.020)	(0.023)	(0.014)	(0.019)	(0.021)
Acresq	−0.001 *	−0.002	−0.001	−0.002 ***	−0.001	−0.002 ***
	(0.001)	(0.001)	(0.001)	(0.001)	(0.001)	(0.001)
Parcel	-	-	-	−0.011	−0.009	−0.009
				(0.007)	(0.010)	(0.011)
Hhedu	-	-	-	−0.043 *	−0.024	−0.065 *
				(0.025)	(0.033)	(0.038)
Hhsizenl	-	-	-	−0.314 ***	−0.309 ***	−0.313 ***
				(0.021)	(0.029)	(0.032)
Hhincomediff	-	-	-	0.004 ***	0.004 ***	0.006 ***
				(0.001)	(0.001)	(0.001)
Constant	0.090	0.054	0.151 *	0.148	0.141	0.114
	(0.055)	(0.073)	(0.084)	(0.092)	(0.126)	(0.134)
Observations	736	416	320	736	416	320
Pseudo R^2	0.001	0.001	0.002	0.045	0.038	0.056

Note: Robust standard errors in parentheses. *** $p < 0.01$; ** $p < 0.05$; * $p < 0.1$.

Table 6 shows the results of a formal test of the U-shaped relationship. The results based on the whole sample show a positive slope at the lower bound and a negative slope at the upper bound. The t-value for the test of an inverse U-shaped relationship is 1.91 and suggests the rejection of the null hypothesis of a U shape. The extremum point for the overall sample is 8.191 mu, and the 90% confidence interval is about 3 mu to 10 mu. The group with unsecured land tenure expectations shows a similar pattern of an inverse U-shaped relationship to the one identified for the overall sample (although the relationship at the lower bound is weak). The extreme point for the group with unsecured land tenures is 7.32 mu. The lower and upper bounds are set as the minimum and maximum values in the data range. However, the conditions of a positive slope at the lower bound and a negative slope at the upper bound still hold when the lower and upper bounds change within a narrow range. The inverse U-shaped relationship is therefore empirically confirmed.

Table 6. The test results for a U-shaped relationship.

Item	Overall	Secure Tenure	Unsecure Tenure
ACRE	0.033 **	0.030	0.038 *
	(0.014)	(0.019)	(0.021)
ACRESQ	−0.002 ***	−0.001	−0.002 ***
	(0.001)	(0.001)	(0.001)
Slope at lower bound	0.025 **	0.024 *	0.030 *
Slope at upper bound	−0.048 ***	−0.035 *	−0.069 ***
Appropriate for an inverse U shape (t-value)	1.91 **	1.35 *	1.53 *
Extremum point (*mu*)	8.191	9.695	7.332
90% confidence interval, Fieller method	[3.332, 10.547]	[-, -]	[-, -]

Notes: The numbers of coefficients and robust standard errors in parentheses come from the estimation in Table 5. The Lind and Mehlum (2010) test provides the slope, test, extremum point, and confidence interval [16]. *** $p < 0.01$; ** $p < 0.05$; * $p < 0.1$.

Given their inverse U-shaped relationships, Figure 2 shows the plotted predictions for land endowment and the share of migrating household members for both the overall sample and the unsecure tenure group. Both lines display clear inverse U-shaped relationships (Figure 2). That is, households tend to allocate about half of their labor to jobs in urban areas when the land area they operate is relatively small. The motivation to migrate is driven by the income difference between urban and rural areas. With the increase in land endowment, farm households tend to allocate more labor to jobs in urban areas, and when the land endowment is abundant, households are more likely to retain labor for farming. The group with unsecured land tenure expectations has a similar pattern of an inverse U-shaped relationship. However, the group with unsecured land tenures is associated with a curve that has a much steeper slope on the right side (Figure 2), which reinforces the negative relationship between land endowment and the share of household members with off-farm urban jobs.

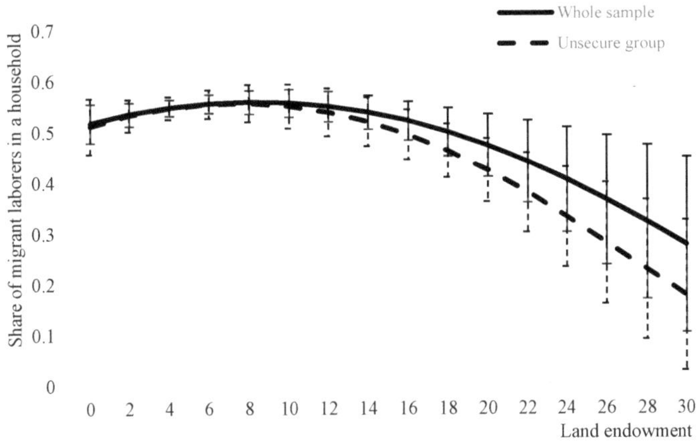

Figure 2. The prediction of the relationship between land endowment and the share of migrating household members using the overall sample and the group with unsecure land tenure expectations. Notes: The continuous vertical lines show the predictive margins with 95% confidence intervals for the whole sample (n = 736). The dashed vertical lines show the predictive margins with 95% confidence intervals for the group with unsecure land tenure expectations (n = 320).

Additionally, the household income difference between migrant employment in urban areas and farming has a significant positive effect on migration to urban areas in both the full sample and the subgroup with unsecure land tenure expectations. This finding is consistent with the Todaro model, which argues that income disparity between urban and rural areas drives migration [23]. Conversely, the presence of non-working household members—specifically children and elderly dependents—exerts a significant negative influence on migration likelihood. The number of non-working members in a household prevents migration. Children and the elderly require care, which could limit labor mobility.

5.2. Individual Migration Decisions

This section examines how land endowment affects individuals' decision to migrate to urban areas. The dependent variable is the choice between remaining in an urban area or returning to a farm. Table 7 shows the results of two relationships, estimated with and without control variables. The discussion focuses on the results of the former specification.

Land endowment and its squared term do not significantly influence individuals' decisions to migrate under specifications with or without control variables. This finding confirms that distinct mechanisms underpin individual and household migration decisions. Migrants perceiving their land tenure as secure are more likely to stay in urban areas. The income difference between urban jobs and farming positively affects migrant workers' decisions, but the measure is statistically insignificant. Notably, gender disparities emerge:

female migrants are more likely to remain in urban areas as opposed to males (Table 7). Since there are fewer women among the rural migrants employed in urban areas, this effect is differently experienced by those remaining on farms. The differences across gender for migration will benefit from additional future research regarding cultural aspects. Age is strongly negatively associated with individual migration decisions, whereas education helps migrant workers to compete in the urban labor market and increases their likelihood of remaining in urban areas. These findings are further supported by the statistically significant variable indicating integration into city life. The higher the level of integration into urban life, the higher the likelihood that migrant workers choose to stay in urban areas.

Table 7. Estimation results of land endowment effects on individuals.

Variables	PLAN	PLAN
Acreind	0.016	0.041
	(0.118)	(0.128)
Acreindsq	−0.006	−0.005
	(0.014)	(0.014)
Parcel	-	−0.034
	-	(0.030)
Landtenure	-	0.394 **
		(0.156)
Incomediff	-	0.005
		(0.004)
Male	-	−0.408 *
		(0.223)
Age		−0.030 ***
		(0.009)
Educ	-	0.323 *
		(0.174)
Integration	-	0.823 ***
		(0.170)
Hhedu		−0.246
		(0.170)
Hhsizenl		−0.034
		(0.094)
Constant	−0.121	0.400
	(0.167)	(0.575)
Pseudo R^2	0.001	0.062

Notes: n = 736. Robust standard errors in parentheses. *** $p < 0.01$; ** $p < 0.05$; * $p < 0.1$.

5.3. The Land Rental Market

The assumption of the above model is that the land rental market is not active. Although land transfer in rural areas across the country has begun to take shape, land transfer is not strongly market-dominated and, in practice, is still dominated by the government; at the same time, due to the high degree of government financial bondage [50], land transfers from small farmers to other operators tend to be relatively limited, which is at variance with national statistics. We also add that, in China, farmland ownership is retained by rural collectives, and only the management right can be circulated as an institutional arrangement [51,52]; meanwhile, while studies have explored the impact of farmland-use right transfer on collective action on public land from an irrigation perspective [53], the unique perspective of this study better reflects the relevant themes. Moreover, many studies have confirmed the enhancing effect of land tenure security on land transactions [54]. On the contrary, unsecured land tenure reduces the activity of the land rental market. Therefore, the inactive land market and unsafe land rights discussed in this paper are meaningful.

6. Conclusions

Land endowment, labor migration, and their relationship are key issues in many developing countries. Understanding those relationship helps policymakers formulate effective policies to stimulate migration processes and, eventually, the permanent urbanization of rural jobseekers. The previous studies on the effects of land endowment on labor allocation provide checkered support for their U-shaped relationship. This study contributes to the literature by examining the relationship between land endowment, labor migration, and land tenure perceptions among rural–urban migrants in China.

The current paper recognizes the heterogeneity of land tenure security perceptions among rural migrants employed in urban areas as the key motivating factor in decisions to undergo rural–urban migration. By classifying migrant workers into secured land tenure and unsecured land tenure groups, we examine how land endowment affects two groups of migrants with differing perceptions of land tenure security, revealing subtle institutional influences on migration behavior. The chosen empirical approach controls directly for the perceptions of land ownership and administrative land adjustment schemes and identifies how the amount of acreage affects household labor distribution as well as individuals' decisions to migrate to urban areas in search of jobs. Though rural land is constitutionally defined as being owned by village collectives in China, rural migrants exhibit heterogeneous perceptions of their land tenures, reflecting strong attachments to the land.

The estimation results confirm an inverse U-shaped relationship between land endowment and household labor distribution. Moreover, this study distinguishes between the effects for households who perceive the security of their land tenures differently. Specifically, the operated land area only hinders labor allocation primarily among rural households who perceive their land tenure to be unsecured, where it displays an inverse U-shaped relationship. Conversely, land endowment does not constrain labor allocation if the household perceives their land tenure to be secured. Notably, there are limits to the land endowment that can be allocated to farm households by village collectives in China. Within the collective framework, increases in land endowment to individual farms induce outmigration due to perceptions of land tenure security. The study further distinguishes between household-level and individual migration decisions. The individual decision to migrate to urban areas to seek a job is more likely if land tenure is perceived as secure. Additionally, another finding from this study is that the level of integration into urban life encourages migrant workers to remain in the cities, underscoring the dual role of land tenure and urban adaptation in shaping migration outcomes. Collectively, these methodologies not only serve to ameliorate the limitations inherent in earlier research, but also enhance the empirical rigor of labor allocation dynamics analysis, thereby advancing the field's capacity for reliable inquiry.

Securing land tenure should be prioritized in rural policy design if more rural migrants are needed to fill jobs in urban areas. Rural labor outmigration, encouraged by secure land tenure perceptions, may stimulate the development of local land renting markets. The land titling program, conducted at the end of 2018 in response to widespread perceptions of unsecured land tenure, may have to be augmented by a program of compensation for titled land. These measures would strengthen tenure security, reducing rural households' reliance on land as a sole safety net and encouraging labor mobility. Additionally, the expansion of social welfare benefits accessible to rural migrants in urban areas could encourage them to remain in cities permanently. As shown during the recent financial crisis and COVID-19 pandemic, rural migrants are most likely to lose urban jobs, forcing them to return to farms. Such unforeseen events prompt the return of millions of people back to rural areas, potentially slowing agricultural productivity growth and economic development in rural and urban areas. To mitigate these risks, policy framework design must clarify whether land donations serve as a transitional buffer or a permanent fallback. Land endowment is viewed by rural migrants and their households as a safety net unless the urban hukou sustains urban residency. Therefore, the ongoing second-round contracting renewal policy, as well as rural revitalization initiatives, should secure land tenure for migrant workers. In this context, the rural social security system reform and the development of the land leasing market are essential to achieve reasonable labor mobility and sustainable rural development. These recommendations align with current policy trajectories, such as land tenure confirmation and urban–rural integration initiatives, while offering measures to address systemic barriers to equitable migration outcomes.

While this study enriches our understanding of land tenure's role in migration dynamics, its limitations are its focus on a specific institutional context and reliance on cross-sectional data. Future research could extend this inquiry to diverse institutional settings to test the robustness of the observed relationships. Also, longitudinal research or panel data would offer deeper insights into how evolving tenure security influences migration decisions.

Author Contributions: Conceptualization, H.C. and Z.L.; methodology, Z.L. and H.C.; software, Z.L.; validation, Z.L., W.J.F. and H.C.; formal analysis, Z.L.; investigation, H.C.; resources, H.C.; data curation, Z.L. and H.C.; writing—original draft preparation, Z.L. and H.C.; writing—review and editing, W.J.F. and H.C.; supervision, W.J.F. and H.C.; project administration, H.C.; funding acquisition, H.C. All authors have read and agreed to the published version of the manuscript.

Funding: This research was funded by the National Natural Science Foundation of China, grant number 71573131, and the Key Project of National Social Science Found of China, grant number 18AZD023. It was also supported by the Social Science Foundation of Jiangsu Province, grant number 23ZXZB022, and the Fundamental Research Funds for the Central Universities, grant number SKYZ2024014.

Data Availability Statement: Some or all data, models, or code that support the findings of this study are available from the corresponding author upon reasonable request.

Acknowledgments: The first author acknowledge the students who participated in the survey, such as Hao Chen and Yanan Yang, etc.

Conflicts of Interest: The authors declare no conflicts of interest. The funders had no role in the design of the study; in the collection, analyses, or interpretation of data; in the writing of the manuscript; or in the decision to publish the results.

Appendix A

Table A1. First-stage estimation for IV procedure for household labor allocation.

VARIABLES	HHINCOMEDIFF	SHARE
ACRE	−1.500 ***	0.034 **
	(0.327)	(0.014)
ACRESQ		−0.002 ***
		(0.001)
PARCEL	0.607	−0.011
	(0.619)	(0.008)
HHEDU	2.927 *	−0.046 *
	(1.685)	(0.026)
HHSIZENL	0.218	−0.315 ***
	(1.930)	(0.021)
HHINCOMEDIFF		0.005 ***
		(0.002)
$\hat{\vartheta}$		−0.001
		(0.002)
MONEYGIFT	3.173 ***	
	(0.445)	
Constant	38.704 ***	0.110
	(5.189)	(0.123)
Observations	736	736
R2/Pseudo R2	0.114	0.045

Robust standard errors in parentheses. *** $p < 0.01$, ** $p < 0.05$, * $p < 0.1$.

Table A2. Testing for income difference endogeneity in individuals' migration decisions.

VARIABLES	INCOMEDIFF	PLAN
ACREIND	−1.096 ***	0.067
	(0.399)	(0.130)
ACREINDSQ		−0.005
		(0.014)
PARCEL	0.016	−0.035
	(0.264)	(0.030)
LANDTENURE	0.007	0.408 ***
	(1.496)	(0.157)
MALE	6.360 ***	−0.567 **
	(1.976)	(0.259)
AGE	−0.186 **	−0.026 ***
	(0.077)	(0.009)
EDUC	0.636	0.306 *
	(1.194)	(0.174)
INTEGRATION	3.083 **	0.748 ***
	(1.544)	(0.179)
HHEDU	0.854	−0.274
	(1.205)	(0.171)
HHSIZENL	2.016 *	−0.088
	(1.098)	(0.104)
INCOMEDIFF		0.029
		(0.020)
$\hat{v}$		−0.024
		(0.020)
MONEYGIFT	0.975 ***	
	(0.281)	
Constant	28.230 ***	−0.282
	(5.544)	(0.795)
Observations	736	736
R2/Pseudo R2	0.081	0.063

Robust standard errors in parentheses. *** $p < 0.01$, ** $p < 0.05$, * $p < 0.1$.

References

1. SCRO (The State Council Research Office). *Survey and Research Report on Chinese Migrant Workers*; China Yanshi Press: Beijing, China, 2006; pp. 2–4. (In Chinese)
2. 2023 Monitoring and Investigation Report on Migrant Workers. (In Chinese). Available online: https://www.stats.gov.cn/sj/zxfb/202404/t20240430_1948783.html (accessed on 25 January 2025).
3. Mullan, K.; Grosjean, P.; Kontoleon, A. Land tenure arrangements and rural-urban migration in China. *World Dev.* **2011**, *39*, 123–133. [CrossRef]
4. Chen, Y.; Feng, S. Access to public schools and the education of migrant children in China. *China Econ. Rev.* **2013**, *26*, 75–88. [CrossRef]
5. Liu, S.; Michael, R.C.; Yao, Y. Dimensions and diversity of property rights in rural China: Dilemmas on the road to further reform. *World Dev.* **1998**, *26*, 1789–1806. [CrossRef]
6. Zhang, L.; Huang, J.; Rozelle, S. Employment, emerging labor markets, and the role of education in rural China. *China Econ. Rev.* **2002**, *13*, 313–328. [CrossRef]
7. Li, H.; Chen, K.; Yan, L.; Yu, L.; Zhu, Y. Citizenization of Rural Migrants in China's New Urbanization: The Roles of Hukou System Reform and Rural Land Marketization. *Cities* **2023**, *132*, 103968. [CrossRef]
8. De Janvry, A.; Sadoulet, E. Progress in the modeling of rural households' behavior under market failures. In *Poverty, Inequality and Development: Economic Studies in Inequality, Social Exclusion and Well-Being*; De Janvry, A., Kanbur, R., Eds.; Springer: Boston, MA, USA, 2003; Volume 1, pp. 155–181. [CrossRef]
9. Kung, J.K. Off-Farm Labor Markets and the Emergence of Land Rental Markets in Rural China. *J. Comp. Econ.* **2002**, *30*, 395–414. [CrossRef]
10. Wang, J.; Cao, Y.; Fang, X.; Li, G.; Cao, Y. Does Land Tenure Fragmentation Aggravate Farmland Abandonment? Evidence from Big Survey Data in Rural China. *J. Rural Stud.* **2022**, *91*, 126–135. [CrossRef]
11. Tao, R.; Xu, Z. Urbanization, rural land system and social security for migrants in China. *J. Dev. Stud.* **2007**, *43*, 1301–1320. [CrossRef]
12. VanWey, L.K. Land ownership as a determinant of temporary migration in Nang Rong, Thailand. *Eur. J. Popul.* **2003**, *19*, 121–145. [CrossRef]

13. Mckenzie, D.; Rapoport, H. Network effects and the dynamics of migration and inequality: Theory and evidence from Mexico. *J. Dev. Econ.* **2007**, *84*, 1–24. [CrossRef]
14. Ma, X.; Heerink, N.; van Ierland, E.; Shi, X. Land tenure insecurity and rural-urban migration in rural China. *Pap. Reg. Sci.* **2016**, *95*, 383–406. [CrossRef]
15. Winters, P.; de Janvry, A.; Sadoulet, E. Family and community networks in Mexico-U.S. migration. *J. Hum. Resour.* **2001**, *36*, 159–184. [CrossRef]
16. Bhandari, P. Relative deprivation and migration in an agricultural setting of Nepal. *Popul. Environ.* **2004**, *25*, 475–499. [CrossRef]
17. Feng, S.; Heerink, N. Are farm households' land renting and migration decisions inter-related in rural China? *NJAS-Wagen. J. Life Sci.* **2008**, *55*, 345–362. [CrossRef]
18. Lind, J.T.; Mehlum, H. With or without U? The appropriate test for a U-shaped relationship. *Oxf. Bull. Econ. Stat.* **2010**, *72*, 109–118. [CrossRef]
19. Ge, D.; Long, H.; Qiao, W.; Wang, Z.; Sun, D.; Yang, R. Effects of Rural–Urban Migration on Agricultural Transformation: A Case of Yucheng City, China. *J. Rural Stud.* **2020**, *76*, 85–95. [CrossRef]
20. Field, E. Entitled to work: Urban property rights and labor supply in Peru. *Q. J. Econ.* **2007**, *122*, 1561–1602. [CrossRef]
21. Besley, T.; Ghatak, M. Property rights and economic development. In *Handbook of Development Economics*, 1st ed.; Rodrik, D., Rosenzweig, M., Eds.; Elsevier BV: Amsterdam, The Netherlands, 2010; Volume 5, Chapter 68; pp. 4525–4595. [CrossRef]
22. Qiu, T.; Luo, B.; Choy, S.B.; Li, Y.; He, Q. Do Land Renting-in and Its Marketization Increase Labor Input in Agriculture? Evidence from Rural China. *Land Use Policy* **2020**, *99*, 104820. [CrossRef]
23. Todaro, M.P. A model of labor migration and urban unemployment in less developed Countries. *Am. Econ. Rev.* **1969**, *59*, 138–148.
24. Adamopoulos, T.; Brandt, L.; Leight, J.; Restuccia, D. Misallocation, selection and productivity: A quantitative analysis with panel data from China. *Econometrica* **2022**, *90*, 1261–1282. [CrossRef]
25. Restuccia, D.; Rogerson, R. The causes and costs of misallocation. *J. Econ. Perspect.* **2017**, *31*, 151–174. [CrossRef]
26. Chen, C.; Restuccia, D.; Santaeulalia-llopis, R. Land misallocation and productivity. *Am. Econ. J. Macroecon.* **2023**, *15*, 441–465. [CrossRef]
27. Restuccia, D. Misallocation and aggregate productivity across time and space. *Can. J. Econ.* **2019**, *19*, 5–32. [CrossRef]
28. Harris, J.R.; Todaro, M.P. Migration, unemployment and development: A two-sector analysis. *Am. Econ. Rev.* **1970**, *60*, 126–142. [CrossRef]
29. De Brauw, A. Seasonal migration and agricultural production in Vietnam. *J. Dev. Stud.* **2010**, *46*, 114–139. [CrossRef]
30. Keshri, K.; Bhagat, R.B. Temporary and seasonal migration in India. *Genus* **2010**, *66*, 25–45.
31. Meng, L.; Zhao, M.Q. Permanent and temporary rural–urban migration in China: Evidence from field surveys. *China Econ. Rev.* **2018**, *51*, 228–239. [CrossRef]
32. Hu, X.; Cook, S.; Salazar, M.A. Internal migration and health in China. *Lancet* **2008**, *372*, 1717–1719. [CrossRef]
33. Lin, J.Y. Rural factor markets in China: After the household responsibility system reform. In *Chinese Economic Policy: Economic Reform at Midstream*; Reynolds, B.L., Kim, I.J., Eds.; Paragon House: New York, NY, USA, 1988; pp. 169–203.
34. Bowlus, A.J.; Sicular, T. Moving toward markets? Labor allocation in rural China. *J. Dev. Econ.* **2003**, *71*, 561–583. [CrossRef]
35. Wang, W.; Gong, J.; Wang, Y.; Shen, Y. Exploring the Effects of Rural Site Conditions and Household Livelihood Capitals on Agricultural Land Transfers in China. *Land Use Policy* **2021**, *108*, 105523. [CrossRef]
36. Wang, Y.; Li, X.; Li, W.; Tan, M. Land titling program and farmland rental market participation in China: Evidence from pilot provinces. *Land Use Policy* **2018**, *74*, 281–290. [CrossRef]
37. Chiang, A.C. *Fundamental Methods of Mathematical Economics*, 3rd ed.; McGraw-Hill, Inc.: New York, NY, USA, 1984; pp. 204–225.
38. Statistical Bulletin on National Economic and Social Development of the People's Republic of China in 2023. (In Chinese). Available online: https://www.stats.gov.cn/sj/zxfb/202402/t20240228_1947915.html (accessed on 10 February 2025).
39. Giles, J.; Mu, R. Village political economy, land tenure insecurity, and the rural to urban migration decision: Evidence from China. *Am. J. Agric. Econ.* **2018**, *100*, 521–544. [CrossRef]
40. Kung, J.K.S.; Bai, Y. Induced institutional change or transaction costs? The economic logic of land reallocations in Chinese agriculture. *J. Dev. Stud.* **2011**, *47*, 1510–1528. [CrossRef]
41. Lazarsfeld, P.F.; Henry, N.W. *Latent Structure Analysis*; Houghton Mifflin: Boston, MA, USA, 1968.
42. McLachlan, G.; Peel, D.A. *Finite Mixture Models*; Wiley: New York, NY, USA, 2000.
43. Scarpa, R.; Thiene, M. Destination choice models for rock climbing in the Northeastern Alps: A latent-class approach based on intensity of preferences. *Land Econ.* **2005**, *81*, 426–444. [CrossRef]
44. Klonsky, E.D.; Olino, T.M. Identifying clinically distinct subgroups of self-injurers among young adults: A latent class analysis. *J. Consult. Clin. Psychol.* **2008**, *76*, 22–27. [CrossRef]
45. Barnes, A.P.; Islam, M.M.; Toma, L. Heterogeneity in climate change risk perception amongst dairy farmers: A latent class clustering analysis. *Appl. Geogr.* **2013**, *41*, 105–115. [CrossRef]

46. Kipnis, A.B. *Producing Guanxi: Sentiment, Self, and Subculture in a North China Village*; Duke University Press: Durham, NC, USA, 1997.
47. Hudik, M.; Fang, E.S. Money or in-kind gift? Evidence from red packets in China. *J. Institutional Econ.* **2020**, *16*, 731–746. [CrossRef]
48. Sasabuchi, S. A test of a multivariate normal mean with composite hypotheses determined by linear inequalities. *Biometrika* **1980**, *67*, 429–439. [CrossRef]
49. Fieller, E.C. Some problems in interval estimation. *J. R. Stat. Soc.* **1954**, *16*, 175–185. [CrossRef]
50. Gyourko, J.; Shen, Y.; Wu, J.; Zhang, R. Land Finance in China: Analysis and Review. *China Econ. Rev.* **2022**, *76*, 101868. [CrossRef]
51. Tian, G.; Duan, J.; Yang, L. Spatio-Temporal Pattern and Driving Mechanisms of Cropland Circulation in China. *Land Use Policy* **2021**, *100*, 105118. [CrossRef]
52. Kung, J.K. Common Property Rights and Land Reallocations in Rural China: Evidence from a Village Survey. *World Dev.* **2000**, *28*, 701–719. [CrossRef]
53. Wang, Y.; Wang, H. Effects of Farmland Use Rights Transfer on Collective Action in the Commons: Evidence from Rural China. *Land Use Policy* **2022**, *120*, 106262. [CrossRef]
54. Li, F.; Feng, S.; Lu, H.; Qu, F.; D'Haese, M. How Do Non-Farm Employment and Agricultural Mechanization Impact on Large-Scale Farming? A Spatial Panel Data Analysis from Jiangsu Province, China. *Land Use Policy* **2021**, *107*, 105517. [CrossRef]

MDPI AG
Grosspeteranlage 5
4052 Basel
Switzerland
Tel.: +41 61 683 77 34

Land Editorial Office
E-mail: land@mdpi.com
www.mdpi.com/journal/land

www.ingramcontent.com/pod-product-compliance
Lightning Source LLC
Chambersburg PA
CBHW041415210926
51850CB00013B/3274
* 9 7 8 3 7 2 5 8 4 2 7 3 5 *